International Workshop on Dirac Electrons in Solids 2015

Journal of Physics: Conference Series Volume 603

Tokyo, Japan
14 – 15 January 2015

Editors:

M. Ogata
Y. Suzumura
Y. Fuseya
H. Matsuura

ISBN: 978-1-5108-0263-6
ISSN: 1742-6588

Printed from e-media with permission by:

Curran Associates, Inc.
57 Morehouse Lane
Red Hook, NY 12571

Printed by Curran Associates, Inc. (2015)

Institute of Physics
Dirac House, Temple Back
Bristol BS1 6BE UK

Phone: 44 1 17 929 7481
Fax: 44 1 17 920 0979

techtracking@iop.org

Additional copies of this publication are available from:

Curran Associates, Inc.
57 Morehouse Lane
Red Hook, NY 12571 USA
Phone: 845-758-0400
Fax: 845-758-2634
Email: curran@proceedings.com
Web: www.proceedings.com

TABLE OF CONTENTS

TUNABLE ORBITAL SUSCEPTIBILITY IN α-T_3 TIGHT-BINDING MODELS 1
F Piechon, J-N Fuchs, A Raoux, G Montambaux

CLASSIFICATION OF CRYSTALLINE TOPOLOGICAL SEMIMETALS WITH AN APPLICATION TO NA_3BI 16
Ching-Kai Chiu, Andreas P Schnyder

MULTIBAND K · P THEORY FOR HOLES AT THE T-POINT IN BISMUTH 26
Yuki Fuseya

MEISSNER EFFECT OF DIRAC ELECTRON IN SUPERCONDUCTING STATE 33
Tomonari Mizoguchi, Masao Ogata

EXCITATION SPECTRUM OF CORRELATED DIRAC FERMIONS 40
Z Jalali, S A Jafari

HIGHLY ANISOTROPIC PHYSICS IN PHOSPHORENE 48
Motohiko Ezawa

CHIRALITY EFFECT ON SUPERCONDUCTIVITY 54
Takao Morinari

SPIN TEXTURE OF THE SURFACE STATE OF THREE-DIMENSIONAL DIRAC MATERIAL CA_3PBO 62
Toshikaze Kariyado

LONGITUDINAL CONDUCTIVITY OF A THREE-DIMENSIONAL DIRAC ELECTRON GAS IN MAGNETIC FIELD 68
Igor Proskurin, Masao Ogata, Yoshikazu Suzumura

PLAQUETTE CHIRALITY PATTERNS FOR ROBUST ZERO-GAP STATES IN α-TYPE ORGANIC CONDUCTOR 80
Frederic Piechon, Yoshikazu Suzumura, Takao Morinari

REFLECTANCE OF DIRAC ELECTRONS IN ORGANIC CONDUCTOR 94
Yoshikazu Suzumura, Igor Proskurin, Masao Ogata

ANISOTROPIC SPIN MOTIVE FORCE IN MULTI-LAYERED DIRAC FERMION SYSTEM, α-$(BEDT\text{-}TTF)_2I_3$ 104
K Kubo, T Morinari

DEFECT-INDUCED KONDO EFFECT IN GRAPHENE: ROLE OF LOCALIZED STATE OF π ELECTRONS 110
Taro Kanao, Hiroyasu Matsuura, Masao Ogata

DIRAC FERMIONS AND KONDO EFFECT 120
Takashi Yanagisawa

LOCAL MOMENT FORMATION IN DIRAC ELECTRONS 124
M Mashkoori, I Mahyaeh, S A Jafari

INTRINSIC BROADENING OF THE ENERGY LEVELS OF DIRAC ELECTRONS IN QUANTUM DOTS 131
E Ahmadi, S A Jafari

WEYL SEMIMETAL WITH STRONG LONG-RANGE COULOMB INTERACTIONS 137
Akihiko Sekine, Kentaro Nomura

STUDY OF SPIN TRANSPORT IN DIRAC SYSTEMS 144
Nobuyuki Okuma, Masao Ogata

DISTRIBUTION OF CHERN NUMBER BY LANDAU LEVEL BROADENING IN HOFSTADTER BUTTERFLY 151
Nobuyuki Yoshioka, Hiroyasu Matsuura, Masao Ogata

ANOMALOUS HALL EFFECT IN THE DIRAC ELECTRON SYSTEM WITH A SPLIT TERM 158
Shota Suetsugu, Hiroyasu Matsuura, Masao Ogata

NEGATIVE MAGNETORESISTANCE IN TI-CLEANED SINGLE-LAYER GRAPHENE 164
A Fujimoto, C A Joiner, Y Jiang, D Terasawa, A Fukuda, Z Jiang, E M Vogel

NONTRIVIAL CONTRIBUTIONS TO THE MAGNETOCONDUCTIVITY DUE TO ANOMALOUS G-FACTOR IN THE LUTTINGER HAMILTONIAN 172
Riki Suemasa, Yuki Fuseya

MAGNETOCONDUCTIVITY OF DIRAC ELECTRONS IN BISMUTH 178
Mitsuaki Owada, Yuki Fuseya

FERRIMAGNETISM AND SINGLE-PARTICLE EXCITATIONS IN A PERIODIC ANDERSON MODEL ON THE HONEYCOMB LATTICE 184
Kazuhiro Seki, Tomonori Shirakawa, Qinfang Zhang, Tao Li, Seiji Yunoki
BAND STRUCTURE OF SUPERCONDUCTING DIRAC ELECTRON SYSTEMS 192
Takahiro Kuraya, Yuki Fuseya
Author Index

Preface

It is our pleasure to publish the Proceedings of the International Workshop on Dirac Electrons in Solids held in University of Tokyo, Japan, for January 14-15, 2015. The workshop was organized by the entitled project which lasted from April 2012 to March 2015 with 10 theorists. It has been supported by a Grand-in-Aid for Scientific Research (A) from the Ministry of Education, Culture, Sports, Science, and Technology, Japan.

The subjects discussed in the workshop include bismuth, organic conductors, graphene, topological insulators, new materials including Ca_3PbO, and new directions in theory (superconductivity, orbital susceptibility, etc). The number of participants was about 70 and the papers presented in the workshop include four invited talks, 16 oral presentations, and 23 poster presentations.

Dirac electron systems appear in various systems, such as graphene, quasi-two-dimensional organic conductors, bismuth, surface states in topological insulators, new materials like Ca_3PbO. In these systems, characteristic transport properties caused by the linear dispersion of Dirac electrons and topological properties, have been extensively discussed. In addition to these, there are many interesting research fields such as Spin-Hall effect, orbital diamagnetism due to interband effects, Landau levels characteristic to Dirac dispersion, anomalous interlayer transport phenomena and magnetoresistance, the effects of spin-orbit interaction, and electron correlation. The workshop focused on recent developments of theory and experiment of Dirac electron systems in the above materials.

We note that all papers published in this volume of Journal of Physics: Conference Series were peer reviewed. Reviews were performed by expert referees with professional knowledge and high scientific standards in this field. Editors made efforts so that the papers may satisfy the criterion of a proceedings journal published by IOP Publishing.

We hope that all the participants of the workshop enjoyed discussions and that these proceedings of the workshop help to extend the international research activities into Dirac Electrons in Solids in the future.

Organizing Committee and Editors
Masao Ogata (University of Tokyo)
Yoshikazu Suzumura (Nagoya University)
Yuki Fuseya (University of Electro-Communications)
Hiroyasu Matsuura (University of Tokyo)

Peer review statement

All papers published in this volume of *Journal of Physics: Conference Series* have been peer reviewed through processes administered by the proceedings Editors. Reviews were conducted by expert referees to the professional and scientific standards expected of a proceedings journal published by IOP Publishing.

Tunable orbital susceptibility in α-$\mathcal{T}_3$ tight-binding models

F Piéchon[1], **J-N Fuchs**[1,2], **A Raoux**[1,3] **and G Montambaux**[1]
[1] Laboratoire de Physique des Solides, CNRS UMR 8502, Université Paris-Sud, F-91405 Orsay Cedex, France
[2] Laboratoire de Physique Théorique de la Matière Condensée, CNRS UMR 7600, Univ. Pierre et Marie Curie 4, place Jussieu, 75252 Paris Cedex 05, France
[3] Département de Physique, École Normale Supérieure, 24 rue Lhomond, 75005 Paris, France

E-mail: piechon@lps.u-psud.fr

Abstract. We study the importance of interband effects on the orbital susceptibility of three bands α-$\mathcal{T}_3$ tight-binding models. The particularity of these models is that the coupling between the three energy bands (which is encoded in the wavefunctions properties) can be tuned (by a parameter α) without any modification of the energy spectrum. Using the gauge-invariant perturbative formalism that we have recently developped [1], we obtain a generic formula of the orbital susceptibility of α-$\mathcal{T}_3$ tight-binding models. Considering then three characteristic examples that exhibit either Dirac, semi-Dirac or quadratic band touching, we show that by varying the parameter α and thus the wavefunctions interband couplings, it is possible to drive a transition from a diamagnetic to a paramagnetic peak of the orbital susceptibility at the band touching. In the presence of a gap separating the dispersive bands, we show that the susceptibility inside the gap exhibits a similar dia to paramagnetic transition.

1. Introduction

The orbital magnetic susceptibility of free electrons was computed long ago by Landau [2] and was found to be diamagnetic. Subsequently, Peierls extended Landau's result to the case of a single band tight-binding model. He found a formula for the orbital susceptibility that only depends on the zero-field band energy spectrum. This result already showed that the band structure can have a strong influence on the magnetic response of crystals [3]. For example, near a saddle point of the dispersion relation, the Peierls orbital susceptibility becomes paramagnetic [4]. However, one important effect was left out, namely the coupling between the bands in the case of several bands. A striking example is the possibility of having a finite orbital susceptibility inside the gap of a band insulator at zero temperature. This was understood as an inter-band effect by Fukuyama and Kubo on the example of bismuth [5]. Fukuyama also provided a compact and quite general linear-response formula that includes interband effects [6]. However, despite its many successes, this formula does not work for non-separable tight-binding models. The aim of this paper is to present orbital susceptibility results obtained from an exact linear response formula that we recently derived for tight-binding models [1]. In order to show the importance of inter-band (or band coupling) effects, we consider here a family of tunable three band tight-binding models constructed on a $\mathcal{T}_3$ lattice (or *dice* lattice) and that we call α-$\mathcal{T}_3$ [7]. The dice model was first introduced [8] as a simple 2-dimensional model that exhibits localized and extended states as the same time. It also presents some peculiar properties under strong magnetic field [9]. These α-$\mathcal{T}_3$ models depend on a real parameter α and have the important property that their zero-field energy spectrum –which is essentially that of graphene with an additional flat band– is independent of α. However, the parameter α has a strong influence on the zero-field eigenstates and therefore on the Berry curvature. This influence is revealed by the orbital susceptibility that changes sign as a function of α.

The paper is organized as follows. In §2 we present the three bands α-$\mathcal{T}_3$ tight-binding models and characterize the key properties of their energy spectrum and wavefunctions. In §3, using the gauge-invariant perturbative formalism that we have recently developped [1], we provide a generic formula for the orbital susceptibility of α-$\mathcal{T}_3$ tight-binding models. In §4 we apply this susceptibity formula to three characteristic examples of α-$\mathcal{T}_3$ tight-binding models that exhibit respectively Dirac, semi-Dirac and quadratic band touching at low-energy. In §5 we summarize the main results of this study.

2. Tight-binding models on the $\mathcal{T}_3$ lattice

Starting from the honeycomb lattice with two sites (A, B) per unit cell, the $\mathcal{T}_3$ or dice lattice is obtained by connecting additional (C) sites at the center of each hexagon to the B sites (see Fig. 1). The dice lattice is thus a triangular Bravais lattice with three sites (A, B, C) per unit cell. We consider tight-binding models that consist of spinless electrons hopping on this lattice. In its simplest form, we allow for a constant onsite potential term $+\Delta$ on sites A, C and $-\Delta$ on sites B and an isotropic nearest-neighbors hopping with amplitude $c_\alpha t$ from A to B and $s_\alpha t$ from C to B with $c_\alpha = \frac{1}{\sqrt{1+\alpha^2}}$, $s_\alpha = \frac{\alpha}{\sqrt{1+\alpha^2}}$ such that $c_\alpha^2 + s_\alpha^2 = 1$. The real space representation of the corresponding Hamiltonian follows as

$$\begin{aligned} h = \textstyle\sum_{\boldsymbol{r}_B} \quad & [c_\alpha t(\delta_{\boldsymbol{r}_B-\boldsymbol{\delta}_1,\boldsymbol{r}_A} + \delta_{\boldsymbol{r}_B-\boldsymbol{\delta}_2,\boldsymbol{r}_A} + \delta_{\boldsymbol{r}_B-\boldsymbol{\delta}_3,\boldsymbol{r}_A})|\boldsymbol{r}_B\rangle\langle\boldsymbol{r}_A| \\ & + s_\alpha t(\delta_{\boldsymbol{r}_B+\boldsymbol{\delta}_1,\boldsymbol{r}_C} + \delta_{\boldsymbol{r}_B+\boldsymbol{\delta}_2,\boldsymbol{r}_C} + \delta_{\boldsymbol{r}_B+\boldsymbol{\delta}_3,\boldsymbol{r}_C})|\boldsymbol{r}_B\rangle\langle\boldsymbol{r}_C|] + \text{h.c}, \end{aligned} \tag{1}$$

where $\boldsymbol{\delta}_1, \boldsymbol{\delta}_2, \boldsymbol{\delta}_3$ are the three vectors connecting nearest-neighbors sites. We assume that the localized orbital basis is orthogonal ($\langle\boldsymbol{r}_{j'}|\boldsymbol{r}_j\rangle = \delta_{\boldsymbol{r}_{j'},\boldsymbol{r}_j}$) such that the position operator is purely diagonal ($\boldsymbol{r} = \sum_{\boldsymbol{r}_{j=A,B,C}} \boldsymbol{r}_j|\boldsymbol{r}_j\rangle\langle\boldsymbol{r}_j|$). Introducing the Bloch states basis $|\boldsymbol{k}_{j=A,B,C}\rangle = \sum_{\boldsymbol{r}_j} e^{i\boldsymbol{k}\boldsymbol{r}_j}|\boldsymbol{r}_j\rangle$, for each wavevector $\boldsymbol{k} = (k_x, k_y)$, the Bloch Hamiltonian matrix associated to this α-$\mathcal{T}_3$ tight-binding model reads

$$h_{\boldsymbol{k}} = \begin{pmatrix} \Delta & c_\alpha f_{\boldsymbol{k}} & 0 \\ c_\alpha f_{\boldsymbol{k}}^* & -\Delta & s_\alpha f_{\boldsymbol{k}} \\ 0 & s_\alpha f_{\boldsymbol{k}}^* & \Delta \end{pmatrix}, \tag{2}$$

where $f_{\boldsymbol{k}} = |f_{\boldsymbol{k}}|e^{-i\theta_{\boldsymbol{k}}} = t(e^{-i\boldsymbol{k}.\boldsymbol{\delta}_1} + e^{-i\boldsymbol{k}.\boldsymbol{\delta}_2} + e^{-i\boldsymbol{k}.\boldsymbol{\delta}_3})$. In the following, we will also consider generalized versions of this model that essentially consist in a modified $f_{\boldsymbol{k}}$. This class of models interpolates between the honeycomb ($\alpha = 0$) and the isotropic dice lattice ($\alpha = 1$).

The remarkable and interesting property of this class of models is that the energy band spectrum does not depend on α (see Fig. 1) whereas the eigenfunctions do. More quantitatively, the energy spectrum consists of two dispersive bands $\epsilon_{\pm,\boldsymbol{k}} = \pm\sqrt{\Delta^2 + |f_{\boldsymbol{k}}|^2}$ and a flat band at energy $\epsilon_0 = \Delta$. For $\Delta > 0$, the corresponding Bloch eigenfunctions $|s_{\alpha,\boldsymbol{k}}\rangle$ $(s = \pm, 0)$ read

$$|+_{\alpha,\boldsymbol{k}}\rangle = \frac{\sqrt{1+c_{\boldsymbol{k}}}}{\sqrt{2}} \begin{pmatrix} c_\alpha e^{-i\theta_{\boldsymbol{k}}} \\ \\ \frac{s_{\boldsymbol{k}}}{1+c_{\boldsymbol{k}}} \\ \\ s_\alpha e^{i\theta_{\boldsymbol{k}}} \end{pmatrix} ; \; |-_{\alpha,\boldsymbol{k}}\rangle = \frac{s_{\boldsymbol{k}}}{\sqrt{2(1+c_{\boldsymbol{k}})}} \begin{pmatrix} c_\alpha e^{-i\theta_{\boldsymbol{k}}} \\ \\ -\frac{1+c_{\boldsymbol{k}}}{s_{\boldsymbol{k}}} \\ \\ s_\alpha e^{i\theta_{\boldsymbol{k}}} \end{pmatrix} ; \; |0_{\alpha,\boldsymbol{k}}\rangle = \begin{pmatrix} s_\alpha e^{-i\theta_{\boldsymbol{k}}} \\ \\ 0 \\ \\ -c_\alpha e^{i\theta_{\boldsymbol{k}}} \end{pmatrix} \tag{3}$$

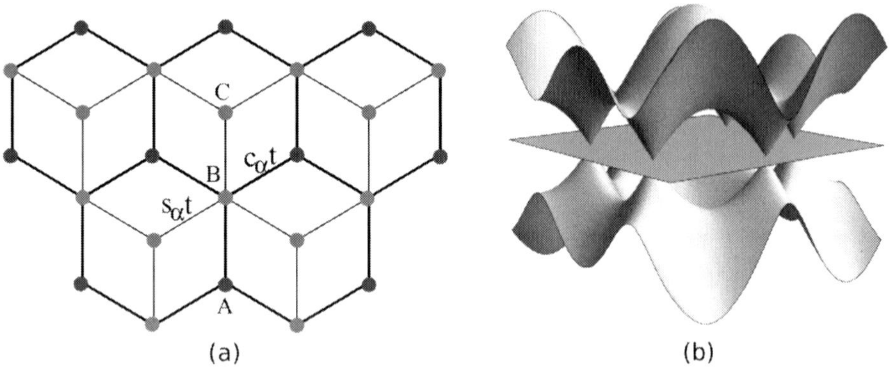

Figure 1. (Color online). α-$\mathcal{T}_3$ model: (a) The dice or $\mathcal{T}_3$ lattice is constituted by two interpenetrating honeycomb lattices. Thick links, nearest-neighbor hoppings $c_\alpha t$ from sites A to sites B ($c_\alpha = \frac{1}{\sqrt{1+\alpha^2}}$). Thin links, nearest-neighbor hoppings $s_\alpha t$ from sites C to sites B ($s_\alpha = \frac{\alpha}{\sqrt{1+\alpha^2}}$). By varying α, the model interpolates between honeycomb ($\alpha = 0$) and dice ($\alpha = 1$). Potential term $+\Delta$ on sites A, C and $-\Delta$ on sites B. (b) Energy bands dispersion in $\boldsymbol{k}$ space, for $\Delta = 0$: two dispersive bands and one flat band. This spectrum does not depend on α.

where $c_{\boldsymbol{k}} = \frac{\Delta}{\sqrt{\Delta^2+|f_{\boldsymbol{k}}|^2}}$ and $s_{\boldsymbol{k}} = \frac{|f_{\boldsymbol{k}}|}{\sqrt{\Delta^2+|f_{\boldsymbol{k}}|^2}}$ with $c_{\boldsymbol{k}}^2 + s_{\boldsymbol{k}}^2 = 1$. The Berry connection $\mathcal{A}_{\boldsymbol{k}}^s = \langle s_{\alpha,\boldsymbol{k}} | i\nabla_{\boldsymbol{k}} | s_{\alpha,\boldsymbol{k}} \rangle$ of each band is then readily obtained as

$$\mathcal{A}_{\boldsymbol{k}}^0 = -\frac{1-\alpha^2}{1+\alpha^2}\nabla_{\boldsymbol{k}}\theta; \quad \mathcal{A}_{\boldsymbol{k}}^+ = -\frac{1+c_{\boldsymbol{k}}}{2}\mathcal{A}_{\boldsymbol{k}}^0; \quad \mathcal{A}_{\boldsymbol{k}}^- = -\frac{s_{\boldsymbol{k}}^2}{2(1+c_{\boldsymbol{k}})}\mathcal{A}_{\boldsymbol{k}}^0, \tag{4}$$

such that quite generally $\mathcal{A}_{\alpha,\boldsymbol{k}}^0 + \mathcal{A}_{\alpha,\boldsymbol{k}}^+ + \mathcal{A}_{\alpha,\boldsymbol{k}}^- = 0$. From this Berry connection it is possible to obtain, for each band, the so-called Berry curvature $\Omega_{\boldsymbol{k}}^s = \nabla_{\boldsymbol{k}} \times \mathcal{A}_{\boldsymbol{k}}^s$. This Berry curvature usually presents some strongly peaked structure near specific $\boldsymbol{k}^*$ points where interband coupling is strong, even if there is a gap separating the bands. Moreover, by considering a k-space closed orbit $\mathscr{C}_{\boldsymbol{k}^*}$ around such a $\boldsymbol{k}^*$ point, it is possible to calculate a Berry phase that measures the winding of the phase $\theta_{\boldsymbol{k}}$ around such points $\boldsymbol{k}^*$. More precisely, considering closed orbits $\mathscr{C}_{\boldsymbol{k}^*}(\varepsilon)$ that corresponds to a constant energy $\varepsilon = \sqrt{\Delta^2 + |f_{\boldsymbol{k}}|^2}$, we deduce that for each band the Berry phase $\Phi_{\boldsymbol{k}^*}^s = \oint_{\mathscr{C}_{\boldsymbol{k}^*}} \mathrm{d}\boldsymbol{k} \cdot \mathcal{A}_{\boldsymbol{k}}^s$ accumulated along such an orbit is given respectively by $\Phi_{\boldsymbol{k}^*}^0 = -\Phi_B W_{\mathscr{C}_{\boldsymbol{k}^*}}$ with $\Phi_{\boldsymbol{k}^*}^\pm = -\frac{1}{2}(1 \pm \frac{\Delta}{\epsilon})\Phi_{\boldsymbol{k}^*}^0$ where $\Phi_B = \pi\frac{1-\alpha^2}{1+\alpha^2}$ and $W_{\mathscr{C}_{\boldsymbol{k}^*}} = \oint_{\mathscr{C}_{\boldsymbol{k}^*}} \mathrm{d}k \cdot \nabla_{\boldsymbol{k}}\theta/2\pi$ is the winding number of the orbit such that $W_{\mathscr{C}} = 0$ for a closed orbit around a regular $\boldsymbol{k}^*$-point and $W_{\mathscr{C}_{\boldsymbol{k}^*}} = \pm 1$ for an orbit encircling a Dirac point $\boldsymbol{k}^*$ [7, 10].

From this perspective, the main interest of the α-$\mathcal{T}_3$ models described by Eq.(2) is that all these Berry quantities (connections, curvature and phase) are proportional to $\Phi_B = \pi\frac{1-\alpha^2}{1+\alpha^2}$ (see Eq.(4)). As a consequence, the interband effects that are encoded by these Berry quantities are reduced upon increasing the value of α from 0 and totally vanish for $\alpha = 1$. Anticipating on what follows, we point out that our results for the orbital susceptibility of α-$\mathcal{T}_3$ models provide clear evidence that all interband effects are not only encoded by the Berry quantities.

3. Orbital susceptibility formula

For time reversal systems which we consider in the following, the orbital magnetization vanishes and the orbital susceptibility $\chi_{\rm orb}(\mu, T)$ is the first measure of the sensitivity of the energy spectrum when it is placed in a uniform static perpendicular magnetic field $\boldsymbol{B} = B\boldsymbol{u}_z$:

$$\chi_{\rm orb}(\mu, T) = -\frac{\mu_0}{S}\frac{\partial^2\Omega}{\partial B^2}\bigg|_{B=0} \tag{5}$$

where S is the area of the system, T is the temperature, μ is the chemical potential, $\mu_0 = 4\pi.10^{-7}$S.I. and $\Omega(T, \mu, B)$ is the grand canonical potential of the non-interacting electrons gas

$$\Omega(T, \mu, B) = -T\int_{-\infty}^{+\infty} \ln\left(1 + e^{-(\omega-\mu)/T}\right)\rho(\omega, B)\,{\rm d}\omega, \tag{6}$$

with $\rho(\omega, B)$ the field dependent density of states (Dos)

$$\rho(\omega, B) = -\frac{1}{\pi}\Im m\,{\rm Tr}\,G(\omega + {\rm i}0^+, B) \tag{7}$$

where $G(\omega, B)$ is the field dependent retarded Green's function (we use $k_{\rm B} = 1$ and $\hbar = 1$).

In order to obtain the susceptibility, it is thus necessary to calculate the density of states $\rho(\omega, B)$ or the Green's function $G(\omega, B) = (\omega{\rm I} - h(B))^{-1}$ associated to the field dependent Hamiltonian $h(B)$. For the tight-binding model (1), $h(B)$ is obtained by multiplying the real space hopping amplitudes by a gauge and position dependent Peierls phase:

$$\begin{aligned} h(B) = \textstyle\sum_{\boldsymbol{r}_B} \quad & [c_\alpha t(\delta_{\boldsymbol{r}_B+\boldsymbol{\delta}_1,\boldsymbol{r}_A} + \delta_{\boldsymbol{r}_B+\boldsymbol{\delta}_2,\boldsymbol{r}_A} + \delta_{\boldsymbol{r}_B+\boldsymbol{\delta}_3,\boldsymbol{r}_A})e^{-i\varphi_{\boldsymbol{r}_A,\boldsymbol{r}_B}}|\boldsymbol{r}_B\rangle\langle\boldsymbol{r}_A| \\ & +s_\alpha t(\delta_{\boldsymbol{r}_B-\boldsymbol{\delta}_1,\boldsymbol{r}_C} + \delta_{\boldsymbol{r}_B-\boldsymbol{\delta}_2,\boldsymbol{r}_C} + \delta_{\boldsymbol{r}_B-\boldsymbol{\delta}_3,\boldsymbol{r}_C})e^{-i\varphi_{\boldsymbol{r}_C,\boldsymbol{r}_B}}|\boldsymbol{r}_B\rangle\langle\boldsymbol{r}_C|] + {\rm h.c}, \end{aligned} \tag{8}$$

where $\varphi_{\boldsymbol{r},\boldsymbol{r}'} = \frac{e}{\hbar}\int_{\boldsymbol{r}}^{\boldsymbol{r}'}\boldsymbol{A}\cdot{\rm d}\boldsymbol{l}$ with $\boldsymbol{A}(\boldsymbol{r})$ the gauge dependent vector potential associated to the uniform magnetic field. Starting from the Hamiltonian $h(B)$ there are essentially two approaches to calculate $\rho(\omega, B)$.

The first approach consists in computing the exact magnetic field dependent energy spectrum $\epsilon_n(B)$ associated to $h(B)$; this leads to the so-called Hofstadter butterfly spectrum. Using such an approach, we have recently studied the orbital susceptibility of the α-$\mathcal{T}_3$ model (2) for the case $\Delta = 0$ [7]. In particular we have shown that the orbital susceptibility $\chi_{\rm orb}(\mu, T)$ strongly varies with the parameter α. More precisely, at the Dirac point, $\chi_{\rm orb}(\mu = 0, T)$ exhibits a continuous transition from a diamagnetic peak for $\alpha = 0$ (honeycomb-graphene) to a paramagnetic peak for $\alpha = 1$ (dice) (see Fig. 2). Away from the Dirac point, an opposite transition from paramagnetism to diamagnetism takes place such that a sum rule $\int {\rm d}\mu\chi_{\rm orb}(\mu, T) = 0$ is preserved for all α [7]. Despite these interesting results, such an approach suffers from being essentially numerical and thus it does not allow to fully understand how the dependence on the parameter α enters in the susceptibility.

In order to better understand to role of the parameter α, we have developped a second approach which consists in calculating the susceptibility $\chi_{\rm orb}(\mu, T)$ using our recently established gauge-invariant perturbative response formula [1]:

$$\chi_{\rm orb} = -\frac{\mu_0 e^2}{12\hbar^2}\frac{\Im m}{\pi S}\int_{-\infty}^{+\infty}{\rm d}\omega\, n_{\rm F}(\omega)\;{\rm Tr}\left\{gh^{xx}gh^{yy} - gh^{xy}gh^{xy} - 4(gh^x gh^x gh^y gh^y - gh^x gh^y gh^x gh^y)\right\}, \tag{9}$$

where $n_{\rm F}(\omega) = 1/(e^{\frac{\omega-\mu}{T}} + 1)$ is the Fermi function, $g(\omega) = (\omega{\rm I} - h)^{-1}$ is the retarded Green's function associated to the zero-field Hamiltonian and $h^x = [x, h]$, $h^{xx} = [x, [x, h]]$ are the single

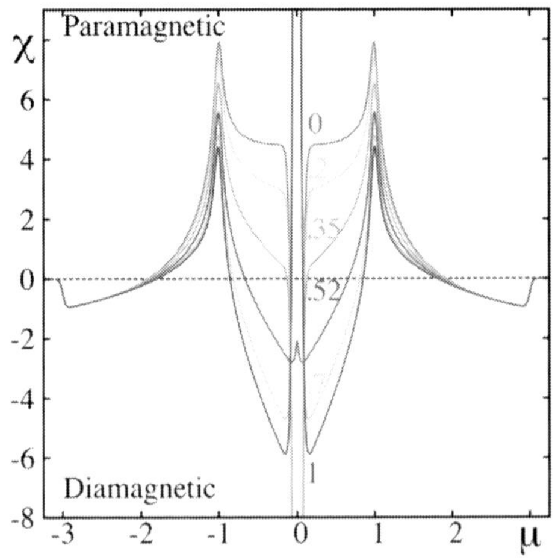

Figure 2. (Color online). Orbital susceptibility $\chi(\mu, T)$ obtained from the numerically computed Hofstadter spectrum [7]. χ (in units of the Landau band edge value $|\chi_L| = (\mu_0/16\pi)(e^2ta^2/\hbar)$) as a function of the chemical potential μ (in units of t) in the whole band for various α as indicated and for a temperature $T = 0.02t$. At $\mu = 0$ there is a transition from a diamagnetic peak for $\alpha = 0$ (red: graphene) to a paramagnetic peak for $\alpha = 1$ (blue: dice). Because of a sum rule, the orbital response at zero doping is systematically compensated by an opposite response at finite doping.

and double commutators of the position operator with the zero-field Hamiltonian. As discussed in [1], the expression (9) is equivalent to a recent formula derived for graphene [11] but it differs from the well-known Fukuyama formula [6]. We stress however that when $h^{xy} \neq 0$, only formula (9) fully agrees with the susceptibility results obtained from direct numerical computation of the corresponding Hofstadter butterfly spectrum. Moreover, we remind that formula (9) is valid not only for Bloch electrons in infinite crystals but it also applies to disordered and finite systems as well.

In the present paper, we restrict to multiband Bloch electrons in infinite crystals. In that situation, the trace operator is rewritten $\mathrm{Tr}(\bullet) = \sum_{\boldsymbol{k}} \mathrm{tr}(\bullet) = S\int \frac{d^2k}{4\pi^2}\mathrm{tr}(\bullet)$ where the integration is performed over the first Brillouin zone (BZ) and $\mathrm{tr}(\bullet)$ is the partial trace operator on the band index. Accordingly, the susceptibility formula follows as

$$\chi_{\mathrm{orb}} = -\frac{\mu_0 e^2}{12}\frac{\Im m}{\pi}\int_{-\infty}^{+\infty} d\omega \int \frac{d^2k}{4\pi^2}\, n_{\mathrm{F}}(\omega)[U_{\boldsymbol{k}}(\omega) - 4V_{\boldsymbol{k}}(\omega)], \tag{10}$$

where

$$\begin{aligned} U_{\boldsymbol{k}}(\omega) &= \mathrm{tr}\left\{(gh^{xx}gh^{yy} - gh^{xy}gh^{xy})_{\boldsymbol{k}}\right\}, \\ V_{\boldsymbol{k}}(\omega) &= \mathrm{tr}\left\{(gh^{x}gh^{x}gh^{y}gh^{y} - gh^{x}gh^{y}gh^{x}gh^{y})_{\boldsymbol{k}}\right\}, \end{aligned} \tag{11}$$

with $g_{\boldsymbol{k}}(\omega) = (\omega\mathrm{I} - h_{\boldsymbol{k}})^{-1}$ the Green's function matrix associated to the zero-field Bloch Hamiltonian matrix and $h^j_{\boldsymbol{k}} = \frac{\partial h_{\boldsymbol{k}}}{\partial k_j}$, $h^{ij}_{\boldsymbol{k}} = \frac{\partial^2 h_{\boldsymbol{k}}}{\partial k_i \partial k_j}$ with $(i,j) \in (x,y)$. As detailed in the appendix, for the class of models described by Eq.(2), by defining the two components vectors ($\vec{x} \equiv (x_1, x_2)$)

$$\vec{f_{\boldsymbol{k}}} = (\Re e f_{\boldsymbol{k}}, \Im m f_{\boldsymbol{k}}), \quad \vec{f^i_{\boldsymbol{k}}} = (\frac{\partial \Re e f_{\boldsymbol{k}}}{\partial k_i}, \frac{\partial \Im m f_{\boldsymbol{k}}}{\partial k_i}), \quad \vec{f^{ij}_{\boldsymbol{k}}} = (\frac{\partial^2 \Re e f_{\boldsymbol{k}}}{\partial k_i \partial k_j}, \frac{\partial^2 \Im m f_{\boldsymbol{k}}}{\partial k_i \partial k_j}), \tag{12}$$

and the quantities

$$u^j_{\boldsymbol{k}} = \frac{\vec{f^j_{\boldsymbol{k}}} \cdot \vec{f_{\boldsymbol{k}}}}{|f_{\boldsymbol{k}}|}, \quad v^j_{\boldsymbol{k}} = \frac{(\vec{f^j_{\boldsymbol{k}}} \times \vec{f_{\boldsymbol{k}}})}{|f_{\boldsymbol{k}}|}, \quad u^{ij}_{\boldsymbol{k}} = \frac{\vec{f^{ij}_{\boldsymbol{k}}} \cdot \vec{f_{\boldsymbol{k}}}}{|f_{\boldsymbol{k}}|}, \quad v^{ij}_{\boldsymbol{k}} = \frac{(\vec{f^{ij}_{\boldsymbol{k}}} \times \vec{f_{\boldsymbol{k}}})}{|f_{\boldsymbol{k}}|}, \tag{13}$$

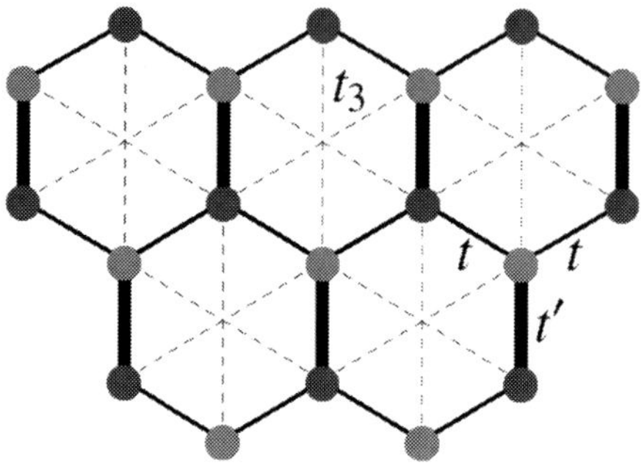

Figure 3. (color online) Honeycomb lattice model that determines the form $f_{\boldsymbol{k}}$: t (full lines) and t_3 (dashed lines) respectively first and third nearest-neighbor hopping amplitudes. The two sites in the unit cell are called A and B and are shown as blue and red dots. In a uniaxially compressed honeycomb lattice, there are two different values for the nearest-neighbor hopping amplitudes: t for thin (non–vertical) lines and $t' \geq t$ for thick (vertical) lines.

it is possible to obtain the following compact expressions for $U_{\boldsymbol{k}}(\omega)$ and $V_{\boldsymbol{k}}(\omega)$:

$$\begin{aligned} U_{\boldsymbol{k}}(\omega) &= 2g_+g_-[(u^{xx}_{\boldsymbol{k}}u^{yy}_{\boldsymbol{k}} - u^{xy}_{\boldsymbol{k}}u^{xy}_{\boldsymbol{k}}) + (v^{xx}_{\boldsymbol{k}}v^{yy}_{\boldsymbol{k}} - v^{xy}_{\boldsymbol{k}}v^{xy}_{\boldsymbol{k}})] + 4g_+^2g_-^2(u^{xx}_{\boldsymbol{k}}u^{yy}_{\boldsymbol{k}} - u^{xy}_{\boldsymbol{k}}u^{xy}_{\boldsymbol{k}})|f_{\boldsymbol{k}}|^2, \\ V_{\boldsymbol{k}}(\omega) &= (u^x_{\boldsymbol{k}}v^y_{\boldsymbol{k}} - u^y_{\boldsymbol{k}}v^x_{\boldsymbol{k}})^2\left[g_+^2g_-^2[1+3\left(\tfrac{1-\alpha^2}{1+\alpha^2}\right)^2] + 4g_+^3g_-^3|f_{\boldsymbol{k}}|^2\right]. \end{aligned} \tag{14}$$

where $g_\pm(\omega) = 1/(\omega - \epsilon_{\pm,\boldsymbol{k}})$. Interestingly, these last expressions show that $U_{\boldsymbol{k}}(\omega)$ and $V_{\boldsymbol{k}}(\omega)$ have poles only on the two dispersive bands as if the flat band did not play any role. In fact, the implicit effect of the flat band is essentially encoded in the α dependent term appearing in $V_{\boldsymbol{k}}(\omega)$. However it is quite remarkable that $U_{\boldsymbol{k}}(\omega)$ is totally independent of α.

4. Orbital susceptibility of low-energy α-$\mathcal{T}_3$ models

In this part of the paper we use equations (13,14) to calculate the susceptibility of α-$\mathcal{T}_3$ model for three different $f_{\boldsymbol{k}}$. In order to allow for a full analytical calculation, for each example we consider the low-energy effective model Hamiltonian that correctly describes the energy spectrum near the minimum of the corresponding $|f_{\boldsymbol{k}}|$. The three distinct forms of $f_{\boldsymbol{k}}$ that we consider are defined on the honeycomb lattice model illustrated on Fig. 3, that corresponds to the $\alpha = 0$ limit in which the C sites are completely decoupled from A, B sites. The C sites are not shown on Fig. 3 for simplicity.

4.1. α-$\mathcal{T}_3$ graphene model

We first consider the usual isotropic model of graphene which corresponds to $t' = t$ and $t_3 = 0$ in Fig. 3. In this situation,

$$f_{\boldsymbol{k}} = t(e^{-i\boldsymbol{k}.\boldsymbol{\delta}_1} + e^{-i\boldsymbol{k}.\boldsymbol{\delta}_2} + e^{-i\boldsymbol{k}.\boldsymbol{\delta}_3}) = t(e^{-ik_ya} + 2e^{ik_ya/2}\cos(\sqrt{3}k_xa/2)) \tag{15}$$

For $\Delta = 0$, the corresponding α-$\mathcal{T}_3$ energy spectrum (see Fig. 1(b)) exhibits linear band touching at the two inequivalent corners $\pm\boldsymbol{K}$ of the Brillouin zone. Introducing the valley index $\xi = \pm$, the effective low-energy model is obtained by expanding $f_{\boldsymbol{k}}$ near $\xi\boldsymbol{K}$: $f_{\xi\boldsymbol{K}+\boldsymbol{k}} \simeq v(\xi k_x - ik_y)$ with the velocity $v = \frac{3ta}{2}$ where a is the nearest-neighbor distance. Hereafter to simplify the notations of most equations we define the pseudo wavevector $\kappa_{x,y} = v k_{x,y}$. For each valley, the low-energy α-$\mathcal{T}_3$ Hamiltonian reads:

$$h_{\boldsymbol{k}} = \begin{pmatrix} \Delta & c_\alpha(\xi\kappa_x - i\kappa_y) & 0 \\ c_\alpha(\xi\kappa_x + i\kappa_y) & -\Delta & s_\alpha(\xi\kappa_x - i\kappa_y) \\ 0 & s_\alpha(\xi\kappa_x + i\kappa_y) & \Delta \end{pmatrix}. \tag{16}$$

It is then immediate to obtain

$$\vec{f} = (\xi\kappa_x, -\kappa_y), \quad \vec{f^x} = (\xi, 0), \quad \vec{f^y} = (0, -1), \quad \vec{f^{ij}} = 0. \tag{17}$$

and

$$u^x = \xi v_y = \frac{\kappa_x}{\sqrt{\kappa_x^2+\kappa_y^2}}, \quad u^y = -\xi v_x = -\frac{\kappa_y}{\sqrt{\kappa_x^2+\kappa_y^2}}, \quad u^{ij} = v^{ij} = 0, \tag{18}$$

from which we deduce

$$\begin{aligned} U_{\boldsymbol{k}}(\omega) &= 0, \\ V_{\boldsymbol{k}}(\omega) &= \left[\frac{1}{(\omega^2-(\Delta^2+\kappa_x^2+\kappa_y^2))^2}[1+3\left(\frac{1-\alpha^2}{1+\alpha^2}\right)^2] + \frac{4(\kappa_x^2+\kappa_y^2)}{(\omega^2-(\Delta^2+\kappa_x^2+\kappa_y^2))^3}\right] \end{aligned} \tag{19}$$

Substituting these expressions in (10), summing over the two valleys and noting $\kappa_x = \varepsilon\cos\theta$, $\kappa_y = \varepsilon\sin\theta$, we obtain

$$\begin{aligned} \chi_{\rm orb}(\mu) &= \frac{2\mu_0 e^2}{12}\left(\frac{4}{4\pi^2}\right)\frac{\Im m}{\pi}\int_{-\infty}^{+\infty} \mathrm{d}\omega\, n_{\rm F}(\omega)\int_0^{2\pi}\mathrm{d}\theta \\ &\times \int_0^\infty \mathrm{d}\varepsilon\, \varepsilon\left[\frac{1}{(\omega^2-(\Delta^2+\varepsilon^2))^2}[1+3\left(\frac{1-\alpha^2}{1+\alpha^2}\right)^2] + \frac{4\varepsilon^2}{(\omega^2-(\Delta^2+\varepsilon^2))^3}\right]. \end{aligned} \tag{20}$$

Integrating over variables and following the procedure described in appendix B, we finally obtain

$$\chi_{\rm orb}(\mu=0) = \frac{\mu_0 e^2 v^2}{2\pi}\left[\left(\frac{1-\alpha^2}{1+\alpha^2}\right)^2 - \frac{1}{3}\right]\frac{n_F(\Delta)-n_F(-\Delta)}{2\Delta}. \tag{21}$$

which gives the correct behaviour for $\Delta = 0$. For $\alpha = 0$, Eq.(21) with $\Delta = 0$ recovers the famous diamagnetic peak originally found by McClure [12], whereas Eq.(21) with $\Delta \neq 0$ reproduces the in-gap diamagnetic *plateau* first obtained by Koshino and Ando [13]. Note however that these authors obtained their results by performing a low-field expansion of the grand potential calculated from direct summation over the effective Landau levels spectrum $\epsilon_n(B) = \pm\epsilon_B\sqrt{|n|}$ [12] or $\epsilon_n(B) = \pm\sqrt{\Delta^2+\epsilon_B^2|n|}$ [13] where $\epsilon_B = v\sqrt{2eB}$. For $\alpha \neq 0$, Eq.(21) is also coherent with the results of such methods, albeit now with α-dependent effective Landau levels spectrum $\epsilon_n(B) = \pm\epsilon_B\sqrt{|n+\gamma|}$ or $\epsilon_n(B) = \pm\sqrt{\Delta^2+\epsilon_B^2|n+\gamma|}$ where $n \in \mathbb{Z}$ and $\gamma = \frac{\alpha^2}{1+\alpha^2}$ [7]. In this picture, using the relation $\gamma = \frac{1}{2} - \Phi_B/2\pi$ where Φ_B is the α-dependent Berry phase of a Dirac cone [7, 10], the variation of the susceptibility with α is interpreted as a consequence of the variation of the corresponding Berry phase. As we already pointed out, for $\alpha = 1$ the Berry phase vanishes and so do interband effects encoded by the Berry curvature. Nevertheless, for $\alpha = 1$ there is still a paramagnetic susceptibility *plateau* inside the gap which is an indication that interband effects are still present. This result implies that some important interband effects are not encoded by the Berry curvature. More quantitatively we note that the α dependent prefactor of the susceptibility (21) changes sign at $\alpha_c = \frac{\sqrt{3}-1}{\sqrt{2}} \simeq 0.518$. This signals a transition from a diamagnetic peak/*plateau* for $\alpha < \alpha_c$ to a paramagnetic peak/*plateau* for $\alpha > \alpha_c$. This value $\alpha_c = 0.518$ coincides with the numerical results obtained from computing the susceptibility with the Hofstadter spectrum [7]. As a final remark, we note that for the low-energy Hamiltonian (16) which is linear in $\boldsymbol{k}$ and thus separable (i.e. $h_{\boldsymbol{k}}^{xy} = 0$), the Fukuyama formula [6] would have given the same results.

4.2. α-$\mathcal{T}_3$ semi-Dirac model

We now consider a model with $t' = 2t$ and $t_3 = 0$ (see Fig. 3). In that situation

$$f_{\boldsymbol{k}} = t(2e^{-ik_ya} + 2e^{ik_ya/2}\cos(\sqrt{3}k_xa/2)) \tag{22}$$

For $\Delta = 0$, the corresponding α-$\mathcal{T}_3$ energy spectrum (see Fig. 4) exhibits a semi-Dirac band touching at one of the M points of the Brillouin zone [14]. The low-energy model obtained by expanding near this point reads $f_{M+\boldsymbol{k}} \simeq \frac{k_x^2}{2m_*} - ivk_y$ with the effective mass $m_* = \frac{1}{ta^2}$ velocity $v = \frac{ta}{\hbar}$ and it features a linear-quadratic spectrum $\varepsilon_{\boldsymbol{k}} = \pm\sqrt{(\frac{k_x^2}{2m_*})^2 + (vk_y)^2}$. As in previous example, to simplify the notations of most equations we define the pseudo wavevectors $\kappa_x = \frac{k_x}{\sqrt{2m_*}}$ and $\kappa_y = vk_y$. The low-energy α-$\mathcal{T}_3$ Hamiltonian is:

$$h_{\boldsymbol{k}} = \begin{pmatrix} \Delta & c_\alpha(\kappa_x^2 - i\kappa_y) & 0 \\ c_\alpha(\kappa_x^2 + i\kappa_y) & -\Delta & s_\alpha(\kappa_x^2 - i\kappa_y) \\ 0 & s_\alpha(\kappa_x^2 + i\kappa_y) & \Delta \end{pmatrix}, \tag{23}$$

such that

$$\vec{f} = (\kappa_x^2, -\kappa_y), \quad \vec{f}^x = 2(\kappa_x, 0), \quad \vec{f}^y = (0, -1) \tag{24}$$

and

$$u^x = \frac{2\kappa_x^3}{\sqrt{\kappa_x^4+\kappa_y^2}}, \quad u^y = \frac{\kappa_y}{\sqrt{\kappa_x^4+\kappa_y^2}}, \quad v^x = \frac{-2\kappa_x\kappa_y}{\sqrt{\kappa_x^4+\kappa_y^2}}, \quad v^y = \frac{\kappa_x^2}{\sqrt{\kappa_x^4+\kappa_y^2}}, \tag{25}$$

from which we deduce

$$\begin{aligned} U_{\boldsymbol{k}}(\omega) &= 0, \\ V_{\boldsymbol{k}}(\omega) &= 4\kappa_x^2\left[\frac{1}{(\omega^2-(\Delta^2+\kappa_x^4+\kappa_y^2))^2}[1+3\left(\frac{1-\alpha^2}{1+\alpha^2}\right)^2] + \frac{4(\kappa_x^4+\kappa_y^2)}{(\omega^2-(\Delta^2+\kappa_x^4+\kappa_y^2))^3}\right]. \end{aligned} \tag{26}$$

Writing now $\kappa_x^2 = \varepsilon\cos\theta$ and $\kappa_y = \varepsilon\sin\theta$ with $\theta \in [0, \pi/2]$ we deduce

$$\begin{aligned} \chi_{\mathrm{orb}}(\mu) &= \frac{\mu_0 e^2}{12}\left(\frac{64}{4\pi^2}\right)\frac{3m}{\pi}\int_{-\infty}^{+\infty} \mathrm{d}\omega\; n_{\mathrm{F}}(\omega)\int_0^{\pi/2}\mathrm{d}\theta\sqrt{\cos\theta} \\ &\times \int_0^\infty \mathrm{d}\varepsilon\; \varepsilon^{3/2}\left[\frac{1}{(\omega^2-(\Delta^2+\varepsilon^2))^2}\left[1+3\left(\frac{1-\alpha^2}{1+\alpha^2}\right)^2\right] + \frac{4\varepsilon^2}{(\omega^2-(\Delta^2+\varepsilon^2))^3}\right] \end{aligned} \tag{27}$$

Quite interestingly, apart from some prefactor, the main noticable change between equations (20) and (27) is the effective *density of states* that appears in the integral: for the α-$\mathcal{T}_3$ semi-Dirac model it is $\propto \varepsilon^{3/2}$ whereas it is $\propto \varepsilon$ in the α-$\mathcal{T}_3$ graphene model. From formulae (B.4,B.9) given in appendix B, we finally obtain:

$$\chi_{\mathrm{orb}}(\mu) = -\frac{\mu_0 e^2 v}{\sqrt{2m_*}}\frac{\Gamma(\frac{3}{4})^2}{\pi^2\sqrt{2\pi}}\left[\left(\frac{1-\alpha^2}{1+\alpha^2}\right)^2 - \frac{1}{2}\right] \times \begin{cases} \frac{1}{\sqrt{|\mu|}} & \Delta = 0, T = 0 \\ \left(\frac{4\Gamma(\frac{5}{4})^2}{\sqrt{\pi}}\right)\frac{1}{\sqrt{\Delta}} & |\mu| < \Delta, T = 0. \end{cases} \tag{28}$$

For $\alpha = 0$ and $\Delta = 0$, the first expression coincides with the result recently obtained in [1] for a two band model at the semi-Dirac point. For any α it predicts a diamagnetic peak that scales like $1/\sqrt{\min(\mu, T)}$ in the gapless case $\Delta = 0$. In the presence of a gap, the second expression shows that the susceptibility *plateau* in the gap scales as $1/\sqrt{\Delta}$. As in the previous example, Eq.(28) predicts a diamagnetic to paramagnetic transition for $\alpha \geq \alpha_c$ at a critical value $\alpha_c = \sqrt{2} - 1 \simeq 0.414$ which is smaller than in then previous example. We have verified that this value $\alpha_c = 0.414$ agrees with the numerical results obtained from computing the susceptibility

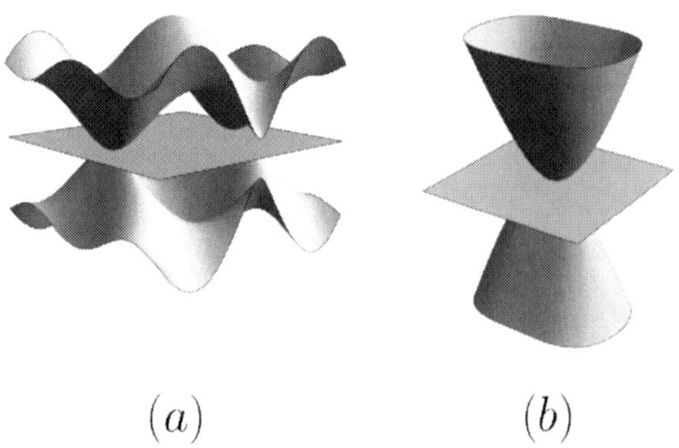

Figure 4. (Color online). α-$\mathcal{T}_3$ semi-Dirac model energy bands dispersion in $\boldsymbol{k}$ space, for $\Delta = 0$. (a) Full spectrum with semi-Dirac band touching of the two dispersive bands. (b) low-energy spectrum.

with the Hofstadter spectrum. We stress that for the α-$\mathcal{T}_3$ semi-Dirac model, as yet, there is no exact analytical expression for the Landau levels $\epsilon_n(B)$ of the low-energy model. The Landau levels are related to the eigenvalues of a modified quartic oscillator [14]. However by computing the Hofstadter spectrum of this α-$\mathcal{T}_3$ semi-Dirac model for small magnetic field and various α, we have observed that the effective Landau levels $\epsilon_n(B)$ are well described by the approximate form $\epsilon_n \propto [(n+1/2)B]^{2/3}$, first obtained in [14]. Interestingly, for $n \geq 3$ the numerical Landau levels $\epsilon_n(B)$ appear to be almost independent of α. This observation implies that the dia- to paramagnetic transition is essentially driven by the variation of the $n = 0, 1, 2$ Landau levels. We also stress that for this semi-Dirac model, there is no Berry phase around the semi-Dirac point. However the Berry curvature is non-zero (for $\alpha \neq 1$) and it exhibits a double-peak structure near the semi-Dirac point. There is still a lack of a clear physical picture of the origin of the dia to paramagnetic transition for this α-$\mathcal{T}_3$ semi-Dirac model. As a final remark, we note that since the low-energy Hamiltonian (23) is still separable (i.e. $h^{xy}_{\boldsymbol{k}} = 0$), the Fukuyama formula [6] would have given the same result.

4.3. α-$\mathcal{T}_3$ pseudo-bilayer model

The last model we consider has parameters $t' = t$ and $t_3 = 1/2$ in Fig. 5 [15]. In that situation,

$$f_{\boldsymbol{k}} = t(e^{-ik_y a} + 2e^{ik_y a/2}\cos(\sqrt{3}k_x a/2)) + \frac{t}{2}(e^{2ik_y a} + 2e^{-ik_y a}\cos(\sqrt{3}k_x a)) \tag{29}$$

For $\Delta = 0$, the corresponding α-$\mathcal{T}_3$ energy spectrum (see Fig. 5) exhibits quadratic band touching at $\pm\boldsymbol{K}$ similar to a bilayer graphene. The low-energy model obtained by expanding near $\xi\boldsymbol{K}$ reads $f_{\xi\boldsymbol{K}+\boldsymbol{k}} \simeq \frac{1}{2m_*}(\xi k_x - ik_y)^2$ with the effective mass $m_* = \frac{2}{ta^2}$ and the energy spectrum $\varepsilon_{\boldsymbol{k}} = \frac{\boldsymbol{k}^2}{2m_*}$. As in previous sections, to simplify the notations of most equations, we define the pseudo wavevectors $\kappa_{x,y} = \frac{k_{x,y}}{\sqrt{2m_*}}$.

The low-energy α-$\mathcal{T}_3$ Hamiltonian becomes:

$$h_{\boldsymbol{k}} = \begin{pmatrix} \Delta & c_\alpha(\xi\kappa_x - i\kappa_y)^2 & 0 \\ c_\alpha(\xi\kappa_x + i\kappa_y)^2 & -\Delta & s_\alpha(\xi\kappa_x - i\kappa_y)^2 \\ 0 & s_\alpha(\xi\kappa_x + i\kappa_y)^2 & \Delta \end{pmatrix}. \tag{30}$$

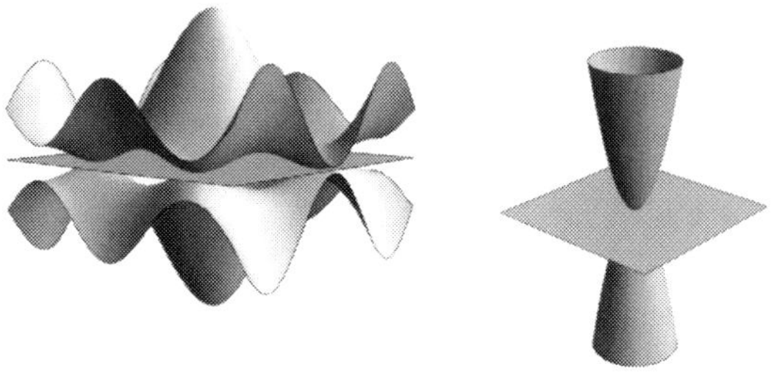

Figure 5. (Color online). α-$\mathcal{T}_3$ *pseudo-bilayer* model energy bands dispersion in $\boldsymbol{k}$ space, for $\Delta = 0$. (a) Full energy spectrum with quadratic band touching. (b) low-energy spectrum.

This effective Hamiltonian is now quadratic in $\kappa_{x,y}$ and it is not separable (i.e. $h_{\boldsymbol{k}}^{x,y} \neq 0$). As a consequence, we do not expect the Fukuyama formula to give the correct result in that situation. Following similar steps as in previous examples we obtain

$$\begin{array}{lll} \vec{f} = (\kappa_x^2 - \kappa_y^2, -2\xi\kappa_x\kappa_y), & \vec{f}^x = 2(\kappa_x, -\xi\kappa_y), & \vec{f}^y = 2(-\kappa_y, -\xi\kappa_x), \\ \vec{f}^{xx} = 2(1,0), & \vec{f}^{yy} = 2(-1,0), & \vec{f}^{xy} = 2(0,-\xi), \end{array} \tag{31}$$

such that by noting $\kappa_x = \sqrt{\varepsilon}\cos\theta$, $\kappa_y = \sqrt{\varepsilon}\sin\theta$ we deduce

$$\begin{array}{l} U_{\boldsymbol{k}}(\omega) = -16[\frac{1}{(\omega^2-(\Delta^2+\varepsilon^2))} + \frac{\varepsilon^2}{(\omega^2-(\Delta^2+\varepsilon^2))^2}], \\ V_{\boldsymbol{k}}(\omega) = 16\varepsilon^2 \left[\frac{1}{(\omega^2-(\Delta^2+\varepsilon^2))^2}[1+3\left(\frac{1-\alpha^2}{1+\alpha^2}\right)^2] + \frac{4\varepsilon^2}{(\omega^2-(\Delta^2+\varepsilon^2))^3}\right]. \end{array} \tag{32}$$

Substituting these expressions in (10), summing over the two valley, integrating over θ we find

$$\begin{array}{rl} \chi_{\mathrm{orb}}(\mu) & = \frac{\mu_0 e^2}{12}\left(\frac{8}{\pi}\right)\frac{\Im m}{\pi}\int_{-\infty}^{+\infty} \mathrm{d}\omega\; n_{\mathrm{F}}(\omega) \\ & \times \int_0^\infty \mathrm{d}\varepsilon \left[\frac{\varepsilon^2}{(\omega^2-(\Delta^2+\varepsilon^2))^2}[5+12\left(\frac{1-\alpha^2}{1+\alpha^2}\right)^2] + \frac{16\varepsilon^4}{(\omega^2-(\Delta^2+\varepsilon^2))^3} + \frac{1}{(\omega^2-(\Delta^2+\varepsilon^2))}\right]. \end{array} \tag{33}$$

From Eqs.(B.4,B.9) given in appendix B, we finally obtain:

$$\chi_{\mathrm{orb}}(\mu) = -\frac{\mu_0 e^2}{\pi m_*} \times \begin{cases} [\left(\frac{1-\alpha^2}{1+\alpha^2}\right)^2 - \frac{3}{4}]\ln\frac{\epsilon_c}{|\mu|} - [\left(\frac{1-\alpha^2}{1+\alpha^2}\right)^2 - \frac{11}{12}] & \Delta = 0, |\mu| < \epsilon_c, T = 0 \\ [\left(\frac{1-\alpha^2}{1+\alpha^2}\right)^2 - \frac{3}{4}]\ln\frac{2\epsilon_c}{\Delta} - [\left(\frac{1-\alpha^2}{1+\alpha^2}\right)^2 - \frac{11}{12}] & |\mu| < \Delta, T = 0 \end{cases} \tag{34}$$

where it is necessary to introduce a cutoff scale ϵ_c [16, 17]. As in previous examples, there is a diamagnetic to paramagnetic transition of the logarithmic term at a critical value $\alpha_c = \sqrt{4} - \sqrt{3} = 0.263$, whereas there is a paramagnetic to diamagnetic transition of the second term at $\alpha_c = \sqrt{12} - \sqrt{11} = 0.147$. For $\alpha = 0$, these expressions coincide with previous results obtained in [16, 17]. In [17], the susceptibility was derived from using the analytical form of the effective Landau levels $\epsilon_n = \epsilon_B\sqrt{n(n-1)}$ ($\epsilon_B = \frac{eB}{m_*}$) that are associated to the low-energy model. A similar derivation for any α seems difficult to achieve even if the Landau level sequence is known. In that perspective, we note that for the gapless α-$\mathcal{T}_3$ *pseudo-bilayer* model it is possible to derive a topological Berry phase that varies with α; this already gives some insight on the modified semiclassical Landau levels spectrum [10]. However as noted in [10], for the bilayer system the semiclassical Landau levels spectrum does not fully agree with the exact quantum Landau levels spectrum and therefore we do not expect the semiclassical Landau levels spectrum to be sufficient to describe correctly the susceptibility –especially the singular behaviour of the susceptibility near half filling $\mu = 0$.

5. Summary

In this work we have studied the importance of interband effects on the orbital susceptibility of three bands α-$\mathcal{T}_3$ tight-binding models that were initiated in [7]. The particularity of the α-$\mathcal{T}_3$ tight-binding models is that the coupling between the three energy bands (which is encoded in the wavefunctions properties) can be tuned (by a parameter α) without any modification of the energy spectrum. To highlight the role of these interbands effects, we have examined the orbital susceptibility of these models. Using the gauge-invariant perturbative formalism that we have recently developped [1], we obtained the generic formula of the orbital susceptibility of the α-$\mathcal{T}_3$ tight-binding models. More quantitatively we have calculated the orbital susceptibility of three distinct α-$\mathcal{T}_3$ tight-binding models: α-$\mathcal{T}_3$ graphene model, the α-$\mathcal{T}_3$ semi-Dirac model and the α-$\mathcal{T}_3$ *pseudo-bilayer* model. To obtain an analytical form of the susceptibility, we have only considered a low-energy Hamiltonian of these models; it correctly describes the energy spectrum near the band touching where it is expected that interband effects are the strongest. The main result of our study is that for each of these models, by varying the parameter α and thus the wavefunctions interband coupling, it is possible to drive a transition from a diamagnetic to a paramagnetic behavior, both in absence or presence of a gap separating the dispersive bands. In particular we emphasize that the in-gap susceptibility does not need to be diamagnetic. Moreover the existence of a finite in-gap (paramagnetic) susceptibility at $\alpha = 1$ (for each model) provides hints that some important interband effects are not encoded in the Berry curvature which vanishes for $\alpha = 1$.

Appendix A. Determination of $U_{\boldsymbol{k}}(\omega)$ and $V_{\boldsymbol{k}}(\omega)$

The aim of this section is to give the key step to find the expressions $U_{\boldsymbol{k}}(\omega)$ and $V_{\boldsymbol{k}}(\omega)$ of Eq.(14). We first slightly rewrite their definitions Eq.(11):

$$\begin{aligned} U_{\boldsymbol{k}}(\omega) &= \mathrm{tr}\left\{(gh^{xx}gh^{yy} - gh^{xy}gh^{xy})_{\boldsymbol{k}}\right\}, \\ V_{\boldsymbol{k}}(\omega) &= \mathrm{tr}\left\{(gh^{x}gh^{y}[gh^{y}, gh^{x}])_{\boldsymbol{k}}\right\}, \end{aligned} \tag{A.1}$$

where $g_{\boldsymbol{k}}(\omega) = (\omega \mathrm{I} - h_{\boldsymbol{k}})^{-1}$ it the Green's function matrix associated to the zero-field Bloch Hamiltonian matrix Eq.(2), $h^i_{\boldsymbol{k}} = \frac{\partial h_{\boldsymbol{k}}}{\partial k_i}$ and $h^{ij}_{\boldsymbol{k}} = \frac{\partial^2 h_{\boldsymbol{k}}}{\partial k_i \partial k_j}$ with $(i, j) = (x, y)$. The calculation of $U_{\boldsymbol{k}}, V_{\boldsymbol{k}}$ is made simple by remarking that the matrices h, g, h^j, h^{ij} can each be written in terms of the following three matrices only:

$$S_\pm = \begin{pmatrix} 0 & \pm c_\alpha e^{-i\theta_{\boldsymbol{k}}} & 0 \\ c_\alpha e^{i\theta_{\boldsymbol{k}}} & 0 & \pm s_\alpha e^{-i\theta_{\boldsymbol{k}}} \\ 0 & s_\alpha e^{i\theta_{\boldsymbol{k}}} & 0 \end{pmatrix} \; ; \; S_0 = \begin{pmatrix} 1 & 0 & 0 \\ 0 & -1 & 0 \\ 0 & 0 & 1 \end{pmatrix}. \tag{A.2}$$

Note that the two matrices $S_\pm$ depend on both parameter α and wavevector $\boldsymbol{k}$; moreover we stress that S_- is an antihermitian matrix whereas $S_{+,0}$ are hermitian. With these definitions, it is immediate to verify the identities

$$\begin{aligned} h &= \Delta S_0 + |f_{\boldsymbol{k}}| S_+, \\ h^j &= u^j_{\boldsymbol{k}} S_+ + i v^j_{\boldsymbol{k}} S_-, \\ h^{ij} &= u^{ij}_{\boldsymbol{k}} S_+ + i v^{ij}_{\boldsymbol{k}} S_-, \end{aligned} \tag{A.3}$$

with

$$\begin{aligned} u^j_{\boldsymbol{k}} &= \frac{\vec{f}^j_{\boldsymbol{k}}\cdot\vec{f}_{\boldsymbol{k}}}{|f_{\boldsymbol{k}}|}, \quad v^j_{\boldsymbol{k}} = \frac{(\vec{f}_{\boldsymbol{k}}\times\vec{f}^j_{\boldsymbol{k}})}{|f_{\boldsymbol{k}}|}, \\ u^{ij}_{\boldsymbol{k}} &= \frac{\vec{f}^{ij}_{\boldsymbol{k}}\cdot\vec{f}_{\boldsymbol{k}}}{|f_{\boldsymbol{k}}|}, \quad v^{ij}_{\boldsymbol{k}} = \frac{(\vec{f}_{\boldsymbol{k}}\times\vec{f}^{ij}_{\boldsymbol{k}})}{|f_{\boldsymbol{k}}|}, \end{aligned} \tag{A.4}$$

where we have defined the two components vectors ($\vec{x} \equiv (x_1, x_2)$)

$$\begin{aligned} \vec{f}_{\boldsymbol{k}} &= (\Re e f_{\boldsymbol{k}}, \Im m f_{\boldsymbol{k}}), \\ \vec{f}^i_{\boldsymbol{k}} &= (\frac{\partial \Re e f_{\boldsymbol{k}}}{\partial k_i}, \frac{\partial \Im m f_{\boldsymbol{k}}}{\partial k_i}), \\ \vec{f}^{ij}_{\boldsymbol{k}} &= (\frac{\partial^2 \Re e f_{\boldsymbol{k}}}{\partial k_i \partial k_j}, \frac{\partial^2 \Im m f_{\boldsymbol{k}}}{\partial k_i \partial k_j}). \end{aligned} \tag{A.5}$$

To obtain a simple form for g we remark that the matrices $S_{\pm,0}$ appear to have very peculiar properties: $S^3_\pm = \pm S_\pm$, $S^2_0 = I$, $[S_\pm, S_0]_+ = 0$. From these properties we deduce that $g = (\omega\mathrm{I} - h_{\boldsymbol{k}})^{-1}$ can be written as $g = a_1 I + a_2 S_0 + a_3 S_+ + a_4 S^2_+$ where $a_{1,2,3,4}$ are determined from using the identity $(\omega - h)g = \mathrm{I}$. We then obtain:

$$g = g_+ g_- (\omega\mathrm{I} - \Delta S_0 - |f_{\boldsymbol{k}}| S_+) - g_+ g_- g_0 |f_{\boldsymbol{k}}|^2 (\mathrm{I} - S^2_+), \tag{A.6}$$

where $g_{\pm\boldsymbol{k}}(\omega) = \frac{1}{\omega - \epsilon_{\pm,\boldsymbol{k}}}$ and $g_0(\omega) = \frac{1}{\omega - \epsilon_0}$. Note that $g(\omega)$ has three poles corresponding to the three bands. Substituting identities Eqs.(A.3,A.6) in Eq.(A.1), we obtain the expressions Eq.(14):

$$\begin{aligned} U_{\boldsymbol{k}}(\omega) &= 2g_+g_-[(u^{xx}_{\boldsymbol{k}}u^{yy}_{\boldsymbol{k}} - u^{xy}_{\boldsymbol{k}}u^{xy}_{\boldsymbol{k}}) + (v^{xx}_{\boldsymbol{k}}v^{yy}_{\boldsymbol{k}} - v^{xy}_{\boldsymbol{k}}v^{xy}_{\boldsymbol{k}})] + 4g^2_+g^2_-(u^{xx}_{\boldsymbol{k}}u^{yy}_{\boldsymbol{k}} - u^{xy}_{\boldsymbol{k}}u^{xy}_{\boldsymbol{k}})|f_{\boldsymbol{k}}|^2, \\ V_{\boldsymbol{k}}(\omega) &= (u^x_{\boldsymbol{k}}v^y_{\boldsymbol{k}} - u^y_{\boldsymbol{k}}v^x_{\boldsymbol{k}})^2 \left[g^2_+g^2_-[1 + 3\left(\frac{1-\alpha^2}{1+\alpha^2}\right)^2] + 4g^3_+g^3_-|f_{\boldsymbol{k}}|^2\right]. \end{aligned} \tag{A.7}$$

Appendix B. Pole decomposition and integration over variables ε and ω

The generic form of the susceptibility of the three different α-$\mathcal{T}_3$ models is:

$$\begin{aligned} \chi_{\mathrm{orb}} &\propto \frac{\Im m}{\pi} \int d\omega\, n_{\mathrm{F}}(\omega) \int_0^\infty d\varepsilon\, \nu(\varepsilon) \left[\frac{A\varepsilon^4}{(\omega^2-(\Delta^2+\varepsilon^2))^3} + \frac{B\varepsilon^2}{(\omega^2-(\Delta^2+\varepsilon^2))^2} + \frac{C}{(\omega^2-(\Delta^2+\varepsilon^2))}\right] \\ &\propto \frac{\Im m}{\pi} \int d\omega\, n_{\mathrm{F}}(\omega) \int_0^\infty d\varepsilon\, \nu(\varepsilon) \left[A\varepsilon^4(g_+g_-)^3 + B\varepsilon^2(g_+g_-)^2 + C(g_+g_-)\right] \end{aligned} \tag{B.1}$$

where $g_s(\omega) = 1/(\omega - \varepsilon_s)$ and with $\varepsilon_s = s\sqrt{\Delta^2 + \varepsilon^2}$ $(s = \pm)$. The effective density of states $\nu(\varepsilon)$ and the parameters A, B, C of the three models (α-graphene, α-semi-Dirac, α-bilayer) are summarized in table B1.

	α-graphene	α-semi-Dirac	α-bilayer
$\nu(\varepsilon)$	$1/\varepsilon$	$1/\sqrt{\varepsilon}$	1
A	4	4	16
B	$1 + 3\left(\frac{1-\alpha^2}{1+\alpha^2}\right)^2$	$1 + 3\left(\frac{1-\alpha^2}{1+\alpha^2}\right)^2$	$5 + 12\left(\frac{1-\alpha^2}{1+\alpha^2}\right)^2$
C	0	0	1

Table B1. Effective density of states $\nu(\varepsilon)$ and parameters A, B, C of the three considered models: α-graphene, α-semi-Dirac and α-bilayer.

We now describe the key steps to perform the explicit integration over variables ε and ω. The first step consists to separate the poles at ε_s:

$$\chi_{\rm orb} \propto \sum_{s=\pm} \frac{\Im m}{\pi} \int d\omega\ n_{\rm F}(\omega) \int_0^\infty d\varepsilon\ \nu(\varepsilon) \left[\frac{A}{8}\frac{\varepsilon^4}{\varepsilon_s^3} g_s^3 + \left(\frac{B}{4}\frac{\varepsilon^2}{\varepsilon_s^2} - 3\frac{A}{16}\frac{\varepsilon^4}{\varepsilon_s^4}\right) g_s^2 + \left(\frac{C}{2\varepsilon_s} - \frac{B}{4}\frac{\varepsilon^2}{\varepsilon_s^3} + 3\frac{A}{16}\frac{\varepsilon^4}{\varepsilon_s^5}\right)) g_s\right] \tag{B.2}$$

The next step consists to use the identity $\frac{\Im m}{\pi} \int d\omega\ n_{\rm F}(\omega) g_s^n(\omega) = -\frac{1}{(n-1)!} n_{\rm F}^{n-1}(\varepsilon_s)$ where $n_{\rm F}^{(n)}$ is the nth derivative; we can the rewrite:

$$\chi_{\rm orb} \propto -\sum_{s=\pm} \int_0^\infty d\varepsilon\ \nu(\varepsilon) \left[\frac{A}{16}\frac{\varepsilon^4}{\varepsilon_s^3} n_{\rm F}''(\varepsilon_s) + \left(\frac{B}{4}\frac{\varepsilon^2}{\varepsilon_s^2} - 3\frac{A}{16}\frac{\varepsilon^4}{\varepsilon_s^4}\right) n_{\rm F}'(\varepsilon_s) + \left(\frac{C}{2\varepsilon_s} - \frac{B}{4}\frac{\varepsilon^2}{\varepsilon_s^3} + 3\frac{A}{16}\frac{\varepsilon^4}{\varepsilon_s^5}\right) n_{\rm F}(\varepsilon_s)\right] \tag{B.3}$$

From this point, we separate the study of case $\Delta = 0$ from case $\Delta \neq 0$.

Gapless models, $\Delta = 0$;

For $\Delta = 0$, since $\varepsilon_s = s\varepsilon$ we first rewrite:

$$\chi_{\rm orb} \propto -\int_0^\infty d\varepsilon\ \nu(\varepsilon) \left[\frac{A}{16}\varepsilon(n_{\rm F}''(\varepsilon) - n_{\rm F}''(-\varepsilon)) + \left(\frac{B}{4} - 3\frac{A}{16}\right)(n_{\rm F}'(\varepsilon) + n_{\rm F}'(-\varepsilon)) + \left(\frac{C}{2} - \frac{B}{4} + 3\frac{A}{16}\right)\frac{n_{\rm F}(\varepsilon) - n_{\rm F}(-\varepsilon)}{\varepsilon}\right] \tag{B.4}$$

We now note that for the three considered cases, $\nu(\varepsilon) \sim \varepsilon^p$ with $-1 \leq p \leq 0$; this property permits an integration by part of the term proportionnal to n_F'' and n_F:

$$\begin{aligned}
\int_0^\infty d\varepsilon \varepsilon^{p+1}(n_{\rm F}''(\varepsilon) - n_{\rm F}''(-\varepsilon)) &= \left[\varepsilon^{p+1}(n_{\rm F}'(\varepsilon) + n_{\rm F}'(-\varepsilon))\right]_0^\infty - (p+1)\int_0^\infty d\varepsilon \varepsilon^p (n_{\rm F}'(\varepsilon) + n_{\rm F}'(-\varepsilon)), \\
\int_0^\infty d\varepsilon \varepsilon^{p-1}(n_{\rm F}(\varepsilon) - n_{\rm F}(-\varepsilon)) &= \left[\frac{\varepsilon^p}{p}(n_{\rm F}(\varepsilon) - n_{\rm F}(-\varepsilon))\right]_0^\infty - \int_0^\infty d\varepsilon \frac{\varepsilon^p}{p}(n_{\rm F}'(\varepsilon) + n_{\rm F}'(-\varepsilon)),
\end{aligned} \tag{B.5}$$

where the last line requires $p < 0$. Using these identities for the three considered cases $(p = -1, -1/2, 0)$ and using the parameters of table B1 we deduce:

$$\begin{aligned}
&\chi_{\rm orb} \propto \frac{3}{2}\left[\left(\frac{1-\alpha^2}{1+\alpha^2}\right)^2 - \frac{1}{3}\right] n_{\rm F}'(0), && \alpha-\text{graphene} \\
&\chi_{\rm orb} \propto \frac{3}{4}\left[\left(\frac{1-\alpha^2}{1+\alpha^2}\right)^2 - \frac{1}{2}\right] \int_0^\infty d\varepsilon \frac{n_{\rm F}'(\varepsilon) + n_{\rm F}'(-\varepsilon)}{\sqrt{\varepsilon}} && \alpha-\text{semi}-\text{Dirac} \\
&\chi_{\rm orb} \propto 3\left[\left[\left(\frac{1-\alpha^2}{1+\alpha^2}\right)^2 - \frac{3}{4}\right] \int_0^\infty d\varepsilon\ \frac{n_{\rm F}(\varepsilon) - n_{\rm F}(-\varepsilon)}{\varepsilon} + \left[\left(\frac{1-\alpha^2}{1+\alpha^2}\right)^2 - \frac{11}{12}\right]\right] && \alpha-\text{bilayer}
\end{aligned} \tag{B.6}$$

Gapped models, $\Delta \neq 0$:

For the gapped case, we can also perform some integration by part of terms proportionnal to n_F'' and n_F. More precisely using the identities $\frac{\partial n_F^n}{\partial \varepsilon} = \frac{\varepsilon}{\varepsilon_s} n_F^{n+1}$ and $\frac{\partial \varepsilon_s}{\partial \varepsilon} = \frac{\varepsilon}{\varepsilon_s}$ we first obtain

$$\begin{aligned}
\int_0^\infty d\varepsilon \frac{\varepsilon^{p+4}}{\varepsilon_s^3} n_F'' &= -\int_0^\infty d\varepsilon \frac{\varepsilon^{p+2}}{\varepsilon_s^2}\left(p + 3 - 2\frac{\varepsilon^2}{\varepsilon_s^2}\right) n_F', \\
\int_0^\infty d\varepsilon \frac{\varepsilon^{p+2}}{\varepsilon_s^3} n_F &= \left[-\frac{\varepsilon^{p+1}}{\varepsilon_s} n_F\right]_0^\infty + \int_0^\infty d\varepsilon \left[\frac{(p+1)\varepsilon^p}{\varepsilon_s} n_F + \frac{\varepsilon^{p+2}}{\varepsilon_s^2} n_F'\right], \\
\int_0^\infty d\varepsilon \frac{\varepsilon^{p+4}}{\varepsilon_s^5} n_F &= \frac{1}{3}\left[-\frac{\varepsilon^{p+3}}{\varepsilon_s^3} n_F\right]_0^\infty + \frac{1}{3}\int_0^\infty d\varepsilon \left[\frac{(p+3)\varepsilon^{p+2}}{\varepsilon_s^3} n_F + \frac{\varepsilon^{p+4}}{\varepsilon_s^4} n_F'\right], \\
&= \left[-\frac{\varepsilon^{p+3}}{\varepsilon_s^3} n_F\right]_0^\infty + \frac{p+3}{3}\left[-\frac{\varepsilon^{p+1}}{\varepsilon_s} n_F\right]_0^\infty + \frac{1}{3}\int_0^\infty d\varepsilon \left[\frac{(p+3)(p+1)\varepsilon^p}{\varepsilon_s} n_F + \frac{(p+3)\varepsilon^{p+2}}{\varepsilon_s^2} n_F' + \frac{\varepsilon^{p+4}}{\varepsilon_s^4} n_F'\right]
\end{aligned} \tag{B.7}$$

from which we deduce:

$$\begin{aligned}\chi_{\mathrm{orb}} \propto -\sum_{s=\pm} \quad & \left[\frac{A}{16}\left[-\frac{\varepsilon^{p+3}}{\varepsilon_s^3} n_F\right]_0^\infty + \left[\frac{B}{4} - \frac{A}{16}(p+3)\right]\left[\frac{\varepsilon^{p+1}}{\varepsilon_s} n_F\right]_0^\infty \right. \\ & \left. +\left(\frac{C}{2} - (p+1)\left[\frac{B}{4} - \frac{A}{16}(p+3)\right]\right) \int_0^\infty \mathrm{d}\varepsilon\, \frac{\varepsilon^p}{\varepsilon_s} n_{\mathrm{F}}(\varepsilon_s)\right].\end{aligned} \tag{B.8}$$

For the three considered cases ($p = -1, -1/2, 0$) with the parameters of table B1 we finally obtain:

$$\begin{aligned}
\chi_{\mathrm{orb}} &\propto \frac{3}{4}\left[\left(\frac{1-\alpha^2}{1+\alpha^2}\right)^2 - \frac{1}{3}\right]\frac{n_F(\Delta)-n_F(-\Delta)}{\Delta} && \alpha\text{-graphene}, \\
\chi_{\mathrm{orb}} &\propto \frac{3}{8}\left[\left(\frac{1-\alpha^2}{1+\alpha^2}\right)^2 - \frac{1}{2}\right]\int_\Delta^\infty \mathrm{d}\varepsilon\, \frac{n_{\mathrm{F}}(\sqrt{\Delta^2+\varepsilon^2})-n_{\mathrm{F}}(-\sqrt{\Delta^2+\varepsilon^2})}{\sqrt{\varepsilon}\sqrt{\Delta^2+\varepsilon^2}} && \alpha\text{-semi}-\text{Dirac} \\
\chi_{\mathrm{orb}} &\propto 3\left[\left[\left(\frac{1-\alpha^2}{1+\alpha^2}\right)^2 - \frac{3}{4}\right]\int_\Delta^\infty \mathrm{d}\varepsilon\, \frac{n_{\mathrm{F}}(\sqrt{\Delta^2+\varepsilon})-n_{\mathrm{F}}(-\sqrt{\Delta^2+\varepsilon})}{\sqrt{\Delta^2+\varepsilon^2}} - \left[\left(\frac{1-\alpha^2}{1+\alpha^2}\right)^2 - \frac{11}{12}\right]\right], && \alpha\text{-bilayer}
\end{aligned} \tag{B.9}$$

[1] Raoux A Piéchon F Fuchs J-N and Montambaux G 2014 *arXiv*:1411.5940, accepted in Phys. Rev. B.
[2] Landau L D 1930 *Z. Phys.* **64**, 629 .
[3] Peierls R 1933 *Z. Phys.* **80**, 763.
[4] Vignale G 1991 Phys. Rev. Lett **67**, 358.
[5] Fukuyama H and Kubo R 1970 *J. Phys. Soc. Jap.* **28**, 570.
[6] Fukuyama H 1970 *Phys. Lett.* **32A**, 111 ; Fukuyama H 1971 *Prog. Theor. Phys.* **45**, 704.
[7] Raoux A Morigi M Fuchs J-N Piéchon F and Montambaux G 2014 *Phys. Rev. Lett.* **112**, 026402.
[8] Sutherland B 1986 *Phys. Rev. B* **34**, 5208.
[9] Vidal J Mosseri R and Doucot B, 1998 *Phys. Rev. Lett.* **81**, 5888.
[10] Fuchs J-N Piéchon F Goerbig M-O and Montambaux G 2010 *Eur. Phys. J.* B **77** 351.
[11] Gómez-Santos G and Stauber T 2011 *Phys. Rev. Lett.* **106**, 045504.
[12] McClure J W 1956 *Phys. Rev.* **104**, 666.
[13] Koshino M and Ando T 2007 *Phys. Rev. B* **76**, 085425.
[14] Dietl P Piéchon F and Montambaux G 2008 *Phys. Rev. Lett.* **100**, 236405.
[15] Montambaux G 2012 *Eur. Phys. J.* B **85**, 375.
[16] Safran S A 1984 *Phys. Rev. B* **30**, 421.
[17] Koshino M 2011 *Orbital Magnetism of Graphenes*, in *Physics and Applications of Graphene - Theory*, Mikhailov S (Ed.), InTech.

Classification of crystalline topological semimetals with an application to Na_3Bi

Ching-Kai Chiu

Department of Physics and Astronomy, University of British Columbia, Vancouver, BC, Canada V6T 1Z1

Quantum Matter Institute, University of British Columbia, Vancouver BC, Canada V6T 1Z4

E-mail: chiu7@phas.ubc.ca

Andreas P. Schnyder

Max-Planck-Institut für Festkörperforschung, Heißenbergstrasse 1, D-70569 Stuttgart, Germany

E-mail: a.schnyder@fkf.mpg.de

Abstract. Topological phases can not only be protected by internal symmetries (e.g., time-reversal symmetry), but also by crystalline symmetries, such as reflection or rotation symmetry. Recently a complete topological classification of reflection symmetric insulators, superconductors, nodal semimetals, and nodal superconductors has been established. In this article, after a brief review of the classification of reflection-symmetry-protected semimetals and nodal superconductors, we discuss an example of a three-dimensional topological Dirac semimetal, which exhibits time-reversal symmetry as well as reflection and rotation symmetries. We compute the surface state spectrum of this Dirac semimetal and identify the crystal lattice symmetries that lead to the protection of the surface states. We discuss the implications of our findings for the stability of the Fermi arc surface states of the Dirac material Na_3Bi. Our analysis suggests that the Fermi arc of Na_3Bi is gapped except at time-reversal invariant surface momenta, which is in agreement with recent photoemission measurements.

1. Introduction

Finding new topological materials beyond the Bi-based topological insulators has become an important goal in solid state research. In the past few years, it has been realized that not only internal symmetries [1, 2, 3, 4, 5], such as time-reversal, but also crystal symmetries, for example crystal reflection or rotation, can lead to the protection of topological states [6, 7, 8, 9, 10, 11, 12, 13, 14, 15, 16, 17]. In fact, a complete classification of reflection-symmetry-protected insulators and fully gapped superconductors was recently presented in Refs. [15, 16, 17]. The topological character of these crystalline topological insulators and superconductors gives rise to gapless states at the boundary. Importantly, since the surface of a material has lower crystal symmetry than its bulk, not all surfaces of a crystalline topological insulator (or superconductor) exhibit gapless boundary states. Only those surfaces that are invariant under the crystal symmetry can support topological surface states. Experimentally, SnTe has been identified to be a crystalline topological insulator with Dirac cone surface states, that are protected against gap opening by reflection symmetry together with time-reversal symmetry [18, 19, 20, 21]. Other candidate materials for reflection-symmetry-protected topological insulators are the anti-perovskites Ca_3PbO and Sr_3PbO [22, 23, 24].

The concept of topological band theory can also be applied to nodal systems, such as semimetals and nodal superconductors. The stability of the Fermi surfaces (superconducting nodes) of these materials is ensured by the conservation of a topological invariant which is defined in terms of an integral along a contour enclosing (encircling) the Fermi point (Fermi line). Topological nodal systems can be protected by global symmetries [25, 26, 27, 28, 29, 30, 31, 32, 33] (i.e., internal symmetries, such as time-reversal) as well as crystal lattice symmetries, or a combination of the two [34, 35, 36, 37, 38]. The nontrivial topology of the electronic band structure of these gapless materials manifests itself at the boundary in terms of protected surface states. Depending on the symmetry properties and the dimension of the Fermi surface, these surface states form Dirac cones, flat bands, or Fermi arc states [30]. Recently, several materials have been experimentally identified to be crystal-symmetry-protected topological semimetals. Among them are the Dirac materials Cd_3As_2 [39, 40, 41, 42, 43, 44, 45] and Na_3Bi [46, 47, 48, 49], whose Dirac points are protected by rotation symmetry. The Fermi arc states of Na_3Bi have recently been observed in angle-resolved photoemission measurements [48, 49].

In this paper, we first present in Sec. 2 a short review of the topological classification of reflection symmetric semimetals and nodal superconductors [34]. For brevity, we focus on the case where the Fermi surfaces (superconducting nodes) are located within the reflection plane. Second, we discuss in Sec. 3 a four-band tight-binding model on the cubic lattice describing a three-dimensional Dirac semimetal, which is protected by C_4 rotation symmetry. We discuss the surface states of this topological semimetal and determine the symmetries that protect the gapless surface states. Our tight-binding model has similar symmetry properties and topological features as the Dirac semimetal Na_3Bi. Hence, from our topological analysis we can obtain information about the stability of the Fermi arc surface states of Na_3Bi, see Sec. 3.4.

2. Topological classification of reflection symmetric semimetals and nodal superconductors

Before discussing a concrete example of a crystalline topological semimetal, let us briefly review the topological classification of reflection symmetric semimetals and nodal superconductors [34]. The topological properties of reflection symmetric semimetals (nodal superconductors) can be classified in a similar manner as those of reflection symmetric insulators (fully gapped superconductors) [15, 16, 17]. The classification depends on the internal symmetry properties of the nodal system as well as on whether the reflection symmetry commutes or anticommutes with the internal symmetries. There are three fundamental internal symmetries, which act locally in position space, namely time-reversal symmetry (TRS), particle-hole symmetry (PHS), and chiral or sublattice symmetry (SLS). In momentum space, time-reversal and particle-hole symmetry act on the Bloch Hamiltonian $H(\mathbf{k})$ as

$$T^{-1}H(-\mathbf{k})T = +H(\mathbf{k}) \tag{1a}$$

and

$$C^{-1}H(-\mathbf{k})C = -H(\mathbf{k}), \tag{1b}$$

respectively, where $T = \mathcal{K}U_T$ and $C = \mathcal{K}U_C$ are antiunitary operators and $\mathcal{K}$ is the complex conjugation operator. Chiral symmetry is implemented in terms of a unitary matrix S which anti-commutes with the Hamiltonian, i.e., $SH(\mathbf{k}) + H(\mathbf{k})S = 0$.

Crystal reflection symmetry acts non-locally in position space. In the following, we consider a $d_{\rm BZ}$-dimensional Hamiltonian $H(\mathbf{k})$ which is symmetric under reflection in the first direction.

Hence, $H(\mathbf{k})$ satisfies

$$R^{-1}H(-k_1, \tilde{\mathbf{k}})R = H(k_1, \tilde{\mathbf{k}}), \tag{2}$$

where $\tilde{\mathbf{k}} = (k_2, k_3, \ldots, k_{d_{\rm BZ}})$ and R is a unitary reflection operator. Since a phase factor can be absorbed in the definition of the electron creation and annihilation operators, we can assume that R is Hermitian, i.e., $R^\dagger = R$. As a consequence, the eigenvalues of R are either $+1$ or -1 and all commutation and anti-commutation relations between R and the internal symmetries are defined unambiguously, i.e.,

$$TRT^{-1} = \eta_T R, \quad CRC^{-1} = \eta_C R \quad \text{and} \quad SRS^{-1} = \eta_S R, \tag{3}$$

where the three indices η_T, η_C, and η_S specify whether R commutes $(+1)$ or anti-commutes (-1) with the corresponding internal symmetry operator. For the symmetry classes AI, AII, AIII, C, and D these different possibilities are labeled in Table 1 as R_{η_T}, R_{η_C}, and R_{η_S}, respectively. For the symmetry classes that contain two internal symmetries (i.e., classes BDI, CI, CII, and DIII), the different (anti-)commutation relations are denoted by $R_{\eta_T \eta_C}$. Hence, reflection symmetries together with the three internal symmetries define a total of 27 symmetry classes, see Table 1.

The classification of reflection-symmetry-protected semimetals and nodal superconductors depends on the symmetry class and on the codimension of the Fermi surface (superconducting nodes) $p = d_{\rm BZ} - d_{\rm FS}$, where $d_{\rm BZ}$ denotes the spatial dimension (i.e., the dimension of the Brillouin zone) and $d_{\rm FS}$ is the dimension of the Fermi surface. Furthermore, the classification depends on how the Fermi surfaces (superconducting nodes) transform under the reflection symmetry and the internal symmetries. In general, one can distinguish three different situations: (i) Each Fermi surface is left invariant by both reflection and internal symmetries, (ii) Fermi surfaces are invariant under reflection symmetry, but are pairwise related to each other by the internal symmetries, and (iii) different Fermi surfaces are pairwise related to each other by both reflection and in internal symmetry operations. In case (i) and case (ii) the Fermi surfaces are located within the refection plane, whereas in case (iii) the Fermi surfaces are positioned away from the reflection plane. For brevity we focus here on only case (i) and case (ii). Case (iii) has been discussed extensively in Refs. [34] and [36].

2.1. Fermi surfaces at high-symmetry points within mirror plane

First we consider case (i), i.e., Fermi surfaces that are invariant under both reflection and internal symmetries. These Fermi surfaces are located within the reflection plane and at high-symmetry points in the Brillouin zone, that is, at time-reversal invariant momenta. For this case the classification of reflection symmetric Fermi points can be inferred from the classification of reflection-symmetry-protected insulators by means of a dimensional reduction procedure [34]. Namely, the surface states of reflection symmetric $d_{\rm BZ}$-dimensional topological insulators can be viewed as reflection-symmetry-protected Fermi points in $d_{\rm BZ}-1$ dimensions. From this it follows that the topological classification of reflection symmetric Fermi points at time-reversal invariant momenta with $d_{\rm FS} = 0$ is obtained from the classification of reflection symmetric insulators by the dimensional shift $d_{\rm BZ} \to d_{\rm BZ} - 1$, see Table 1. This logic also works for Femi surfaces with $d_{\rm FS} > 0$, if their stability is guaranteed by an $M\mathbb{Z}$ or $2M\mathbb{Z}$ topological number. However, $\mathbb{Z}_2$ and $M\mathbb{Z}_2$ topological numbers ensure only the stability of Femi points, i.e., Fermi surfaces with $d_{\rm FS} = 0$. Derivations based on Clifford analysis and K theory [17, 34] corroborate these findings. The classification of reflection-symmetry-protected Fermi surfaces at high symmetry points of the Brillouin zone is presented in Table 1, where the first row indicates the codimension p of the Fermi surface. We note that the classification is eight-fold periodic in p.

Table 1. Topological classification of reflection-symmetry-protected semimetals and nodal superconductors. The first and second rows specify the codimension $p = d_{\mathrm{BZ}} - d_{\mathrm{FS}}$ of the reflection symmetric Fermi surfaces (superconducting nodes) at high-symmetry points and away from high-symmetry points of the Brillouin zone, respectively. The third row shows the classification of reflection symmetric insulators and fully gapped superconductors [15, 16]. The first column indicates whether the reflection operator R commutes ("R_+", "R_{++}") or anti-commutes ("R_-", "R_{--}") with the global symmetries. "R_{+-}" and "R_{-+}" denote the case where R commutes with one of the global symmetries but anti-commutes with the other one. The second column lists the global symmetry classes (using the Cartan nomenclature [3]), which are distinguished by the presence or absence of time-reversal symmetry, particle-hole symmetry, and chiral symmetry.

Reflection	FS in mirror plane at high-sym. point	p=7	p=8	p=1	p=2	p=3	p=4	p=5	p=6
	FS in mirror plane off high-sym. point	p=1	p=2	p=3	p=4	p=5	p=6	p=7	p=8
	top. insul. and top. SC	d_{BZ}=8	d_{BZ}=1	d_{BZ}=2	d_{BZ}=3	d_{BZ}=4	d_{BZ}=5	d_{BZ}=6	d_{BZ}=7
R	A	0	$M\mathbb{Z}$	0	$M\mathbb{Z}$	0	$M\mathbb{Z}$	0	$M\mathbb{Z}$
R_+	AIII	$M\mathbb{Z}$	0	$M\mathbb{Z}$	0	$M\mathbb{Z}$	0	$M\mathbb{Z}$	0
R_-	AIII	0	$M\mathbb{Z}\oplus\mathbb{Z}$	0	$M\mathbb{Z}\oplus\mathbb{Z}$	0	$M\mathbb{Z}\oplus\mathbb{Z}$	0	$M\mathbb{Z}\oplus\mathbb{Z}$
	AI	$M\mathbb{Z}_2^{\mathrm{a,b}}$	$M\mathbb{Z}$	0	0	0	$2M\mathbb{Z}$	0	$M\mathbb{Z}_2^{\mathrm{a,b}}$
	BDI	$M\mathbb{Z}_2^{\mathrm{a,b}}$	$M\mathbb{Z}_2^{\mathrm{a,b}}$	$M\mathbb{Z}$	0	0	0	$2M\mathbb{Z}$	0
	D	0	$M\mathbb{Z}_2^{\mathrm{a,b}}$	$M\mathbb{Z}_2^{\mathrm{a,b}}$	$M\mathbb{Z}$	0	0	0	$2M\mathbb{Z}$
R_+, R_{++}	DIII	$2M\mathbb{Z}$	0	$M\mathbb{Z}_2^{\mathrm{a,b}}$	$M\mathbb{Z}_2^{\mathrm{a,b}}$	$M\mathbb{Z}$	0	0	0
	AII	0	$2M\mathbb{Z}$	0	$M\mathbb{Z}_2^{\mathrm{a,b}}$	$M\mathbb{Z}_2^{\mathrm{a,b}}$	$M\mathbb{Z}$	0	0
	CII	0	0	$2M\mathbb{Z}$	0	$M\mathbb{Z}_2^{\mathrm{a,b}}$	$M\mathbb{Z}_2^{\mathrm{a,b}}$	$M\mathbb{Z}$	0
	C	0	0	0	$2M\mathbb{Z}$	0	$M\mathbb{Z}_2^{\mathrm{a,b}}$	$M\mathbb{Z}_2^{\mathrm{a,b}}$	$M\mathbb{Z}$
	CI	$M\mathbb{Z}$	0	0	0	$2M\mathbb{Z}$	0	$M\mathbb{Z}_2^{\mathrm{a,b}}$	$M\mathbb{Z}_2^{\mathrm{a,b}}$
	AI	0	0	0	$2M\mathbb{Z}$	0	$T\mathbb{Z}_2^{\mathrm{a,b,c}}$	$\mathbb{Z}_2^{\mathrm{a,b}}$	$M\mathbb{Z}$
	BDI	$M\mathbb{Z}$	0	0	0	$2M\mathbb{Z}$	0	$T\mathbb{Z}_2^{\mathrm{a,b,c}}$	$\mathbb{Z}_2^{\mathrm{a,b}}$
	D	$\mathbb{Z}_2^{\mathrm{a,b}}$	$M\mathbb{Z}$	0	0	0	$2M\mathbb{Z}$	0	$T\mathbb{Z}_2^{\mathrm{a,b,c}}$
R_-, R_{--}	DIII	$T\mathbb{Z}_2^{\mathrm{a,b,c}}$	$\mathbb{Z}_2^{\mathrm{a,b}}$	$M\mathbb{Z}$	0	0	0	$2M\mathbb{Z}$	0
	AII	0	$T\mathbb{Z}_2^{\mathrm{a,b,c}}$	$\mathbb{Z}_2^{\mathrm{a,b}}$	$M\mathbb{Z}$	0	0	0	$2M\mathbb{Z}$
	CII	$2M\mathbb{Z}$	0	$T\mathbb{Z}_2^{\mathrm{a,b,c}}$	$\mathbb{Z}_2^{\mathrm{a,b}}$	$M\mathbb{Z}$	0	0	0
	C	0	$2M\mathbb{Z}$	0	$T\mathbb{Z}_2^{\mathrm{a,b,c}}$	$\mathbb{Z}_2^{\mathrm{a,b}}$	$M\mathbb{Z}$	0	0
	CI	0	0	$2M\mathbb{Z}$	0	$T\mathbb{Z}_2^{\mathrm{a,b,c}}$	$\mathbb{Z}_2^{\mathrm{a,b}}$	$M\mathbb{Z}$	0
R_{-+}	BDI, CII	0	$2\mathbb{Z}$	0	$2M\mathbb{Z}$	0	$2\mathbb{Z}$	0	$2M\mathbb{Z}$
R_{+-}	DIII, CI	0	$2M\mathbb{Z}$	0	$2\mathbb{Z}$	0	$2M\mathbb{Z}$	0	$2\mathbb{Z}$
R_{+-}	BDI	$M\mathbb{Z}_2\oplus\mathbb{Z}_2^{\mathrm{a,b}}$	$M\mathbb{Z}\oplus\mathbb{Z}$	0	0	0	$2M\mathbb{Z}\oplus 2\mathbb{Z}$	0	$M\mathbb{Z}_2\oplus\mathbb{Z}_2^{\mathrm{a,b}}$
R_{-+}	DIII	0	$M\mathbb{Z}_2\oplus\mathbb{Z}_2^{\mathrm{a,b}}$	$M\mathbb{Z}_2\oplus\mathbb{Z}_2^{\mathrm{a,b}}$	$M\mathbb{Z}\oplus\mathbb{Z}$	0	0	0	$2M\mathbb{Z}\oplus 2\mathbb{Z}$
R_{+-}	CII	0	$2M\mathbb{Z}\oplus 2\mathbb{Z}$	0	$M\mathbb{Z}_2\oplus\mathbb{Z}_2^{\mathrm{a,b}}$	$M\mathbb{Z}_2\oplus\mathbb{Z}_2^{\mathrm{a,b}}$	$M\mathbb{Z}\oplus\mathbb{Z}$	0	0
R_{-+}	CI	0	0	0	$2M\mathbb{Z}\oplus 2\mathbb{Z}$	0	$M\mathbb{Z}_2\oplus\mathbb{Z}_2^{\mathrm{a,b}}$	$M\mathbb{Z}_2\oplus\mathbb{Z}_2^{\mathrm{a,b}}$	$M\mathbb{Z}\oplus\mathbb{Z}$

[a] $\mathbb{Z}_2$ and $M\mathbb{Z}_2$ invariants only protect Fermi surfaces of dimension zero (i.e., Fermi points with $d_{\mathrm{FS}} = 0$) at high-symmetry points of the Brillouin zone.
[b] Fermi surfaces located within the mirror plane but away from high symmetry points cannot be protected by a $\mathbb{Z}_2$ or $M\mathbb{Z}_2$ topological number. However, the nodal system can exhibit zero-energy surface states that are protected by a $\mathbb{Z}_2$ or $M\mathbb{Z}_2$ topological number.
[c] For topological semimetals (nodal topological superconductors) the presence of translation symmetry is always assumed. Therefore, one does not need to distinguish between $T\mathbb{Z}_2$ and $\mathbb{Z}_2$ invariants for these gapless topological materials.

2.2. Fermi surfaces within mirror plane but off high-symmetry points

Second, we review case (ii), i.e., the classification of reflection symmetric Fermi surfaces that are located away from high-symmetry points of the Brillouin zone. These Fermi surfaces transform pairwise into each other by the global (i.e., internal) symmetries, which relate $\mathbf{k} \to -\mathbf{k}$. Using an analysis based on the minimal-Dirac-Hamiltonian method, it was show in Ref. [34] that only $M\mathbb{Z}$ and $2M\mathbb{Z}$ topological numbers can ensure the stability of reflection symmetric Fermi surfaces off high-symmetry points. $\mathbb{Z}_2$ and $M\mathbb{Z}_2$ invariants, on the other hand, do not give rise to stable Fermi surfaces. However, as illustrated in Sec. 3, $\mathbb{Z}_2$ or $M\mathbb{Z}_2$ invariants may nevertheless lead to protected zero-energy surface states at time-reversal-invariant momenta of the surface Brillouin zone. The classification of reflection symmetric Fermi surfaces located away from high-symmetry points is summarized in Table 1. As above, we note that due to Bott's periodicity theorem [50], the classification scheme exhibits an eight-fold periodicity in p. We observe that the classification of reflection-symmetric Fermi surfaces located away from high symmetry points with codimension p is related to the classification of reflection-symmetric insulators with spatial dimension $d_{\mathrm{BZ}} = p - 1$.

3. Topological Dirac semimetal with reflection and rotation symmetry

In this section we discuss a tight-binding model describing a topological semimetal which exhibits time-reversal symmetry as well as reflection and rotation symmetries. This model is inspired by the Dirac material Na_3Bi, whose three-dimensional bulk Dirac cones are protected by reflection symmetry. Using angle-resolved photoemission experiments, topological surface modes have recently been observed on the (100) surface of Na_3Bi [49].

3.1. Model Hamiltonian

We consider the following cubic-lattice Hamiltonian describing a four-band semimetal with two Dirac points

$$\hat{H} = \frac{1}{2}\sum_{\vec{x}} \left[C^{\dagger}_{\vec{x}+\hat{x}}(\tau_z s_0 + i\tau_x s_x)C_{\vec{x}} + C^{\dagger}_{\vec{x}+\hat{y}}(\tau_z s_0 + i\tau_x s_y)C_{\vec{x}} + C^{\dagger}_{\vec{x}+\hat{z}}\tau_z s_0 C_{\vec{x}} + mC^{\dagger}_{\vec{x}}\tau_z s_0 C_{\vec{x}} \right] + \text{h.c.}, \quad (4a)$$

with the spinor $C_{\vec{x}} = [c_{1\uparrow}(\vec{x}), c_{1\downarrow}(\vec{x}), c_{2\uparrow}(\vec{x}), c_{2\downarrow}(\vec{x})]$, where $c_{n\sigma}(\vec{x})$ denotes the electron annihilation operator with orbital index n and spin σ. The two sets of Pauli matrices τ_α and s_α operate in spin space and orbital space, respectively. In momentum space Hamiltonian (4a) takes diagonal form $\mathcal{H} = \sum_{\mathbf{k}} C^{\dagger}_{\mathbf{k}} H(\mathbf{k}) C_{\mathbf{k}}$, with $C_{\mathbf{k}}$ the Fourier-transformed spinor, and

$$H(\mathbf{k}) = \sin k_x \tau_x s_x + \sin k_y \tau_x s_y + (\cos k_x + \cos k_y + \cos k_z - m)\tau_z s_0. \quad (4b)$$

For concreteness we set $m = 2$. With this choice, the bulk Dirac cones of $H(\mathbf{k})$ are positioned at $\mathbf{k}_\pm = (0, 0, \pm\pi/2)$, i.e., they are located on the k_z axis just as in the Dirac material Na_3Bi. The cubic-lattice Hamiltonian (4b) possesses a rotation symmetry of order four about the z axis. This C_4 rotation symmetry acts on $H(\mathbf{k})$ as

$$R^{-1}_{C_4} H(-k_y, k_x, k_z) R_{C_4} = H(k_x, k_y, k_z), \quad (5)$$

where $R_{C_4} = \tau_z(s_0 + is_z)/\sqrt{2}$ denotes the C_4 rotation operator. The Dirac semimetal (4) also preserves time reversal symmetry, Eq. (1a), with the time-reversal operator $T = \tau_0 s_y \mathcal{K}$. Since $T^2 = -\mathbb{1}$, Hamiltonian (4) belongs to symmetry class AII. The codimension of the Dirac cones of $H(\mathbf{k})$, which are located away from the high symmetry points in the Brillouin zone, is $p = 3$. Hence, according to the classification of Ref. [34] (see Table 1 in Ref. [34]), the Dirac points are not protected by time-reversal symmetry, even though a binary $\mathbb{Z}_2$ invariant can be defined for this symmetry class. (As we will show below, this $\mathbb{Z}_2$ invariant leads to gapless surface states at time-reversal-invariant momenta of the surface Brillouin zone.) Indeed, we find that the bulk Dirac cones can be gapped out by the mass term $\sin k_z \tau_y s_0$, which preserves time reversal symmetry. However, the mass term $\sin k_z \tau_y s_0$ breaks the C_4 rotation symmetry (5). In general, we find that the only possible gap opening terms are $f_1(k_z)\tau_y s_0$ and $f_2(k_z)\tau_x s_z$, since these are the terms that anti-commute with $H(\mathbf{k})$. Here, $f_1(k_z)$ and $f_2(k_z)$ represent k_z dependent masses. We find that these two gap opening terms break the C_4 rotation symmetry (5), since they anti-commute with R_{C_4}. As a consequence, the two Dirac cones of Hamiltonian (4) are protected against gap opening by the C_4 rotation symmetry (5).

3.2. Surface states

As discussed above and in Ref. [34], class AII Dirac points with codimension $p = 3$ that are located away from high-symmetry points of the Brillouin zone are unstable in the absence of

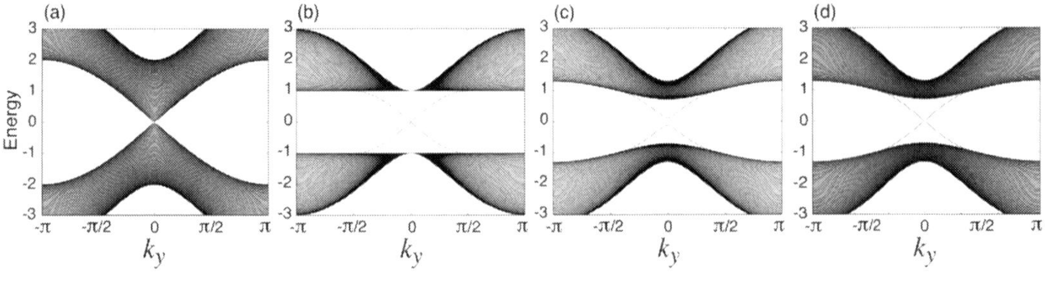

Figure 1. (a) Surface band structure of the Dirac semimetal (4) for a (100) slab as a function of surface momentum k_y with fixed $k_z = \pi/2$. (b) Same as panel (a) but for $k_z = 0$. The gapless surface state at $k_y = 0$ is protected by a $\mathbb{Z}_2$ invariant. (c) Band structure at the (100) surface of the Dirac semimetal (8) as a function of k_y with $k_z = \pi/4$. In the absence of reflection symmetry (6) and chiral symmetry (7) the surface state (red trace) is gapped. (d) Surface band structure of the Dirac semimetal (4) (i.e., with reflection symmetry and chiral symmetry) as a function of k_y with $k_z = \pi/4$.

crystal lattice symmetries, even though a binary $\mathbb{Z}_2$ index can be defined for this symmetry class. This $\mathbb{Z}_2$ invariant does not guarantee the stability of Dirac points, however it gives rise to protected gapless surface states at time-reversal invariant momenta of the surface Brillouin zone. For the Dirac semimetal (4) these gapless modes appear at $(k_y, k_z) = (0, 0)$ of the (100) surface Brillouin zone, see Fig. 1(b). For other surface momenta, the surface states are in general gapped. Similarly, rotation symmetry (5) can protect surface states only at time-reversal-invariant momenta of the surface Brillouin zone, but not away from these points.

To illustrate this, we consider Hamiltonian (4) with periodic boundary conditions along the z-axis and open boundary conditions along the x- and y-directions. We observe that model Hamiltonian (4) exhibits besides rotation symmetry (5) an *accidental* reflection symmetry

$$R_y^{-1} H(k_x, -k_y, k_z) R_y = H(k_x, k_y, k_z), \tag{6}$$

with $R_y = \tau_z s_y$, and an *accidental* chiral symmetry

$$H(\mathbf{k}) S = -S H(\mathbf{k}), \tag{7}$$

with $S = \tau_x s_z$. In order to break these accidental symmetries we need to add to the Hamiltonian the term $+g \sin k_z \tau_x s_z$ on the (100) and $(\bar{1}00)$ surfaces, and the term $-g \sin k_z \tau_x s_z$ on the (010) and $(0\bar{1}0)$ surfaces. That is, we consider

$$\tilde{H}(k_z) = \hat{H}(k_z) + g \sum_{\vec{x}_\perp} C^\dagger_{\vec{x}_\perp} \sin k_z \tau_x s_z \left[\delta_{\vec{x}_\perp, \vec{x}^{(100)}_\perp} + \delta_{\vec{x}_\perp, \vec{x}^{(\bar{1}00)}_\perp} - \delta_{\vec{x}_\perp, \vec{x}^{(010)}_\perp} - \delta_{\vec{x}_\perp, \vec{x}^{(0\bar{1}0)}_\perp} \right] C_{\vec{x}_\perp}, \tag{8}$$

where $\hat{H}(k_z)$ is obtained from Eq. (4a) by Fourier transforming the z-coordinate and the sum in Eq. (8) is over a collection of one-dimensional chains which are oriented along the z-axis and labeled by $\vec{x}_\perp$. Hamiltonian (8) is symmetric under time-reversal symmetry and rotation symmetry (5), but breaks the accidental symmetries (6) and (7). Fig. 1(c) shows the (100) surface spectrum of Hamiltonian (8) with $g = 0.2$ as a function of surface momentum k_y for $k_z = \pi/4$. Fig. 2(a) displays the (100) surface spectrum as a function of both surface momenta k_y and k_z. We observe that the surface state is gapped, except at $(k_y, k_z) = (0, 0)$, where it is protected by time-reversal symmetry (i.e., the aforementioned $\mathbb{Z}_2$ invariant).

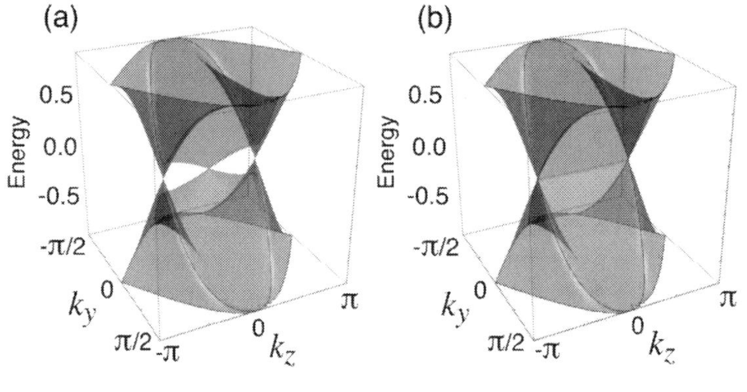

Figure 2. (a) Surface spectrum of the Dirac semimetal (8) for the (100) face as a function of surface momenta k_y and k_z. The surface states and bulk states are colored in green and gray, respectively. In the absence of reflection symmetry (6) and chiral symmetry (7) the surface state is gapped except at $(k_y, k_z) = (0, 0)$. (b) Surface spectrum of the Dirac semimetal (4), i.e., in the presence of reflection symmetry and chiral symmetry. The gaplessness of the Fermi arc is guaranteed by the mirror winding number (9).

On the other hand, if we consider a semimetal which possesses also symmetries (6) and (7) [i.e., Hamiltonian (4) instead of Hamiltonian (8)], then the surface states are gapless along a one-dimensional arc that connects the two projected Dirac nodes $\mathbf{k}_\pm = (0, 0, \pm\pi/2)$, see Fig. 1(d) and Fig. 2(b). Since the reflection operator R_y anti-commutes with the time-reversal operator $T = \tau_0 s_y \mathcal{K}$ and with the effective particle-hole operator $C = T^{-1}S = i\tau_x s_x \mathcal{K}$, Hamiltonian (4) can be viewed as a member of symmetry class DIII with R_{--} in Table 1. Hence, according to the classification scheme of Sec. 2.2 the Dirac nodes of Eq. (4) are protected by an $M\mathbb{Z}$ invariant, i.e., a mirror winding number. Moreover, Hamiltonian (4) exhibits Fermi arc surface states, which are protected by this mirror winding number. This mirror invariant is defined in terms of a one-dimensional integral along a contour that lies within the mirror plane $k_y = 0$. Within the $k_y = 0$ mirror plane the Hamiltonian can be block diagonalized with respect to R_y. For the block with mirror eigenvalue $R_y = +1$ the mirror winding number takes the form

$$n_{\mathcal{C}}^{+} = \frac{i}{2\pi} \int_{\mathcal{C}} \mathrm{Tr}(q^* dq), \quad \text{with} \quad q = \frac{\sin k_x - iM(k_x)}{\sin^2 k_x + M^2(k_x)} \tag{9}$$

and $M(k_x) = \cos k_x + \cos k_z - 1$, where $\mathcal{C}$ denotes a contour that lies within the mirror plane. Choosing the contour $\mathcal{C}$ to be parallel to the k_x-axis with $k_y = 0$ and k_z a fixed parameter we find that

$$n^{+}(k_z) = \begin{cases} -1, & |k_z| < \pi/2 \\ 0, & |k_z| > \pi/2 \end{cases}. \tag{10}$$

This indicates that there exists a gapless Fermi arc state at $k_y = 0$ on the (100) surface, which is in agreement with the spectrum shown in Fig. 2(b).

3.3. The role of translation symmetry

While each Dirac point of a topological semimetal is stable against deformations that are local in momentum space, the Dirac points are not protected against commensurate perturbations, such as charge-density wave modulations, which connect Dirac points with opposite momenta.

However, these deformations are forbidden, if we impose besides internal and crystal point-group symmetries, also translation symmetry.

To illustrate the instability of the Dirac points against charge-density wave perturbations, let us consider a low-energy description of semimetal (4). Expanding Eq. (4) around the two Dirac points $\mathbf{k}_{\pm} = (0, 0, \pm\pi/2)$, we obtain the following low-energy effective Hamiltonian

$$\hat{H}_{\text{eff}} = \sum_{k_x, k_y, k_\Delta} \begin{pmatrix} C^{\dagger}_{\mathbf{K}_+} & C^{\dagger}_{\mathbf{K}_-} \end{pmatrix} \begin{pmatrix} h_+ & 0 \\ 0 & h_- \end{pmatrix} \begin{pmatrix} C_{\mathbf{K}_+} \\ C_{\mathbf{K}_-} \end{pmatrix}, \tag{11}$$

where $\mathbf{K}_{\pm} = (k_x, k_y, k_\Delta \pm \pi/2)$ and $h_{\pm} = k_x \tau_x s_x + k_y \tau_x s_y \mp k_\Delta \tau_z s_0$. Hamiltonian $\hat{H}_{\text{eff}}$ is symmetric under the C_4 rotation symmetry (5), which transforms $C_{\mathbf{K}_{\pm}} \to \tau_z (s_0 + i s_z) C_{\tilde{\mathbf{K}}_{\pm}} / \sqrt{2}$, and the time reversal symmetry (1a), which transforms $C_{\mathbf{K}_{\pm}} \to \tau_0 s_y \mathcal{K} C_{-\mathbf{K}_{\pm}}$. We find that the two Dirac cones can be gapped out by the charge-density wave modulation

$$\hat{D} = \sum_{k_x, k_y, k_\Delta} \begin{pmatrix} C^{\dagger}_{\mathbf{K}_+} & C^{\dagger}_{\mathbf{K}_-} \end{pmatrix} \begin{pmatrix} 0 & \tau_z s_0 \\ \tau_z s_0 & 0 \end{pmatrix} \begin{pmatrix} C_{\mathbf{K}_+} \\ C_{\mathbf{K}_-} \end{pmatrix}, \tag{12}$$

which preserves rotation symmetry (5), but breaks translation symmetry. We conclude that in the absence of translation symmetry, the Dirac cones of semimetals are unstable against charge-density wave-type perturbations.

3.4. Implications for Na_3Bi

Albeit Na_3Bi has hexagonal $P6_3/mmc$ crystal structure rather than cubic one, its topological and symmetry properties are similar to the model Hamiltonian (8). Instead of C_4 rotation symmetry, Na_3Bi exhibits a C_3 rotation symmetry, which protects the two gapless bulk Dirac cones. Using a downfolding procedure, it is possible to derive from DFT band structure calculations a low-energy hexagonal-lattice tight-binding model for Na_3Bi. This tight-binding model is expected to exhibit similar symmetry properties and the same topological features as our tight-binding Hamiltonian (8). Hence, this suggests that Na_3Bi supports Fermi arc surface states, which are gapped except at time-reversal invariant points of the surface Brillouin zone. This expectation is confirmed by recent angle-resolved photoemission experiments [48, 49].

4. Summary and discussion

In this paper, we have reviewed the topological classification of reflection-symmetry-protected semimetals and nodal superconductors. This classification scheme depends on (i) the internal symmetry properties of the Hamiltonian, (ii) whether the reflection symmetry commutes or anti-commutes with the internal symmetries, (iii) the codimension of the Fermi surface (superconducting node) $p = d_{\text{BZ}} - d_{\text{FS}}$, and (iv) the transformation properties of the Fermi surface (superconducting node) under the action of the reflection and internal symmetries. Reflection symmetry together with the internal (i.e., non-spatial) symmetries define a total of 27 symmetry classes. The result of the classification scheme is presented in Table 1. The stability of the Fermi surface is ensured by the conservation of a topological invariant that is defined as an integral along a contour surrounding the Fermi surface. These topological numbers can be Chern or winding numbers (indicated by "$\mathbb{Z}$" in Table 1), binary invariants (indicated by "$\mathbb{Z}_2$"), mirror Chern or mirror winding numbers (indicated by "$M\mathbb{Z}$"), or mirror binary invariants (indicated by "$M\mathbb{Z}_2$").

To illustrate the usefulness of the classification Table 1, we have discussed in Sec. 3 an example of a topological semimetal with time-reversal symmetry and rotation symmetry. This model Hamiltonian represents a cubic lattice version of the hexagonal Dirac material Na_3Bi. We determined the surface state spectrum of this Dirac model system and discussed its topological properties. We found that the (100) surface supports Fermi arc surface states, which are gapped except at the $\bar{\Gamma}$ point of the surface Brillouin zone, see Fig. 2(a). This is in agreement with recent angular-resolved photoemission experiments on the (100) surface of Na_3Bi [48, 49].

Acknowledgments

The authors thank Wei-Feng Tsai for useful discussions. The support of the Max-Planck-UBC Centre for Quantum Materials is gratefully acknowledged.

References

[1] Schnyder A P, Ryu S, Furusaki A and Ludwig A W W 2008 *Phys. Rev. B* **78** 195125
[2] Kitaev A 2009 *AIP Conf. Proc.* **1134** 22
[3] Schnyder A P, Ryu S, Furusaki A and Ludwig A W W 2009 *AIP Conf. Proc.* **1134** 10
[4] Ryu S, Schnyder A P, Furusaki A and Ludwig A W W 2010 *New J. Phys.* **12** 065010
[5] Chiu C K 2014 *ArXiv e-prints* (*Preprint* 1410.1117)
[6] Teo J C Y, Fu L and Kane C L 2008 *Phys. Rev. B* **78** 045426
[7] Fu L 2011 *Phys. Rev. Lett.* **106** 106802
[8] Slager R J, Mesaros A, Juricic V and Zaanen J 2013 *Nat. Phys.* **9** 98
[9] Ueno Y, Yamakage A, Tanaka Y and Sato M 2013 *Phys. Rev. Lett.* **111** 087002
[10] Zhang F, Kane C L and Mele E J 2013 *Phys. Rev. Lett.* **111** 056403
[11] Benalcazar W A, Teo J C Y and Hughes T L 2014 *Phys. Rev. B* **89**(22) 224503
[12] Teo J C Y and Hughes T L 2013 *Phys. Rev. Lett.* **111**(4) 047006
[13] Turner A M, Zhang Y, Mong R S K and Vishwanath A 2012 *Phys. Rev. B* **85** 165120
[14] Hughes T L, Prodan E and Bernevig B A 2011 *Phys. Rev. B* **83** 245132
[15] Chiu C K, Yao H and Ryu S 2013 *Phys. Rev. B* **88**(7) 075142
[16] Morimoto T and Furusaki A 2013 *Phys. Rev. B* **88**(12) 125129
[17] Shiozaki K and Sato M 2014 *Phys. Rev. B* **90**(16) 165114
[18] Tanaka Y, Ren Z, Sato T, Nakayama K, Souma S, Takahashi T, Segawa K and Ando Y 2012 *Nat Phys* **8** 800
[19] Hsieh T H, Lin H, Liu J, Duan W, Bansil A and Fu L 2012 *Nat. Commun.* **3** 982
[20] Xu S Y, Liu C, Alidoust N, Neupane M, Qian D, Belopolski I, Denlinger J D, Wang Y J, Lin H, Wray L A, Landolt G, Slomski B, Dil J H, Marcinkova A, Morosan E, Gibson Q, Sankar R, Chou F C, Cava R J, Bansil A and Hasan M Z 2012 *Nat. Commun.* **3** 1192
[21] Dziawa *et al* P 2012 *Nat. Mater.* **11** 1023
[22] Kariyado T and Ogata M 2011 *Journal of the Physical Society of Japan* **80** 083704
[23] Kariyado T and Ogata M 2012 *Journal of the Physical Society of Japan* **81** 064701
[24] Hsieh T H, Liu J W and Fu L 2014 *Phys. Rev. B* **90** 08112
[25] Béri B 2010 *Phys. Rev. B* **81**(13) 134515
[26] Wan X, Turner A M, Vishwanath A and Savrasov S Y 2011 *Phys. Rev. B* **83**(20) 205101
[27] Schnyder A P and Ryu S 2011 *Phys. Rev. B* **84**(6) 060504
[28] Brydon P M R, Schnyder A P and Timm C 2011 *Phys. Rev. B* **84**(2) 020501
[29] Schnyder A P, Brydon P M R and Timm C 2011 *Phys. Rev. B* **85** 24522
[30] Matsuura S, Chang P Y, Schnyder A P and Ryu S 2013 *New J. Phys.* **15** 065001
[31] Zhao Y X and Wang Z D 2013 *Phys. Rev. Lett.* **110**(24) 240404
[32] Zhao Y X and Wang Z D 2014 *Phys. Rev. B* **89**(7) 075111
[33] Turner A M and Vishwanath A 2013 *ArXiv e-prints* (*Preprint* 1301.0330)
[34] Chiu C K and Schnyder A P 2014 *Phys. Rev. B* **90**(20) 205136
[35] Yang B J and Nagaosa N 2014 *Nat. Commun.* **5** 4898
[36] Morimoto T and Furusaki A 2014 *Phys. Rev. B* **89**(23) 235127
[37] Chang P Y, Matsuura S, Schnyder A P and Ryu S 2014 *Phys. Rev. B* **90**(17) 174504
[38] Chen Y, Lu Y M and Kee H Y 2014 *ArXiv e-prints* (*Preprint* 1410.5830)
[39] Jeon S, Zhou B B, Gyenis A, Feldman B E, Kimchi I, Potter A C, Gibson Q D, Cava R J, Vishwanath A and Yazdani A 2014 *Nat. Mater.* **13** 851
[40] Wang Z, Weng H, Wu Q, Dai X and Fang Z 2013 *Phys. Rev. B* **88**(12) 125427
[41] Neupane M, Xu S Y, Sankar R, Alidoust N, Bian G, Liu C, Belopolski I, Chang T R, Jeng H T, Lin H, Bansil A, Chou F and Hasan M Z 2014 *Nat. Commun.* **5** 3786

[42] Borisenko S, Gibson Q, Evtushinsky D, Zabolotnyy V, Büchner B and Cava R J 2014 *Phys. Rev. Lett.* **113**(2) 027603

[43] Liu Z K, Jiang J, Zhou B, Wang Z J, Zhang Y, Weng H M, Prabhakaran D, Mo S K, Peng H, Dudin P, Kim T, Hoesch M, Fang Z, Dai X, Shen Z X, Feng D L, Hussain Z and Chen Y L 2014 *Nat. Mater.* **13** 677

[44] He L P, Hong X C, Dong J K, Pan J, Zhang Z, Zhang J and Li S Y 2014 *Phys. Rev. Lett.* **113**(24) 246402

[45] Liang T, Gibson Q, Ali M N, Liu M, Cava R J and Ong N P 2014 *Nature Materials* (*Preprint* `http://nmat/journal/vaop/ncurrent/full/nmat4143.html`)

[46] Liu Z K, Zhou B, Zhang Y, Wang Z J, Weng H M, Prabhakaran D, Mo S K, Shen Z X, Fang Z, Dai X, Hussain Z and Chen Y L 2014 *Science* **343** 864

[47] Wang Z, Sun Y, Chen X Q, Franchini C, Xu G, Weng H, Dai X and Fang Z 2012 *Phys. Rev. B* **85**(19) 195320

[48] Xu S Y, Liu C, Kushwaha S K, Chang T R, Krizan J W, Sankar R, Polley C M, Adell J, Balasubramanian T, Miyamoto K, Alidoust N, Bian G, Neupane M, Belopolski I, Jeng H T, Huang C Y, Tsai W F, Lin H, Chou F C, Okuda T, Bansil A, Cava R J and Hasan M Z 2013 *ArXiv e-prints* (*Preprint* `1312.7624`)

[49] Xu S Y, Liu C, Kushwaha S K, Sankar R, Krizan J W, Belopolski I, Neupan M, e, Bian G, Alidoust N, Chang T R, Jeng H T, Huang C Y, Tsai W F, Lin H, Shibayev P P, Chou F, Cava R J and Hasan M Z 2014 *Science* (*Preprint* `http://www.sciencemag.org/content/early/2014/12/17/science.1256742.full.pdf`)

[50] Bott R 1970 *Advances in Mathematics* **4** 353

Journal of Physics: Conference Series **603** (2015) 012003 doi:10.1088/1742-6596/603/1/012003

Multiband $\boldsymbol{k} \cdot \boldsymbol{p}$ theory for holes at the T-point in bismuth

Yuki Fuseya

Department of Engineering Science, University of Electro-Communications, Chofu, Tokyo 182-8585, Japan

E-mail: `fuseya@pc.uec.ac.jp`

Abstract. Effective g-factor for multiband system with a strong spin-orbit interaction is investigated. A simple, general, and gauge-invariant formula for the effective g-factor is obtained on the basis of $\boldsymbol{k} \cdot \boldsymbol{p}$ theory. By taking into account the symmetry for the T-point in Brillouin zone of bismuth, the formula of the effective g-factor is further simplified. The obtained results give a reasonable explanation for the longstanding puzzle about the large anisotropic g-factor of holes in bismuth.

1. Introduction

Bismuth has been studied from the dawn of solid state physics. Diamagnetism, Seebeck, Nernst, Shubnikov-de Haas and de Haas-van Alfven effects were all discovered in elemental solid of bismuth. The reason why such discoveries happened in succession is closely related to its specific electronic structure [1–4]. The crystal structure of bismuth, the rhombohedral structure, is obtained from two interpenetrating fcc lattices by adding two types of lattice distortions along the body diagonal direction: shift one sublattice relative to the other and stretch both sublattices. By these distortions, energy gaps open at the L- and T-point in the Brillouin zone and the lowest conduction band at the L-point overlaps with the highest valence band at the T-point, namely, the crystal becomes semimetallic (Fig. 1).

The effective Hamiltonian of bismuth has been studied mainly for the electrons at the L-point. A two-band model was introduced by Cohen and Blount on the basis of $\boldsymbol{k} \cdot \boldsymbol{p}$ theory with large spin-orbit interaction (SOI) [6], which succeeded to give an interpretation of the anomalously large g-factor at the L-point observed in bismuth. Soon later, Wolff pointed out that Cohen-Blount's two-band model is essentially equivalent to the Dirac Hamiltonian in the relativistic quantum mechanics [7] by applying a suitable unitary transformation to the spin space [8]. Hence, the electrons at the L-point in bismuth can be described as the Dirac electrons in solids. Nowadays this two-band model becomes the "standard model" for bismuth since it has been succeeded in interpreting various physical properties of bismuth, such as magneto reflection [9–11], diamagnetism [12, 13], and angular dependent quantum oscillations [14–16]. The properties of the Dirac electrons in solids has attracted modern interests, such as interband effects on Hall conductivity [17–19], spin polarized electric current [20], and spin Hall effect [21, 22].

On the other hand, for holes at the T-point of bismuth, it has been *believed* that its effective Hamiltonian is given by a much simpler form than that of electrons [1, 23], since the angle resolved quantum oscillation measurements show that the T-point hole has a (free electron like)

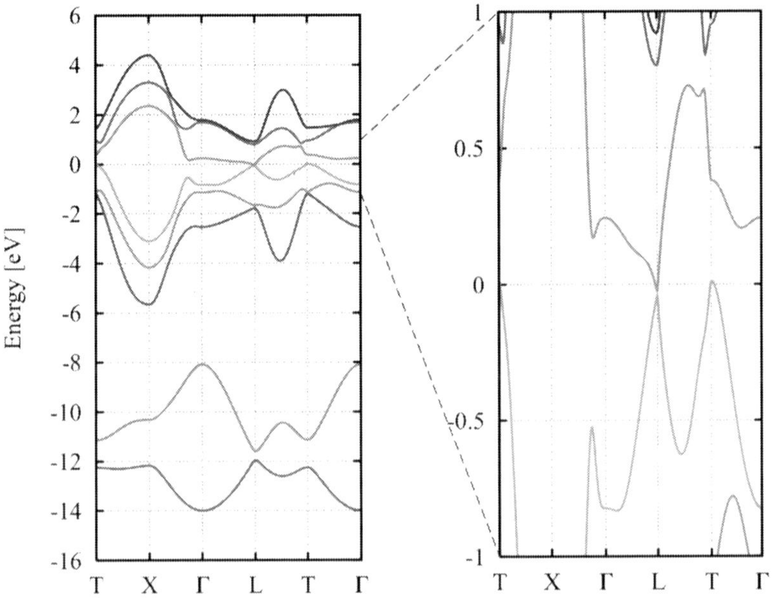

Figure 1. Band structure of bismuth obtained by the tight-binding approximation [5].

parabolic dispersion with an anisotropic effective mass [24–26]. In order to explain the energy spectrum under a magnetic field, Smith, Baraff and Rowell phenomenologically introduced a spin mass tensor $\boldsymbol{m}_s$ defined by $g^{*2} = 4m^2\boldsymbol{h}\cdot\boldsymbol{m}_s\cdot\boldsymbol{h}/\det\boldsymbol{m}_s$, where g^* is the effective g-factor, m is the bare electron mass, and $\boldsymbol{h}$ is a unit vector along the magnetic field direction. Their model with the spin mass can fit well with the angle resolved quantum oscillation results [15, 16, 23]. However, there still remains some mysteries about the g-factor of holes in bismuth: (i) the microscopic origin of the effective g-factor (or spin mass) is unknown; (ii) there is no reasonable explanation for the vanishingly small effective g-factor for $\boldsymbol{B} \perp$ trigonal axis [15]; (iii) why the ratio of the Zeeman splitting to the cyclotron energy, M, can greatly exceed 1 ($M = 2.12$ for $\boldsymbol{B} \parallel$ trigonal axis), in spite of the fact that the two-band model predicts that M is independent from the field direction and its maximum is $M = 1$ [6, 8].

In this paper, we derive a general theory of the effective g-factor on the basis of the multiband $\boldsymbol{k}\cdot\boldsymbol{p}$ theory. By applying the obtained formulae to the T-point holes in bismuth, we show that the longstanding puzzles (more than half a century) can be solved in a unified picture.

2. Multiband $\boldsymbol{k}\cdot\boldsymbol{p}$ Hamiltonian

We start from the one electron Hamiltonian with a SOI,

$$\mathscr{H} = \frac{p^2}{2m} + V(\boldsymbol{r}) + \frac{\hbar}{4m^2c^2}\boldsymbol{\sigma}\cdot\nabla V(\boldsymbol{r})\times\boldsymbol{p}, \tag{1}$$

where $V(\boldsymbol{r})$ is the periodic potential arising from the crystal, $\boldsymbol{\sigma}$ is the Pauli spin matrix vector, and c is the light velocity. By applying $\boldsymbol{k}\cdot\boldsymbol{p}$ theory, we obtain the eigenvalue equation

$$\sum_{n'}\left[\left(\epsilon_{n\boldsymbol{k}_0} + \frac{\hbar^2k^2}{2m}\right)\delta_{nn'}\delta_{\sigma\sigma'} + \hbar\boldsymbol{k}\cdot\boldsymbol{v}_{nn'}^{\sigma\sigma'}\right]c_{n'\sigma'}(\boldsymbol{k}) = \epsilon c_{n\sigma}(\boldsymbol{k}), \tag{2}$$

where $\epsilon_{n\boldsymbol{k}_0}$ is the energy of nth Bloch band at a band extremum $\boldsymbol{k}_0$, $\boldsymbol{v}_{nn'}^{\sigma\sigma'}$ is a matrix element of the velocity operator $\boldsymbol{v} = \boldsymbol{p}/m + (\hbar/4m^2c^2)\boldsymbol{\sigma} \times \boldsymbol{\nabla}V(\boldsymbol{r})$ between nth band with σ spin and n'th band with σ' spin states, $c_{n\sigma}(\boldsymbol{k})$ is an expansion coefficient for the eigenfunction, and the wave vector $\boldsymbol{k}$ is measured from $\boldsymbol{k}_0$. If we consider an n-band systems with a large SOI, Eq. (2) becomes

$$\begin{pmatrix} \epsilon_0 & 0 & \hbar\boldsymbol{k}\cdot\boldsymbol{v}_{01}^{\uparrow\uparrow} & \hbar\boldsymbol{k}\cdot\boldsymbol{v}_{01}^{\uparrow\downarrow} & \hbar\boldsymbol{k}\cdot\boldsymbol{v}_{02}^{\uparrow\uparrow} & \hbar\boldsymbol{k}\cdot\boldsymbol{v}_{02}^{\uparrow\downarrow} & \cdots \\ 0 & \epsilon_0 & \hbar\boldsymbol{k}\cdot\boldsymbol{v}_{01}^{\downarrow\uparrow} & \hbar\boldsymbol{k}\cdot\boldsymbol{v}_{01}^{\downarrow\downarrow} & \hbar\boldsymbol{k}\cdot\boldsymbol{v}_{02}^{\downarrow\uparrow} & \hbar\boldsymbol{k}\cdot\boldsymbol{v}_{02}^{\downarrow\downarrow} & \cdots \\ \hbar\boldsymbol{k}\cdot\boldsymbol{v}_{10}^{\uparrow\uparrow} & \hbar\boldsymbol{k}\cdot\boldsymbol{v}_{10}^{\uparrow\downarrow} & \epsilon_1 & 0 & \hbar\boldsymbol{k}\cdot\boldsymbol{v}_{12}^{\uparrow\uparrow} & \hbar\boldsymbol{k}\cdot\boldsymbol{v}_{12}^{\uparrow\downarrow} & \cdots \\ \hbar\boldsymbol{k}\cdot\boldsymbol{v}_{10}^{\downarrow\uparrow} & \hbar\boldsymbol{k}\cdot\boldsymbol{v}_{10}^{\downarrow\downarrow} & 0 & \epsilon_1 & \hbar\boldsymbol{k}\cdot\boldsymbol{v}_{12}^{\downarrow\uparrow} & \hbar\boldsymbol{k}\cdot\boldsymbol{v}_{12}^{\downarrow\downarrow} & \cdots \\ \hbar\boldsymbol{k}\cdot\boldsymbol{v}_{20}^{\uparrow\uparrow} & \hbar\boldsymbol{k}\cdot\boldsymbol{v}_{20}^{\uparrow\downarrow} & \hbar\boldsymbol{k}\cdot\boldsymbol{v}_{21}^{\uparrow\uparrow} & \hbar\boldsymbol{k}\cdot\boldsymbol{v}_{21}^{\uparrow\downarrow} & \epsilon_2 & 0 & \cdots \\ \hbar\boldsymbol{k}\cdot\boldsymbol{v}_{20}^{\downarrow\uparrow} & \hbar\boldsymbol{k}\cdot\boldsymbol{v}_{20}^{\downarrow\downarrow} & \hbar\boldsymbol{k}\cdot\boldsymbol{v}_{21}^{\downarrow\uparrow} & \hbar\boldsymbol{k}\cdot\boldsymbol{v}_{21}^{\downarrow\downarrow} & 0 & \epsilon_2 & \cdots \\ \vdots & \vdots & \vdots & \vdots & \vdots & \vdots & \ddots \end{pmatrix} \begin{pmatrix} c_{0\uparrow} \\ c_{0\downarrow} \\ c_{1\uparrow} \\ c_{1\downarrow} \\ c_{2\uparrow} \\ c_{2\downarrow} \\ \vdots \end{pmatrix} = \epsilon \begin{pmatrix} c_{0\uparrow} \\ c_{0\downarrow} \\ c_{1\uparrow} \\ c_{1\downarrow} \\ c_{2\uparrow} \\ c_{2\downarrow} \\ \vdots \end{pmatrix}, \quad (3)$$

where $\uparrow$ and $\downarrow$ denote the degree of freedom of Kramers doublet. The quadratic term $\hbar^2k^2/2m$ in Eq. (2) is discarded from ϵ since its contribution is much smaller than the $\hbar\boldsymbol{k}\cdot\boldsymbol{v}_{nn'}^{\sigma\sigma'}$ terms. By expanding the Cohen-Blount's notation [6], we have relations

$$\boldsymbol{t}_n = v_{0n}^{\uparrow\uparrow} = v_{n0}^{\downarrow\downarrow}, \quad (4)$$

$$\boldsymbol{u}_n = v_{0n}^{\uparrow\downarrow} = -v_{n0}^{\uparrow\downarrow}, \quad (5)$$

when the system is symmetric with respect to the time reversal and space inversion operation [27]. Then, we have a multiband $\boldsymbol{k}\cdot\boldsymbol{p}$ Hamiltonian in the form

$$\mathcal{H} = \begin{pmatrix} \epsilon_0 & 0 & \hbar\boldsymbol{k}\cdot\boldsymbol{t}_1 & \hbar\boldsymbol{k}\cdot\boldsymbol{u}_1 & \hbar\boldsymbol{k}\cdot\boldsymbol{t}_2 & \hbar\boldsymbol{k}\cdot\boldsymbol{u}_2 & \cdots \\ 0 & \epsilon_0 & -\hbar\boldsymbol{k}\cdot\boldsymbol{u}_1^* & \hbar\boldsymbol{k}\cdot\boldsymbol{t}_1^* & -\hbar\boldsymbol{k}\cdot\boldsymbol{u}_2^* & \hbar\boldsymbol{k}\cdot\boldsymbol{t}_2^* & \cdots \\ \hbar\boldsymbol{k}\cdot\boldsymbol{t}_1^* & -\hbar\boldsymbol{k}\cdot\boldsymbol{u}_1 & \epsilon_1 & 0 & \hbar\boldsymbol{k}\cdot\boldsymbol{s}_1 & \hbar\boldsymbol{k}\cdot\boldsymbol{w}_1 & \cdots \\ \hbar\boldsymbol{k}\cdot\boldsymbol{u}_1^* & \hbar\boldsymbol{k}\cdot\boldsymbol{t}_1 & 0 & \epsilon_1 & -\hbar\boldsymbol{k}\cdot\boldsymbol{w}_1^* & \hbar\boldsymbol{k}\cdot\boldsymbol{s}_1 & \cdots \\ \hbar\boldsymbol{k}\cdot\boldsymbol{t}_2^* & -\hbar\boldsymbol{k}\cdot\boldsymbol{u}_2 & \hbar\boldsymbol{k}\cdot\boldsymbol{s}_1^* & -\hbar\boldsymbol{k}\cdot\boldsymbol{w}_1 & \epsilon_2 & 0 & \cdots \\ \hbar\boldsymbol{k}\cdot\boldsymbol{u}_2^* & \hbar\boldsymbol{k}\cdot\boldsymbol{t}_2 & \hbar\boldsymbol{k}\cdot\boldsymbol{w}_1^* & \hbar\boldsymbol{k}\cdot\boldsymbol{s}_1 & 0 & \epsilon_2 & \cdots \\ \vdots & \vdots & \vdots & \vdots & \vdots & \vdots & \ddots \end{pmatrix}. \quad (6)$$

If we consider the two band system ($n = 0, 1$), Eq. (6) is equivalent to the Cohen-Blount's two-band model.

3. Effective g-factor of multiband $\boldsymbol{k}\cdot\boldsymbol{p}$ Hamiltonian

In order to obtain a general formula for the g-factor, we need to derive an effective Hamiltonian of a particular band under a magnetic field in terms of 2×2 matrix including the spin degree of freedom. This can be systematically achieved by using the Löwdin partitioning, which is a canonical transformation[28, 29]. The canonical transformed Hamiltonian is generally expressed in a perturbative form as

$$\mathcal{H} = \mathcal{H}^{(0)} + \mathcal{H}^{(1)} + \mathcal{H}^{(2)} + \cdots, \quad (7)$$

$$\mathcal{H}_{mm'}^{(0)} = \mathcal{H}_{mm'}^{0}, \quad (8)$$

$$\mathcal{H}_{mm'}^{(1)} = \mathcal{H}_{mm'}^{\prime}, \quad (9)$$

$$\mathcal{H}_{mm'}^{(2)} = \frac{1}{2}\sum_{\ell}\mathcal{H}_{m\ell}^{\prime}\mathcal{H}_{\ell m'}^{\prime}\left(\frac{1}{E_m - E_\ell} + \frac{1}{E_{m'} - E_\ell}\right), \quad (10)$$

where $\mathcal{H}^0$ is a Hamiltonian with known eigenvalues and $\mathcal{H} = \mathcal{H}^0 + \mathcal{H}'$. In the present case, we get

$$\mathcal{H} = \frac{\boldsymbol{\pi}\cdot\boldsymbol{\alpha}\cdot\boldsymbol{\pi}}{2m} + \sum_{n\neq 0}\left[\frac{im\mu_{\rm B}}{\epsilon_0-\epsilon_n}\begin{pmatrix} -(\boldsymbol{t}_n\times\boldsymbol{t}_n^* + \boldsymbol{u}_n\times\boldsymbol{u}_n^*) & 2\boldsymbol{t}_n\times\boldsymbol{u}_n \\ -2\boldsymbol{t}_n^*\times\boldsymbol{u}_n^* & \boldsymbol{t}_n\times\boldsymbol{t}_n^* + \boldsymbol{u}_n\times\boldsymbol{u}_n^* \end{pmatrix}\cdot\boldsymbol{B}\right], \quad (11)$$

up to the second order, where $\mu_{\rm B}$ is the Bohr magneton, $\boldsymbol{\alpha} = \sum_{n\neq 0}\boldsymbol{\alpha}_n$ is the inverse mass tensor of $n = 0$ band, and

$$\alpha_{n,ij} = \frac{t_{ni}t_{nj}^* + t_{nj}t_{ni}^* + u_{ni}u_{nj}^* + u_{nj}u_{ni}^*}{E_0 - E_n}. \quad (12)$$

We made the replacement $\hbar\boldsymbol{k} \to i\boldsymbol{\nabla} + (e/c)\boldsymbol{A} \equiv \boldsymbol{\pi}$ in the multiband $\boldsymbol{k}\cdot\boldsymbol{p}$ Hamiltonian (6) and used the commutation relation $\boldsymbol{\pi}\times\boldsymbol{\pi} = -i(\hbar e/c)\boldsymbol{B}$, where e is the elementary charge, $\boldsymbol{B}$ is the external magnetic field, and $\boldsymbol{A}$ is the vector potential satisfying $\boldsymbol{B} = \boldsymbol{\nabla}\times\boldsymbol{A}$. By using this commutation relation, the resultant formulae are gauge invariant.

The first term of Eq. (11) gives the cyclotron energy

$$\hbar\omega_{\rm c} = \frac{\hbar eB}{m_{\rm c}^* c}, \quad (13)$$

$$m_{\rm c}^* = \sqrt{\det\boldsymbol{\alpha}\,(\boldsymbol{\alpha}^{-1})_{ii}}, \quad (14)$$

for $\boldsymbol{B} \parallel i$. The second term of Eq. (11) gives the Zeeman energy

$$\begin{aligned} E_{\rm Z} &= \pm m\mu_{\rm B}\sqrt{4\left|\sum_{n\neq 0}\frac{(\boldsymbol{t}_n\times\boldsymbol{u}_n)\cdot\boldsymbol{B}}{E_0-E_n}\right|^2 - \left\{\sum_{n\neq 0}\frac{(\boldsymbol{t}_n\times\boldsymbol{t}_n^* + \boldsymbol{u}_n\times\boldsymbol{u}_n^*)\cdot\boldsymbol{B}}{E_0-E_n}\right\}^2} \\ &= \pm m\mu_{\rm B}\sqrt{\boldsymbol{B}\cdot\boldsymbol{G}\cdot\boldsymbol{B}}, \end{aligned} \quad (15)$$

$$\begin{aligned} G_{ij} &= 4\left(\sum_{n\neq 0}\frac{\boldsymbol{t}_n\times\boldsymbol{u}_n}{E_0-E_n}\right)_i\left(\sum_{n\neq 0}\frac{\boldsymbol{t}_n^*\times\boldsymbol{u}_n^*}{E_0-E_n}\right)_j \\ &\quad - \left(\sum_{n\neq 0}\frac{\boldsymbol{t}_n\times\boldsymbol{t}_n^* + \boldsymbol{u}_n\times\boldsymbol{u}_n^*}{E_0-E_n}\right)_i\left(\sum_{n\neq 0}\frac{\boldsymbol{t}_n\times\boldsymbol{t}_n^* + \boldsymbol{u}_n\times\boldsymbol{u}_n^*}{E_0-E_n}\right)_j. \end{aligned} \quad (16)$$

The effective g-factor is defined by the Zeeman energy as $E_{\rm Z} = \pm g^*\mu_{\rm B}B_i/2$, so we have a general formula of the effective g-factor in the form

$$g^* = 2m\sqrt{G_{ii}}. \quad (17)$$

This is the answer to the question (i). Then the ratio of the Zeeman splitting to the orbital splitting, which characterizes the effect of crystalline SOI, for $B \parallel i$ is given by

$$M = \frac{\Delta E_{\rm Z}}{\hbar\omega_{\rm c}} = \sqrt{\frac{G_{ii}}{\alpha_{jj}\alpha_{kk} - \alpha_{jk}^2}}, \quad (18)$$

where i, j, and k are perpendicular to each other.

Journal of Physics: Conference Series **603** (2015) 012003 doi:10.1088/1742-6596/603/1/012003

When we apply Eq. (17) to the two-band model ($n = 0, 1$), we have

$$\begin{aligned} G_{zz} &= \frac{1}{\Delta_1^2}\Big[4\left(|t_{1x}|^2|u_{1y}|^2 + |t_{1y}|^2|u_{1x}|^2\right) + 2\left(|t_{1x}|^2|t_{1y}|^2 + |u_{1y}|^2|u_{1x}|^2\right) \\ &\quad - \left(t_{1x}^2 t_{1y}^{*2} + t_{1y}^2 t_{1x}^{*2} + u_{1x}^2 u_{1y}^{*2} + u_{1y}^2 u_{1x}^{*2}\right) \\ &\quad -2\left(t_{1x}t_{1y}^* u_{1y}u_{1x}^* + t_{1y}t_{1x}^* u_{1x}u_{1y}^* - t_{1x}t_{1y}^* u_{1x}u_{1y}^* - t_{1y}t_{1x}^* u_{1y}u_{1x}^*\right)\Big] \\ &= \alpha_{xx}\alpha_{yy} - \alpha_{xy}^2 = (\hbar\omega_c)^2 , \end{aligned} \tag{19}$$

so $M = 1$, which, of course, agrees with the result of Cohen-Blount.

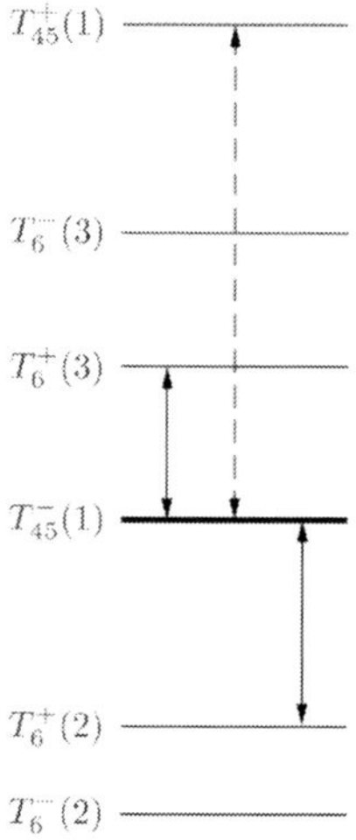

Figure 2. Energy levels at the T-point in bismuth. The selection rules for the hole band (T_{45}^-) are indicated by the arrows: the solid arrows are for the transitions of x and y components and the dashed arrow is for that of z component [30]. Here, x-, y-, and z-directions are taken along the binary, bisectrix, and trigonal axes.

4. Holes at the T-point of bismuth

So far, we have derived a general theory of the effective g-factor for multiband systems with a strong SOI. The obtained formulae are applicable for various systems. In the following, we shall obtain a further simplified formula of the g-factor of the holes at the T-point in bismuth by taking into account the specific symmetry of the T-point.

The group theoretical symmetry of band at the T-point has been investigated both experimentally and theoretically [1, 30, 31]. The symmetry of each band are indicated in Fig. 2. The matrix elements related to the hole band (T_{45}^-) are finite only for

$$\begin{aligned} \boldsymbol{t}_n^{(6)} &= \langle T_{45}^-(1)|\boldsymbol{v}|T_6^+(n)\rangle = (-a_n, ia_n, 0) && (20) \\ \boldsymbol{u}_n^{(6)} &= \langle T_{45}^-(1)|\boldsymbol{v}|CT_6^+(n)\rangle = (-a_n, -ia_n, 0), && (21) \\ \boldsymbol{u}_n^{(45)} &= \langle T_{45}^-(1)|\boldsymbol{v}|CT_{45}^+(n)\rangle = (0, 0, b_n), && (22) \end{aligned}$$

where operator C reverses the spin and a_n, b_n are complex numbers [30, 31]. The x-, y-, and z-directions are taken along the binary, bisectrix, and trigonal axes. By substituting Eqs. (20)–(22) into Eq. (16), only G_{zz} is finite as

$$G_{zz} = 4\left|\sum_{n\neq 0}\frac{2a_n^2}{\epsilon_0 - \epsilon_n}\right|^2 . \tag{23}$$

and the other ($G_{xx} = G_{yy} = G_{xy} = G_{yz} = G_{zx}$) are zero since the effective g-factor is given by the outer products of $\boldsymbol{t}_n$ and $\boldsymbol{u}_n$. Consequently, the Zeeman splitting as a function of the field direction is given by $\Delta E_Z(\theta) = 2m\mu_B B|\cos\theta|\sqrt{G_{zz}}$, where θ is an angle from the trigonal axis. Therefore, G_{zz} is finite for $\boldsymbol{B} \parallel$ trigonal and is zero for $\boldsymbol{B} \perp$ trigonal, which gives a possible solution to the mystery (ii).

The ratio of the Zeeman splitting to the cyclotron energy becomes

$$M = \sqrt{\frac{\left|\sum_{n\neq 0} \frac{a_n^2}{\epsilon_0 - \epsilon_n}\right|^2}{\left(\sum_{n\neq 0} \frac{|a_n|^2}{\epsilon_0 - \epsilon_n}\right)^2}}. \tag{24}$$

This analytic result tells us how the ratio M behaves. Let's consider a simple example with three band model ($n = 0, 1, 2$). If $\epsilon_0 < \epsilon_1 < \epsilon_2$,

$$M = \frac{|\tilde{a}_1^2 + \tilde{a}_2^2|}{|\tilde{a}_1|^2 + |\tilde{a}_2|^2} \leq 1, \tag{25}$$

where $\tilde{a}_n = a_n/\sqrt{|\epsilon_0 - \epsilon_n|}$. If $\epsilon_2 < \epsilon_0 < \epsilon_1$,

$$M = \frac{|\tilde{a}_1^2 - \tilde{a}_2^2|}{|\tilde{a}_1|^2 - |\tilde{a}_2|^2} \geq 1. \tag{26}$$

Consequently, we can obtain a simple understanding about the ratio M from these examples. When the contribution from the lower (higher) energy band is dominant, the ratio M will increase (decrease). This gives a reasonable explanation to the mystery (iii). Moreover, by evaluating M from the tight-binding model of Liu and Allen [5] with Eqs. (12), (16), and (17), we obtain $M = 2.08$, which agrees well with the experimental value of $M = 2.12$ [15].

5. Conclusion

In this paper, we have investigated the g-factor on the basis of the multiband $\boldsymbol{k}\cdot\boldsymbol{p}$ Hamiltonian with the strong SOI. The general formula of the effective g-factor was derived in the gauge invariant form by applying the Löwdin partitioning to the multiband $\boldsymbol{k}\cdot\boldsymbol{p}$ Hamiltonian. The effective g-factor of holes at the T-point in bismuth was obtained by considering the specific symmetries for the T-point. It was found that the effective g-factor for holes is highly anisotropic; g^* is finite for the magnetic field along the trigonal axis, while $g^* = 0$ for the field perpendicular to the trigonal axis. Further more, it was shown that the ratio of the Zeeman splitting to the cyclotron energy becomes larger than $M = 1$, which is the maximum value of the two-band model, when the contributions from the lower energy band is taken into account. $M = 2.08$ was numerically obtained by applying the theoretical results to the tight-binding calculation. These results gives possible solutions to the longstanding problem for the holes at the T-point in bismuth in a unified picture.

References

[1] Dresselhaus M S 1971 *J. Phys. Chem. Solids* **32** 3
[2] Édel'man V S 1976 *Adv. Phys.* **25** 555
[3] Issi J P 1979 *Aust. J. Phys.* **32** 585
[4] Fuseya Y, Ogata M and Fukuyama H 2015 *J. Phys. Soc. Jpn.* **84** 012001

Journal of Physics: Conference Series **603** (2015) 012003 doi:10.1088/1742-6596/603/1/012003

[5] Liu Y and Allen R E 1995 *Phys. Rev. B* **52** 1566
[6] Cohen M H and Blount E I 1960 *Phil. Mag.* **5** 115
[7] Dirac P A M 1928 *Proc. Roy. Soc. Lond. A* **117** 610
[8] Wolff P A 1964 *J. Phys. Chem. Solids* **25** 1057
[9] Maltz M and Dresselhaus M S 1970 *Phys. Rev. B* **2** 2877
[10] Vecchi M P and Dresselhaus M S 1974 *Phys. Rev. B* **9** 3257
[11] Vecchi M P, Pereira J R and Dresselhaus M S 1976 *Phys. Rev. B* **14** 298
[12] Fukuyama H and Kubo R 1970 *J. Phys. Soc. Jpn.* **28** 570
[13] Mikitik G P and Sharlai Y V 2000 *Low Temp. Phys.* **26** 39–46
[14] Sharlai Y V and Mikitik G P 2009 *Phys. Rev. B* **79** 081102(R)
[15] Zhu Z, Fauqué B, Fuseya Y and Behnia K 2011 *Phys. Rev. B* **84** 115137
[16] Zhu Z, Fauqué B, Malone L, Antunes A B, Fuseya Y and Behnia K 2012 *Proc. Natl. Acad. Sci. U.S.A.* **109** 14813
[17] Fuseya Y, Ogata M and Fukuyama H 2009 *Phys. Rev. Lett.* **102** 066601
[18] Fukuyama H, Fuseya Y, Ogata M, Kobayashi A and Suzumura Y 2012 *Physica B,* **407** 1943
[19] Kajita K, Nishio Y, Tajima N, Suzumura Y and Kobayashi A 2014 *J. Phys. Soc. Jpn.* **83** 072002
[20] Fuseya Y, Ogata M and Fukuyama H 2012 *J. Phys. Soc. Jpn* **81** 013704
[21] Fuseya Y, Ogata M and Fukuyama H 2012 *J. Phys. Soc. Jpn* **81** 093704
[22] Fuseya Y, Ogata M and Fukuyama H 2014 *J. Phys. Soc. Jpn.* **83** 074702
[23] Smith G E, Baraff G A and Rowell J M 1964 *Phys. Rev.* **135** A1118
[24] Kao Y H 1963 *Phys. Rev.* **129** 1122
[25] Brandt N B, Dolgolenko T F and Stuposhenko N N 1964 *Sov. Phys. JETP* **18** 908
[26] Bhargava R N 1967 *Phys. Rev.* **156** 785
[27] Yafet Y 1963 *Solid State Phys.* **14** 1 – 98 ISSN 0081-1947
[28] Löwdin P O 1951 *J. Chem. Phys.* **19**(11) 1396–1401
[29] Winkler R 2003 *Spin-Orbit Coupling Effects in Two-Dimensional Electron and Hole Systems* (Springer-Verlag)
[30] Golin S 1968 *Phys. Rev.* **166**(3) 643–651
[31] Bate R T, Einspruch N G and May P J 1969 *Phys. Rev.* **186**(3) 599–608

Meissner effect of Dirac electron in superconducting state

Tomonari Mizoguchi and Masao Ogata

Department of Physics, The University of Tokyo, 7-3-1 Hongo, Bunkyo, Tokyo 113-0033, Japan

E-mail: `mizoguchi@hosi.phys.s.u-tokyo.ac.jp`

Abstract. We investigate the Meissner effect (ME) of Dirac electrons in a superconducting state on the basis of Kubo formula, and clarify that Meissner kernel becomes finite by use of the inter-band contribution. This mechanism of the ME for Dirac electron is completely different from that for the electron in the usual metals. We also derive the result of the electron gas by taking the non-relativistic limit of Dirac Hamiltonian, and clarify that the diamagnetic term of the Meissner kernel can be regarded as the inter-band contribution between electrons and positrons in terms of the Dirac model.

1. Introduction

Recently, Dirac electron systems have been attracting a great interest. Dirac electron is characterized by the linear dispersion relation, and its low-energy physics is described by relativistic Dirac equation. It has been found that Dirac electron appears in such materials as graphene [1], Bi [2], α-(BEDT-TTF)$_2$I$_3$ [3], and so on. Up to now, many works have been carried out to investigate the properties in normal state of Dirac electron systems, such as the Hall conductivity [4], magnetoresistance [5], orbital magnetism, [6] and spin Hall conductivity [7]. In these studies, it has been pointed out that the inter-band effect plays important roles, since the conduction and valence bands are close to each other.

However, the properties of Dirac electrons in a superconducting state have not been understood completely. One of the most fundamental properties in a superconducting state is the Meissner effect (ME) in which the system becomes perfectly diamagnetic. In the case of parabolic-dispersion system, the ME arises from the diamagnetic term of the current operator. Note that the diamagnetic current term appears when the Hamiltonian has a kinetic energy proportional to $\boldsymbol{k}^2$ with $\boldsymbol{k}$ being the wave vector. [8]. On the other hand, in the Dirac electron system, the diamagnetic term of the current operator is absent since the Hamiltonian contains only up to k-linear terms [9]. Therefore, it is not trivial whether the Dirac electron in the superconducting state shows the ME.

In this paper, we discuss the ME of the 4×4 massive Dirac electron system in the three-dimensional space. We calculate the Meissner kernel by use of Kubo formula. We clarify that the ME occurs not due to the diamagnetic term of the current operator but due to the inter-band effect. We further study the large band-gap limit, which corresponds to the non-relativistic limit. We clarify that we can reproduce the Meissner kernel of the non-relativistic theory from the Dirac model, and that the paramagnetic and the diamagnetic terms of the kernel correspond to the intra-band and the inter-band terms in Dirac theory.

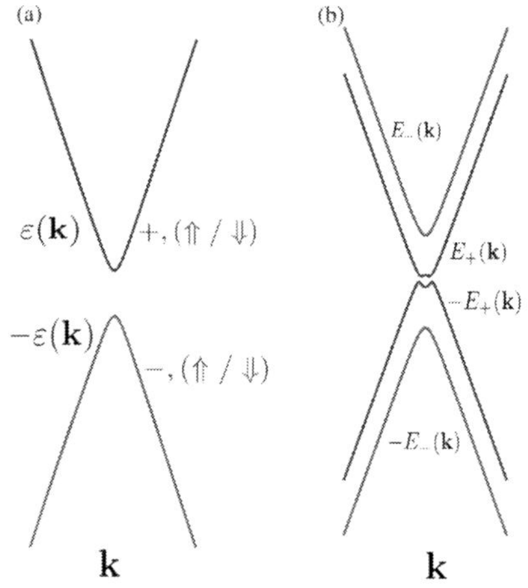

Figure 1. (a) the dispersion relation of the 4×4 Dirac Hamiltonian in normal state. The blue (green) line corresponds to the upper (lower) band. (b) excitation energy of Bogoliubov quasi-particles. The blue (green) lines originate from the upper (lower) band.

2. Formulation

Hamiltonian.- We consider the 4×4 massive Dirac Hamiltonian in the three-dimensional space:

$$H_0(\boldsymbol{k}) = \hat{c}_{\boldsymbol{k}}^{\dagger} \begin{pmatrix} M\hat{I} & iv\boldsymbol{k}\cdot\boldsymbol{\sigma} \\ -iv\boldsymbol{k}\cdot\boldsymbol{\sigma} & -M\hat{I} \end{pmatrix} \hat{c}_{\boldsymbol{k}}, \tag{1}$$

where $\hat{I}$ is the 2×2 unit matrix and $\boldsymbol{\sigma} = (\sigma_x, \sigma_y, \sigma_z)$ are Pauli matrices. Here, M is the band gap, and v is the Fermi velocity. We use the basis $\hat{c}_{\boldsymbol{k}} = (c_{\boldsymbol{k},1,\uparrow}, c_{\boldsymbol{k},1,\downarrow}, c_{\boldsymbol{k},2,\uparrow}, c_{\boldsymbol{k},2,\downarrow})^{\mathrm{T}}$ where $1, 2$ denote the orbital, and $\uparrow, \downarrow$ denote the real spin.

Diagonalized form of this Hamiltonian is

$$H_0(\boldsymbol{k}) = \hat{a}_{\boldsymbol{k}}^{\dagger} \begin{pmatrix} \varepsilon(\boldsymbol{k})\hat{I} & 0 \\ 0 & -\varepsilon(\boldsymbol{k})\hat{I} \end{pmatrix} \hat{a}_{\boldsymbol{k}}, \tag{2}$$

where $\varepsilon(\boldsymbol{k}) = \sqrt{M^2 + (vk)^2}$, and $\hat{a}_{\boldsymbol{k}}$ represents the band-basis which is expressed as $\hat{a}_{\boldsymbol{k}} = (a_{\boldsymbol{k},+,\Uparrow}, a_{\boldsymbol{k},+,\Downarrow}, a_{\boldsymbol{k},-,\Uparrow}, a_{\boldsymbol{k},-,\Downarrow})^{\mathrm{T}}$. Here, the indices $+$ and $-$ denote the upper and the lower bands respectively, and $\Uparrow$ and $\Downarrow$ denote psudo-spins corresponding to the two-fold degeneracy of each band [See Fig. 1(a)] . Two bases, $\hat{c}_{\boldsymbol{k}}$ and $\hat{a}_{\boldsymbol{k}}$ are related by the unitary transformation as $\hat{a}_{\boldsymbol{k}} = U(\boldsymbol{k})\hat{c}_{\boldsymbol{k}}$

.

Superconductivity.- We introduce superconductivity in this system. Here we do not discuss the microscopic origin of attractive interaction. Instead, we simply assume the intra-band s-wave interaction as

$$H_{\mathrm{int}} = -V \sum_{|\boldsymbol{k}|,|\boldsymbol{k}'|<k_c} \sum_{\eta,\eta'=\pm} a_{\boldsymbol{k},\eta,\Uparrow}^{\dagger} a_{-\boldsymbol{k},\eta,\Downarrow}^{\dagger} a_{-\boldsymbol{k}',\eta',\Downarrow} a_{\boldsymbol{k}',\eta',\Uparrow}, \tag{3}$$

where k_c is the cutoff momentum corresponding to the range of energy in which the attractive interaction works. We assume $vk_c \gg M$. Due to this assumption, we obtain the superconducting phase even if the chemical potential is located in the band gap (i.e., there is no density of states at the Fermi level) for sufficiently large V, since electrons with $k < k_c$ can still contribute to the Cooper pairing. Then, we define the superconducting order parameter in the s-wave symmetry,

$$\Delta_\pm = V \sum_{|\boldsymbol{k}'|<k_c} \langle a_{-\boldsymbol{k}',\pm,\Downarrow} a_{\boldsymbol{k}',\pm,\Uparrow} \rangle. \tag{4}$$

By applying the mean field approximation to the Hamiltonian, $H = H_0 + H_{\rm int}$, we obtain the BCS Hamiltonian,

$$H_{\rm BCS} = \sum_{|\boldsymbol{k}|<k_c} \sum_{\eta=\pm} \{ \sum_{\Sigma=\Uparrow,\Downarrow} [\xi_\pm(\boldsymbol{k}) a^\dagger_{\boldsymbol{k},\eta,\Sigma} a^\dagger_{\boldsymbol{k},\eta,\Sigma}] - \Delta[a^\dagger_{\boldsymbol{k},\eta,\Uparrow} a^\dagger_{-\boldsymbol{k},\eta,\Downarrow} + a_{-\boldsymbol{k},\eta,\Downarrow} a_{\boldsymbol{k},\eta,\Uparrow}]\} \tag{5}$$

where $\Delta = \Delta_+ + \Delta_-$ is the sum of the band-dependent order parameters and $\xi_\pm(\boldsymbol{k}) = \pm\varepsilon(\boldsymbol{k}) - \mu$.

Diagonalizing $H_{\rm BCS}$, we obtain the the excitation energy $E_\pm(\boldsymbol{k}) \equiv \sqrt{\xi_\pm(\boldsymbol{k})^2 + \Delta^2}$ [see Fig. 1(b)]. Then, thermal Green's function and the anomalous Green's functions in the band basis are obtained in the form of 4×4 matrices as follows.

$$\mathcal{G}(\boldsymbol{k}, {\rm i}\omega_n) = \begin{pmatrix} \mathcal{G}_+(\boldsymbol{k}, {\rm i}\omega_n)\hat{I} & 0 \\ 0 & \mathcal{G}_-(\boldsymbol{k}, {\rm i}\omega_n)\hat{I} \end{pmatrix}, \tag{6}$$

and

$$\mathcal{F}(\boldsymbol{k}, {\rm i}\omega_n) = -\mathcal{F}^\dagger(\boldsymbol{k}, {\rm i}\omega_n) = \begin{pmatrix} {\rm i}\sigma_y\mathcal{F}_+(\boldsymbol{k}, {\rm i}\omega_n) & 0 \\ 0 & {\rm i}\sigma_y\mathcal{F}_-(\boldsymbol{k}, {\rm i}\omega_n) \end{pmatrix}, \tag{7}$$

where $\mathcal{G}_\pm(\boldsymbol{k}, {\rm i}\omega_n) = \frac{u^2_\pm(\boldsymbol{k})}{i\omega_n - E_\pm(\boldsymbol{k})} + \frac{v^2_\pm(\boldsymbol{k})}{i\omega_n + E_\pm(\boldsymbol{k})}$, and $\mathcal{F}_\pm(\boldsymbol{k}, {\rm i}\omega_n) = -u_\pm(\boldsymbol{k})v_\pm(\boldsymbol{k})(\frac{1}{{\rm i}\omega_n - E_\pm(\boldsymbol{k})} - \frac{1}{{\rm i}\omega_n + E_\pm(\boldsymbol{k})})$, with $u^2_\pm(\boldsymbol{k}) = \frac{1}{2}[1 + \frac{\xi_\pm(\boldsymbol{k})}{E_\pm(\boldsymbol{k})}]$, $v^2_\pm(\boldsymbol{k}) = \frac{1}{2}[1 - \frac{\xi_\pm(\boldsymbol{k})}{E_\pm(\boldsymbol{k})}]$, and $u_\pm(\boldsymbol{k})v_\pm(\boldsymbol{k}) = \frac{\Delta}{2E_\pm(\boldsymbol{k})}$. It should be noted that the upper and lower parts of these Green's functions are decoupled with each other.

Current operator.- To derive the Meissner kernel, we have to define the current operator in the absence and presence of the vector potential. In the absence of vector potential, $\boldsymbol{A}$, the current operator in the momentum space is given by

$$\hat{\boldsymbol{j}}(\boldsymbol{q}) = -e\sum_{\boldsymbol{k}} \hat{c}^\dagger_{\boldsymbol{k}-\boldsymbol{q}} \partial_{\boldsymbol{k}} H_0(\boldsymbol{k}) c^\dagger_{\boldsymbol{k}} = -e\sum_{\boldsymbol{k}} \hat{c}^\dagger_{\boldsymbol{k}-\boldsymbol{q}} \begin{pmatrix} 0 & {\rm i}v\boldsymbol{\sigma} \\ -{\rm i}v\boldsymbol{\sigma} & 0 \end{pmatrix} \hat{c}_{\boldsymbol{k}} \equiv \sum_{\boldsymbol{k}} \hat{c}^\dagger_{\boldsymbol{k}-\boldsymbol{q}} \boldsymbol{j} \hat{c}_{\boldsymbol{k}}. \tag{8}$$

where $e > 0$ is the absolute value of the charge of an electron. The vector potential $\boldsymbol{A}$ is introduced in the Hamiltonian by replacing $\boldsymbol{k}$ by $\boldsymbol{k} + e\boldsymbol{A}$. However, we can readily see that the current operator in Eq. (8) does not change even in the presence of $\boldsymbol{A}$, since $H_0(\boldsymbol{k})$ contains only linear terms with respect to $\boldsymbol{k}$.

This property of the current operator is essentially different from that in jellium model in which the current operator is given by $\boldsymbol{j} = \sum_{\boldsymbol{k},\sigma} c^\dagger_{\boldsymbol{k},\sigma} \frac{e}{m}(\boldsymbol{k} + e\boldsymbol{A})c_{\boldsymbol{k},\sigma}$ where the last term is called as the diamagnetic term. Apparently, the current operator of the Dirac electron systems does not have diamagnetic current. As we will show shortly, the absence of the diamagnetic term is crucial for discussing the mechanism of the ME in Dirac electron in superconducting state.

3. Meissner effect in Dirac electron systems

Small band gap case: Dirac electrons in solids.- Using the formulations derived in the previous section, we calculate Meissner kernel, $K^{(S)}_{xx}$, through Kubo formula. Since we do not have the diamagnetic term of the current operator, $K^{(S)}_{xx}(\mathbf{q}, {\rm i}\omega_\lambda)$ is given only by the current-current correlation function.

$$
\begin{aligned}
K_{xx}^{(S)}(\boldsymbol{q}, \mathrm{i}\omega_\lambda) =& T\sum_{\boldsymbol{k},\omega_n} \mathrm{Tr}[\mathcal{G}(\boldsymbol{k}-\boldsymbol{q}, \omega_n-\omega_\lambda)\tilde{j}_x(\boldsymbol{k},\boldsymbol{q})\mathcal{G}(\boldsymbol{k},\omega_n)\tilde{j}_x(\boldsymbol{k}-\boldsymbol{q},-\boldsymbol{q}) \\
&-\mathcal{F}^\dagger(\boldsymbol{k}-\boldsymbol{q},\omega_n-\omega_\lambda)\tilde{j}_x(\boldsymbol{k},\boldsymbol{q})\mathcal{F}(\boldsymbol{k},\omega_n)\tilde{j}_x^{\mathrm{T}}(-\boldsymbol{k},-\boldsymbol{q})]. \qquad (9)
\end{aligned}
$$

where $\tilde{j}_x(\boldsymbol{k},\boldsymbol{q})$ is given by $\tilde{j}_x(\boldsymbol{k},\boldsymbol{q}) \equiv [U(\boldsymbol{k}-\boldsymbol{q})j_x U^\dagger(\boldsymbol{k})]$, and $\omega_n = \frac{(2n+1)\pi}{\beta}(n = 0, \pm1, \pm2, \cdots)$ denotes a fermionic Matubara frequency. It should be noted that the limit $\omega \to 0$ has to be taken before the limit $\boldsymbol{q} \to 0$.

By taking the trace in Eq. (9), we obtain the Meissner kernel which consists of two parts as $K_{xx}^{(S)}(\boldsymbol{q},0) = K_{xx}^{\mathrm{intra},(\mathrm{S})}(\boldsymbol{q},0) + K_{xx}^{\mathrm{inter},(S)}(\boldsymbol{q},0)$; $K_{xx}^{\mathrm{intra},(\mathrm{S})}(\boldsymbol{q},0)$ comes from the term in which the two Green's functions in Eq. (9), ($\mathcal{G}\mathcal{G}$ or $\mathcal{F}^\dagger\mathcal{F}$) have the same band indix $(+,+)$ or $(-,-)$, and $K_{xx}^{\mathrm{inter},(\mathrm{S})}(\boldsymbol{q},0)$ from the Green's functions with the opposite band index, i.e., $(+,-)$ or $(-,+)$.
The explicit forms of these are obtained as

$$
\begin{aligned}
K_{xx}^{\mathrm{intra},(\mathrm{S})}(\boldsymbol{q},0) =& -\frac{e^2v^2}{2}\sum_{\boldsymbol{k}}\sum_{\eta=\pm} g_{xx}^{\mathrm{intra}}(\boldsymbol{k},\boldsymbol{q})[(1+\frac{\xi_\eta(\boldsymbol{k}-\boldsymbol{q})\xi_\eta(\boldsymbol{k})+\Delta^2}{E_\eta(\boldsymbol{k}-\boldsymbol{q})E_\eta(\boldsymbol{k})})\frac{f(E_\eta(\boldsymbol{k}))-f(E_\eta\boldsymbol{k}-\boldsymbol{q}))}{E_\eta(\boldsymbol{k}-\boldsymbol{q})-E_\eta(\boldsymbol{k})} \\
&-(1-\frac{\xi_\eta(\boldsymbol{k}-\boldsymbol{q})\xi_\eta(\boldsymbol{k})+\Delta^2}{E_\eta(\boldsymbol{k}-\boldsymbol{q})E_\eta(\boldsymbol{k})})\frac{f(E_\eta(\boldsymbol{k}))+f(E_\eta(\boldsymbol{k}-\boldsymbol{q}))-1}{E_\eta(\boldsymbol{k}-\boldsymbol{q})+E_\eta(\boldsymbol{k})}], \qquad (10)
\end{aligned}
$$

and

$$
\begin{aligned}
K_{xx}^{\mathrm{inter},(\mathrm{S})}(\boldsymbol{q},0) =& \frac{e^2v^2}{2}\sum_{\boldsymbol{k}}\sum_{\eta=\pm} g_{xx}^{\mathrm{inter}}(\boldsymbol{k},\boldsymbol{q})[(1+\frac{\xi_+(\boldsymbol{k}-\boldsymbol{q})\xi_-(\boldsymbol{k})-\Delta^2}{E_+(\boldsymbol{k}-\boldsymbol{q})E_-(\boldsymbol{k})})\frac{f(E_{-\eta}(\boldsymbol{k}))-f(E_\eta(\boldsymbol{k}-\boldsymbol{q}))}{E_\eta(\boldsymbol{k}-\boldsymbol{q})-E_{-\eta}(\boldsymbol{k})} \\
&-(1-\frac{\xi_+(\boldsymbol{k}-\boldsymbol{q})\xi_-(\boldsymbol{k})-\Delta^2}{E_+(\boldsymbol{k}-\boldsymbol{q})E_-(\boldsymbol{k})})\frac{f(E_{-\eta}(\boldsymbol{k}))+f(E_\eta(\boldsymbol{k}-\boldsymbol{q}))-1}{E_\eta(\boldsymbol{k}-\boldsymbol{q})+E_{-\eta}(\boldsymbol{k})}]. \qquad (11)
\end{aligned}
$$

Here, $f(x)$ is the Fermi distribution fucntion, and $g_{xx}^{\mathrm{intra}}(\boldsymbol{k},\boldsymbol{q})$, $g_{xx}^{\mathrm{inter}}(\boldsymbol{k},\boldsymbol{q})$ are given by

$$
g_{xx}^{\mathrm{intra}}(\boldsymbol{k},\boldsymbol{q}) = \frac{v^2[k_x(k_x-q_x)-k_y(k_y-q_y)-k_z(k_z-q_z)]}{\epsilon(\boldsymbol{k})\epsilon(\boldsymbol{k}-\boldsymbol{q})} + \frac{\epsilon(\boldsymbol{k})\epsilon(\boldsymbol{k}-\boldsymbol{q})-M^2}{\epsilon(\boldsymbol{k})\epsilon(\boldsymbol{k}-\boldsymbol{q})}, \qquad (12a)
$$

$$
g_{xx}^{\mathrm{inter}}(\boldsymbol{k},\boldsymbol{q}) = \frac{v^2[k_x(k_x-q_x)-k_y(k_y-q_y)-k_z(k_z-q_z)]}{\epsilon(\boldsymbol{k})\epsilon(\boldsymbol{k}-\boldsymbol{q})} - \frac{\epsilon(\boldsymbol{k})\epsilon(\boldsymbol{k}-\boldsymbol{q})+M^2}{\epsilon(\boldsymbol{k})\epsilon(\boldsymbol{k}-\boldsymbol{q})}. \qquad (12b)
$$

In the limit of $\boldsymbol{q} \to \boldsymbol{0}$, we obtain $K_{xx}^{(\mathrm{S})}(0,0) = K_{xx}^{\mathrm{intra},(\mathrm{S})}(0,0) + K_{xx}^{\mathrm{inter},(\mathrm{S})}(0,0)$ with

$$
K_{xx}^{\mathrm{intra},(\mathrm{S})}(0,0) = 2e^2v^4\sum_{\boldsymbol{k}}\frac{k_x^2}{\varepsilon^2(\boldsymbol{k})}[\frac{\partial f(E_+(\boldsymbol{k}))}{\partial E_+(\boldsymbol{k})} + \frac{\partial f(E_-(\boldsymbol{k}))}{\partial E_-(\boldsymbol{k})}], \qquad (13)
$$

and

$$
\begin{aligned}
K_{xx}^{\mathrm{inter},(\mathrm{S})}(0,0) =& -2e^2v^2\sum_{\boldsymbol{k}}(1-\frac{v^2k_x^2}{\varepsilon^2(\boldsymbol{k})})[(1+\frac{\xi_+(\boldsymbol{k})\xi_-(\boldsymbol{k})-\Delta^2}{E_+(\boldsymbol{k})E_-(\boldsymbol{k})})\frac{f(E_-(\boldsymbol{k}))-f(E_+(\boldsymbol{k}))}{E_+(\boldsymbol{k})-E_-(\boldsymbol{k})} \\
&-(1-\frac{\xi_+(\boldsymbol{k})\xi_-(\boldsymbol{k})-\Delta^2}{E_+(\boldsymbol{k})E_-(\boldsymbol{k})})\frac{f(E_-(\boldsymbol{k}))+f(E_+(\boldsymbol{k}))-1}{E_+(\boldsymbol{k})+E_-(\boldsymbol{k})}]. \qquad (14)
\end{aligned}
$$

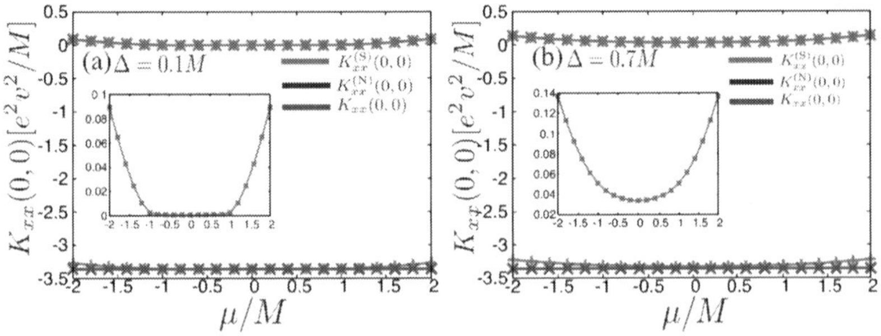

Figure 2. μ dependence of $K_{xx}^{(S)}(0,0)$, $K_{xx}^{(N)}(0,0)$, and $K_{xx}(0,0)$ for (a) $\Delta = 0.1M$ and (b) $\Delta = 0.7M$. μ dependence of K_{xx} in a large scale is in the inset.

It should be noted that for $T = 0$, $K_{xx}^{\text{intra},(S)}(0,0)$ is 0 since the spectrum has a finite gap at the chemical potential. Hence, $K_{xx}^{(S)}$ only consists of the inter-band contribution. This is in sharp contrast to the conventional BCS case, in which the Meissner kernel from the paramagnetic current vanishes at $T = 0$, but the Meissner kernel from the diamagnetic current remains which leads to the ME. In contrast, in the Dirac case, the diamagnetic current does not exist from the beginning, but instead the inter-band contribution exists, which leads to the ME.

However, there is an interesting problem appearing from the inter-band contribution. Assume that $\Delta = 0$, i.e., in the normal state. Even in this case, the kernel $K_{xx}^{(N)}(0,0)$ does not vanish; the explicit form is

$$\begin{aligned} K_{xx}^{\text{intra},(N)}(0,0) =& 2e^2v^2\sum_{\boldsymbol{k}}\frac{v^2k_x^2}{\varepsilon^2(\boldsymbol{k})}\left[\frac{\partial f(\xi_+(\boldsymbol{k}))}{\partial\xi_+(\boldsymbol{k})}+\frac{\partial f(\xi_-(\boldsymbol{k}))}{\partial\xi_-(\boldsymbol{k})}\right] \\ =& -\frac{e^2v^2}{3\pi^2}\frac{(\mu^2-M^2)^{3/2}\Theta(|\mu|-M)}{|\mu|v^3}, \end{aligned} \tag{15}$$

and

$$\begin{aligned} K_{xx}^{\text{inter},(N)}(0,0) =& -2e^2v^2\sum_{\boldsymbol{k}}(1-\frac{v^2k_x^2}{\varepsilon^2(\boldsymbol{k})})\left[\frac{f(\xi_-(\boldsymbol{k}))-f(\xi_+(\boldsymbol{k}))}{\varepsilon(\boldsymbol{k})}\right] \\ =& -\frac{e^2v^2}{3\pi^2}\left[\frac{k_c^3}{\varepsilon(k_c)}-\frac{(\mu^2-M^2)^{3/2}\Theta(|\mu|-M)}{|\mu|v^3}\right], \end{aligned} \tag{16}$$

with $\Theta(x)$ being the step function.

In the usual situations in solids with finite Brillouin zone, $K_{xx}^{(N)} = 0$ hold since the paramagnetic term exactly cancels with the diamagnetic contribution. Therefore we expect that in the Dirac case, there must be some counter-contributions from the vicinity of the Brillouin zone boundary which exactly cancel the contributions from the vicinity of the Dirac point. Therefore, in the following, we calculate the Meissner kernel in superconducting state as

$$K_{xx} \equiv K_{xx}^{(S)} - K_{xx}^{(N)}. \tag{17}$$

The mathematical justification of this subtraction can be carried out by studying the slightly modified model [10].

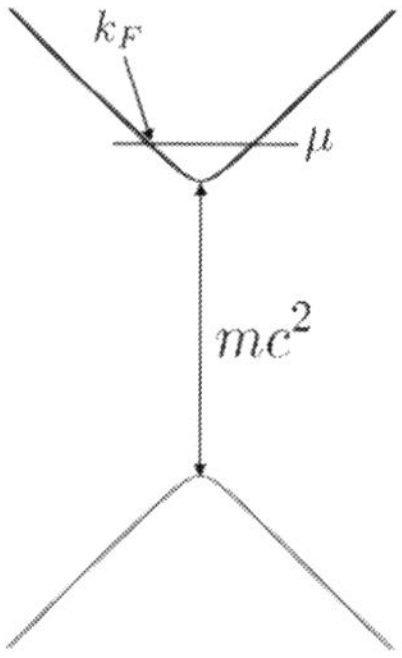

Figure 3. Dispersion relation in the non-relativistic limit.

By using Eq. (17), we numerically calculate $K_{xx}(0,0)$ at $T=0$. In the numerical calculation, we set $k_c = 10M/v$. Figures 2 (a) and 2 (b) show μ dependence of $K_{xx}^{(S)}, K_{xx}^{(N)}$, and K_{xx} for $\Delta = 0.1M$ and $\Delta = 0.7M$ respectively. We can see that although $K_{xx}^{(S)}$ and $K_{xx}^{(N)}$ are both negative, $K_{xx}(0,0)$ is always positive and finite, thus we obtain the Meissner state. (For $\Delta = 0.1M, \mu = 0$, $K_{xx} \sim 0.0007e^2v^2$.) In particular, when $|\mu| < M$, the Meissner kernel purely comes from the inter-band contribution, since the intra-band contributions vanish both in $K_{xx}^{(N)}$ and $K_{xx}^{(S)}$.

We can also see in Figs. 2 (a) and 2 (b) that K_{xx} strongly depends on Δ. This is also different from the result of the conventional BCS case in which K_{xx} does not depend on Δ and depends only on the number and the mass of careers.

Large band gap case: non-relativistic electron.- Up to now, we consider the case in which the band gap is very small ($\sim$ 10 meV). On the other hand, we can also consider the opposite case, i.e., the large band gap case. In particular, it is fascinating to test whether our theory can reproduce the results for non-relativistic theory. To access to the non-relativistic limit, we replace our parameters as M to mc^2, and v to c, with m being the mass of an electron and c the speed of light. Then, we can regard the upper band as the real electron, and the lower band as the positron [see Fig. 3]. We further assume that the chemical potential is on the upper band, and that $\mu - mc^2$ is much smaller than mc^2. Then we obtain the Meissner kernel in the normal state as

$$K_{xx}^{\mathrm{intra},(\mathrm{N})}(0,0) = -\frac{e^2c^2}{3\pi^2}\frac{(\mu^2-(mc^2)^2)^{3/2}}{\mu c^3} \sim -\frac{e^2n}{m}, \tag{18}$$

and

$$K_{xx}^{\mathrm{inter},(\mathrm{N})}(0,0) = -\frac{e^2c^2}{3\pi^2}\left[\frac{k_c^3}{\varepsilon(k_c)} - \frac{(\mu^2-(mc^2)^2)^{3/2}}{\mu c^3}\right] \sim \frac{e^2n}{m} - \frac{e^2c^2}{3\pi^2}\frac{k_c^3}{\varepsilon(k_c)}, \tag{19}$$

where $k_F = \sqrt{2m(\mu - mc^2)}$ is the Fermi momentum and $n = \frac{k_F^3}{3\pi^2}$ is the electron density. Here we use the approximantion $\mu \sim mc^2$.

Comparing Eqs. (18) and (19) with the results for the electron gas [8], we can see that K_{xx}^{intra} equals to the paramagnetic term in the electron gas and that K_{xx}^{inter} equals to the diamagnetic term apart from the divergent part $(-\frac{e^2c^2}{3\pi^2}\frac{k_c^3}{\varepsilon(k_c)})$. Therefore, this result indicates, first, that we can reproduce the result of the electron gas starting from the Dirac model; second, that the diamagnetic term of the kernel in the electron gas theory originates from the inter-band contribution between positrons and electrons.

4. Summary

In summary, we have investigated the ME of three-dimensional massive Dirac electron in superconducting state on the basis of Kubo formula. We have to care about the finite Meissner kernel in the normal state, but this can be eliminated by subtracting the normal component. Although the diamagnetic term of the current operator is absent in Dirac electron system, the Meissner kernel is finite for any value of the chemical potential. This is due to the inter-band contribution. We also have studied the large band-gap limit to discuss the non-relativistic case. We clarify the relation that the paramagnetic term of the kernel originates from the intra-band term and the diamagnetic term from the inter-band term.

References

[1] Castro Neto A H, Guinea F, Peres N M R, Novoselov K S, and Geim A K 2009 *Rev. Mod. Phys.* **81** 109
[2] Fuseya Y, Ogata M, and Fukuyama H 2015 *J. Soc. Phys. Jpn.* **84** 012001
[3] Kajita K, Nishio Y, Tajima N, Suzumura Y, and Kobayashi A 2014 *J. Phys. Soc. Jpn.* **83** 072002
[4] Fuseya Y, Ogata M, and Fukuyama H 2009 *Phys. Rev. Lett.* **102** 066601
[5] Abrikosov A A 1998 *Phys. Rev.* B **58** 2788
[6] Fukuyama H and Kubo R 1970 *J. Phys. Soc. Jpn.* **28** 570
[7] Fuseya Y, Ogata M, and Fukuyama H 2012 *J. Phys. Soc. Jpn.* **81** 093704
[8] See, e.g., J. R. Schrieffer, *Theory of Superconductivity* (Addison-Wesley, Reading, MA, 1988).
[9] The existence of the diamagnetic current in Dirac electron systems is discussed in Uchoa B. and Castro Neto A H 2009 *Phys. Rev. Lett.* **102** 109701, and Kopnin N B and Sonin E B 2009 *Phys. Rev. Lett.* **102** 109702.
[10] Mizoguchi T and Ogata M in preparation.

International Workshop on Dirac Electrons in Solids 2015 IOP Publishing
Journal of Physics: Conference Series **603** (2015) 012005 doi:10.1088/1742-6596/603/1/012005

Excitation spectrum of correlated Dirac fermions

Z Jalali[1], **S A Jafari**[1,2,3]

[1] Department of Physics, Sharif University of Technology, Tehran 11155-9161, Iran
[2] Center of excellence for Complex Systems and Condensed Matter (CSCM), Sharif University of Technology, Tehran 1458889694, Iran
[3] School of Physics, Institute for Research in Fundamental Sciences (IPM), Tehran 19395-5531, Iran

E-mail: jafari@physics.sharif.edu

Abstract. Motivated by the puzzling optical conductivity measurements in graphene, we speculate on the possible role of strong electronic correlations on the two-dimensional Dirac fermions. In this work we employ the slave-particle method to study the excitations of the Hubbard model on honeycomb lattice, away from half-filling. Since the ratio $U/t \approx 3.3$ in graphene is not infinite, double occupancy is not entirely prohibited and hence a finite density of doublons can be generated. We therefore extend the Ioff-Larkin composition rule to include a finite density of doublons. We then investigate the role played by each of these auxiliary particles in the optical absorption of strongly correlated Dirac fermions.

1. Introduction

Strong correlation has remarkable manifestations in properties of metals. Recent emergence of the so called Dirac materials in solids has provided a new playground for condensed matter physicists. One of the most important two-dimensional examples of Dirac materials is graphene the low energy theory of which is described by 2+1 dimensional Dirac Hamiltonian. As far as the non-interacting Dirac theory is concerned, the single-particle eigen-states of this Hamiltonian are chiral which protects them from complete back scattering. Many other amazing properties of graphene follows from the underlying Dirac nature of its free theory. On top of any free theory an interesting theoretical question would be, what happens if some forms of interactions are turned on. In this respect one of the most important forms of interactions are local (on-site) interactions parameterized by the so called Hubbard U. This type of local correlations in ordinary metals leads to interesting spin liquid followed by Mott insulating behavior when the parent metal is at half-filling [1]. Upon doping away from half-filling the on-site Coulomb repulsion would also lead to possible high temperature superconductivity. Given such a background for the short-range Coulomb interactions in two dimensions, it is interesting to ask similar questions about the Dirac matter: i.e. what would be the consequences of strong electronic correlations on the physical properties of Dirac solids?

In the present work, we are interested in examining the role of strong correlations on the physical properties of Dirac electrons. One of the important problems of condensed matter systems is the optical conductivity that couples to the particle-hole channel of excitations, hence providing information not accessible in the single-particle measurements such as angular resolved photo emission spectroscopy (ARPES) or tunneling experiments. Recent *ab-initio* estimates of the Hubbard U in graphene shows that it is remarkably strong on the scale of $\sim$ 10 eV, i.e.

more than 3 times larger than the hopping amplitude t of electrons between the $2p_z$ orbitals of neighboring carbon atoms [2]. Therefore it is timely to approach the problem from strong coupling side, and study the manifestations of strong correlation. In this work we employ slave-boson method to calculate the optical conductivity of Dirac electrons of graphene at zero temperature.

2. Slave Boson Method and Hubbard Model

Hubbard model is the canonical model to describe strong correlations in solids. This model consists of two terms, kinetic term that in the case of graphene has a low-energy Dirac nature; and an on-site Coulomb repulsion term of strength U. Hamiltonian of this model is shown by (1) where t is the hopping amplitude and $\langle i,j\rangle$ implies that sites i,j are nearest neighbors.

$$H = -t\sum_{\langle i,j\rangle,\sigma} c^{\dagger}_{i,\sigma}c_{j,\sigma} + U\sum_{i} n_{i\uparrow}n_{i\downarrow} \tag{1}$$

The hopping takes place on the honeycomb lattice between two different sub-lattices (A,B). In general the Hilbert space of Hubbard model at each site has four states: empty state , two single occupancy states with opposite spin and double occupancy. When U is infinite, the double occupancy is forbidden and the doubly occupied configurations are prohibited by infinitely large U. In the case of graphene, the Hubbard $U \approx 3.3t$ [2] and hence is not infinitely large. Therefore all four states are allowed. A well established technique to study the Hubbard model in this regime is the so called slave-particle method. In this work we use the slave boson formalism to study Hubbard model in the intermediate range of U. Instead of using physical electron operators $c^{\dagger}_{i,\sigma}$ one introduces auxiliary particles $f^{\dagger}_{i,\sigma}$, $h^{\dagger}_{i}$ and $d^{\dagger}_{i}$. The electron creation operator is represented by,

$$c^{\dagger}_{i,\sigma} = f^{\dagger}_{i,\sigma}h_i + \sigma d^{\dagger}_{i} f_{i,-\sigma}\ . \tag{2}$$

The operators $f_{i,\sigma}, h_i, d_i$ are associated with spinons and holons and doublons, respectively. Spinons are particles with spin but no charge. Holons and doublons are particles without spin but they have positive and negative charges, respectively. There is an equivalent between new (unphysical) Hilbert space and old (physical) Hilbert space of real electrons:

$$h^{\dagger}_{i}|vac\rangle \equiv |0\rangle_i \qquad f^{\dagger}_{i,\sigma}|vac\rangle \equiv |\uparrow\rangle_i \quad , \quad f^{\dagger}_{i,-\sigma} \equiv |\downarrow\rangle_i \quad , \quad d^{\dagger}_{i}|vac\rangle \equiv |\uparrow\downarrow\rangle_i$$

. Since each lattice site is either empty, singly occupied or doubly occupied, the number of above auxiliary fields satisfies a local constraint is represented by:

$$y_i \equiv \sum_{\sigma} N^{f}_{i,\sigma} + N^{d}_{i} + N^{h}_{i} - 1 = 0. \tag{3}$$

Substituting the slave-boson representation of electron creation operators in the Hubbard model we obtain the following representation of the Hubbard Hamiltonian in terms of auxiliary particles [3]:

$$H = -t\sum_{\langle i,j\rangle}(\hat{\chi}^{b\dagger}_{i,j}\hat{\chi}^{f}_{i,j} + \hat{\Delta}^{b\dagger}_{i,j}\hat{\Delta}^{f}_{i,j} + h.c.) + U\sum_{i} d^{\dagger}_{i}d_i \tag{4}$$

where bosonic and fermionic bi-particle operators $\hat{\chi}$ and $\hat{\Delta}$ are define as:

$$\hat{\chi}^{f}_{i,j} = \sum_{\sigma} f^{\dagger}_{i,\sigma}f_{j,\sigma}, \qquad \hat{\chi}^{b}_{i,j} = h^{\dagger}_{i}h_j - d^{\dagger}_{i}d_j, \qquad \hat{\Delta}^{f}_{i,j} = \sum_{\sigma} f_{i,-\sigma}f_{j,\sigma}, \qquad \hat{\Delta}^{b}_{i,j} = d_i hj + h_i dj$$

Journal of Physics: Conference Series **603** (2015) 012005 doi:10.1088/1742-6596/603/1/012005

We decouple the above Hamiltonian into bosonic and fermionic hopping ($\hat{\chi}$) and pairing ($\hat{\Delta}$) channels by using Hubbard-Stratonovic transformation and write the partition function as:

$$Z = \int Df^\dagger Df Dh^\dagger Dh Dd^\dagger Dd D\chi D\Delta D\lambda e^{-\int d\tau \mathcal{L}} \tag{5}$$

$$\mathcal{L} = \sum_{i,\sigma} f^\dagger_{i,\sigma}\frac{\partial}{\partial\tau}f_{i,\sigma} + \sum_i h^\dagger_i\frac{\partial}{\partial\tau}h_i + \sum_i d^\dagger_i\frac{\partial}{\partial\tau}d_i + H - \sum_i \lambda_i y_i. \tag{6}$$

The field λ_i is Lagrange multiplier that enforces the local constraint Eq. (3) and in saddle point approximation we replace it by site-independent (mean field value) λ . As can be seen in Eq. (2), in the enlarged Hilbert space of auxiliary particles, the physical electron is invariant under a local $U(1)$ gauge transformation. This internal gauge field is the glue that binds the auxiliary particles back together to form the physical electrons. Therefore it is important to consider the $U(1)$ phase fluctuation of the corresponding order parameters. Within the slave-boson mean field, the order parameters are replaced by their mean field values as $\hat{\Delta}_{i,j} \to \Delta$ and $\hat{\chi}_{i,j} \to \chi$. Of course they may have non-trivial dependence on the direction of the bond connecting sites i and j. Such non-s-wave order parameters can be important in highly doped graphene when it is doped to the M-point of the Brillouin zone [4]. But since in the present work we are concerned with the normal state where superconducting pairing amplitudes will be assumed to be zero, we are not concerned with these possibilities. Ignoring the superconducting order, the normal state Hamiltonian in the slave-boson mean field becomes,

$$\begin{aligned} H_{\text{eff}} &= H_f + H_h + H_d + E_0 \\ H_f &= -\lambda\sum_{i,\sigma} f^\dagger_{i,\sigma}f_{i,\sigma} - t\chi^b \sum_{<i,j>,\sigma} f^\dagger_{i,\sigma}f_{j,\sigma} + h.c \\ H_h &= -(\lambda-\mu_e)\sum_i h^\dagger_i h_i - t\chi^f\sum_{i,j} h^\dagger_i h_j + h.c. \\ H_d &= -(-U+\lambda+\mu_e)\sum_i d^\dagger_i d_i + t\chi^f\sum_{i,j} d^\dagger_i d_j \\ E_0 &= 6tN\chi^f\chi^b + (\lambda-\mu_e)N + \mu_e N_e \end{aligned} \tag{7}$$

In above relation N is the total number of lattice sites half of which are on sub-lattice A, and the other half on sub-lattice B giving total $N/2$ unit cells. The quantity N_e is the total number of electrons. All hopping terms between the nearest neighbors and hence from sub-lattice A to B and vice versa. The chemical potential μ_e controls the total number of electrons with respect to empty lattice, while λ forces the constraint of Eq. (3). The above equation parameterizes the chemical potential $\mu_f = \lambda$, $\mu_h = \lambda - \mu_e$, $\mu_d = \lambda + \mu_e - U$ in terms of two independent Lagrange multipliers μ_e and λ.

3. Extension of Ioffe-Larkin composition rule in presence of doublons

Now that the Hamiltonian in the mean field approximation is decomposed into pieces containing various auxiliary particles, we need a rule that combines the conductivity (i.e. current-current correlation function) in different sectors and gives the conductivity of physical electrons. To this end, we consider phase fluctuation of order parameters around the mean field saddle point values and look at the $U(1)$ gauge theory of the normal state. In this situation we have two order parameters, the phase fluctuations of which are given by a_{ij} link fields as follows,

$$\chi^{f,b}_{ij} = e^{ia_{ij}}\chi^{f,b} \tag{8}$$

Journal of Physics: Conference Series **603** (2015) 012005 doi:10.1088/1742-6596/603/1/012005

In terms of these phase variables the Lagrangian becomes,

$$L = L_f + L_h + L_d \tag{9}$$

$$L_f = \sum_{i,\sigma} f^\dagger_{i,\sigma}(\frac{\partial}{\partial\tau} - \mu_f)f_{i,\sigma} - t\chi^b \sum_{\langle i,j\rangle,\sigma} e^{ia_{ij}} f^\dagger_{i,\sigma} f_{j,\sigma} + h.c.$$

$$L_h = \sum_i h^\dagger_i(\frac{\partial}{\partial\tau} - \mu_h)h_i - t\chi^f \sum_{i,j} e^{ia_{ij}} h^\dagger_i h_j + h.c.$$

$$L_d = \sum_i d^\dagger_i(\frac{\partial}{\partial\tau} - \mu_d)d_i + t\chi^f \sum_{i,j} e^{ia_{ij}} d^\dagger_i d_j + h.c$$

in above equation $a_{i,j}$ is defined as [5, 6]:

$$(\vec{r}_i - \vec{r}_j).\vec{a}[(r_i + r_j)/2] = \int_j^i \vec{a}.d\vec{r}.$$

We are interested to study low energy physics on the Honeycomb lattice where the low energy states is described by massless Dirac fermions [7]:

$$H^\pm_p = v_F\gamma^0\vec{\gamma}.\vec{p}. \tag{10}$$

In above Hamiltonian $\tau_z = \pm$ is associated with two Dirac points (K, K') and v_F is Fermi velocity of electrons and equals $\sqrt{3}ta/2$ and $\vec{p}$ are two dimensional vectors. The matrices $\vec{\gamma}$ are given by:

$$\gamma^0 = -\sigma_z \quad , \quad \gamma^1 = i\sigma_y \quad , \quad \gamma^2 = \pm i\sigma_x \equiv i\tau_z\sigma_x. \tag{11}$$

By going to the continuum limit and considering phase fluctuations of order parameters as internal gauge field, we can write:

$$L_f = \sum_\sigma \int d\vec{r} \;\; \overline{f}(\vec{r},\sigma)\gamma^0(\frac{\partial}{\partial\tau} - \mu_f)f(\vec{r},\sigma) + \hbar v_f \sum_\sigma \int d\vec{r} \;\; \overline{f}(\vec{r},\sigma)\vec{\gamma}.(\frac{\vec{\nabla}}{i} + \vec{a})f(\vec{r},\sigma) \tag{12}$$

$$L_h = \sum_\sigma \int d\vec{r} \;\; \overline{h}(\vec{r},\sigma)\gamma^0(\frac{\partial}{\partial\tau} - \mu_h)h(\vec{r},\sigma) + \hbar v_h \sum_\sigma \int d\vec{r} \;\; \overline{h}(\vec{r},\sigma)\vec{\gamma}.(\frac{\vec{\nabla}}{i} + \vec{a})h(\vec{r},\sigma) \tag{13}$$

$$L_d = \sum_\sigma \int d\vec{r} \;\; \overline{d}(\vec{r},\sigma)\gamma^0(\frac{\partial}{\partial\tau} - \mu_d)h(\vec{r},\sigma) + \hbar v_d \sum_\sigma \int d\vec{r} \;\; \overline{d}(\vec{r},\sigma)\vec{\gamma}.(\frac{\vec{\nabla}}{i} + \vec{a})d(\vec{r},\sigma). \tag{14}$$

Note that $\overline{f}$ is defined as $f^\dagger\gamma^0$ and $f = (f_A, f_B)$ is a spinor encoding the two sub-lattice structure of the honeycomb lattice. Similar definition holds for holons and doublons. Finally the velocity of three auxiliary particles are given by:

$$v_f = \frac{\sqrt{3}ta\chi^b}{2} \quad , \quad v_h = \frac{\sqrt{3}ta\chi^f}{2} \quad , \quad v_d = -\frac{\sqrt{3}ta\chi^f}{2}.$$

The coupling between the auxiliary particles and the internal gauge filed is of the following form,

$$L_{int} = \vec{J}_f.\vec{a} + \vec{J}_h.\vec{a} + \vec{J}_d.\vec{a}. \tag{15}$$

Minimization of the above above term with respect to the internal gauge field $\vec{a}$ gives

$$\vec{J}_f + \vec{J}_h + \vec{J}_h = 0. \tag{16}$$

This relation is indeed very plausible. At every given site the total number of auxiliary particles is conserved. Therefore the at a given site, there is only one of the above auxiliary particles available. Likewise at a neighboring site to which a hopping is going to happen, another particle is available. Therefore always a third particle (among, f, d and h particles) is missing from the nearest neighbor hopping. Hence only two of the above auxiliary particles produce currents which by the very constraint forces the currents to satisfy the above relation.

Now let us couple an external electro-magnetic (EM) gauge field $\vec{A}$. Assuming that spinons have no EM charge, while charge of holons and doublons are $+1$ and -1, respectively, these bosons interact with EM field via the following minimal substitution:

$$(\frac{\vec{\nabla}}{i} + \vec{a}) \longrightarrow (\frac{\vec{\nabla}}{i} + \vec{a} \pm \vec{A}). \tag{17}$$

Having introduced the external EM gauge field $\vec{A}$, expanding up to second order in $\vec{A}$, and performing the integration over auxiliary fields we obtain

$$\begin{aligned} S_{\text{eff}}(a, A) = \quad & \sum_q \Pi^f_{\mu,\nu}(q) a_\mu(q) a_\nu(-q) + \Pi^h_{\mu,\nu}(q)[a(q) - A(q)]_\mu [a(-q) - A(-q)]_\nu \\ & + \Pi^d_{\mu,\nu}[a(q) + A(q)]_\mu [a(-q) + A(-q)]_\nu \end{aligned} \tag{18}$$

where $\Pi_{\mu,\nu}$ is current-current correlation function and is defined by:

$$\Pi^\alpha_{\mu,\nu} = \mathrm{tr}\langle J^\alpha_\mu(q) J^\alpha_\nu(-q)\rangle. \tag{19}$$

Here $\alpha = f, h, d$ and $q \equiv (\vec{q}, i\omega_n)$. If we integrate out internal gauge field a, effective action in terms of $\vec{A}$ becomes,

$$S_{\text{eff}}(A) = \sum_q \Pi_{\mu,\nu} A_\mu A_\nu \tag{20}$$

where

$$\Pi = \frac{4\Pi^d \Pi^h + \Pi^d \Pi^f + \Pi^h \Pi^f}{\Pi^d + \Pi^h + \Pi^f}. \tag{21}$$

This relation arises from the fact that fermions and holons and doublons are coupled to each others via internal gauge fields and hence they can not move independently. Although we employed the linearized dispersion of underlying electrons on a honeycomb lattice to emphasize Dirac nature of the electrons, but the above relation relies on the constraint imposed by the internal gauge field and therefore it also holds when a tight-binding band picture valid over the entire hexagonal Brillouin zone of graphene is used.

The above relation is a generalization of what is known as Ioffe-Larking composition rule to the case where a finite number of doublons is also present. Finite density of doublons arise from the fact that the Hubbard U is not infinitely large. In the following let us repeat the physical argument of Ioffe and Larkin [8] to convince ourselves that the above composition rule in presence of doublons is independent of the details of the band structure of underlying electrons. Assume that the internal gauge fields gives rise to electric field $\vec{e}$. In the presence of external electric field $\vec{E}$, effective electric fields for each particles are:

$$\vec{e}_f = \vec{e}, \qquad \vec{e}_h = \vec{e} - \vec{E}, \qquad \vec{e}_d = \vec{e} + \vec{E}, \tag{22}$$

as we have assumed that the EM charges belong only to holons and doublons. Current due to each particle is given by the following equation:

$$\vec{j}_\alpha = \sigma_\alpha \vec{e}_\alpha \tag{23}$$

where $\alpha = f, h, d$ and the above relation is seen as a tensorial equation where tensor components are not shown for clarity. Using this relation in the constraints on the currents, Eq. (16) gives the following relation between the internal and external electric fields:

$$\vec{e} = \frac{\sigma_h - \sigma_d}{\sigma_f + \sigma_h + \sigma_d} \vec{E}. \tag{24}$$

The physical charge current is defined as difference between doublonic and holonic currents,

$$\vec{j} = \vec{j}_d - \vec{j}_h = \sigma \vec{E}, \quad \Rightarrow \quad \sigma = \frac{4\sigma^d\sigma^h + \sigma^d\sigma^f + \sigma^h\sigma^f}{\sigma^d + \sigma^h + \sigma^f}. \tag{25}$$

This conductivity relation is same as equation that is calculated in Eq. (20) and hence is not dependent on the details of the band dispersion and holds for the low-energy Dirac as well as the full tight-binding kinetic parts. If we consider the situation that there is no doublon in our systems (large limit of U/t), the doublon channel is blocked and hence the conductivity σ^d of doublons vanishes. Therefore in this limit the composition formula for the conductivity reduces to the relation containing only spinons and holons contributions [9, 10].

4. Normal State

Let us now proceed by rewriting Eq (7) in the Fourier space which leads to the following normal state Hamiltonian:

$$H_f = \sum_{k,\sigma} \begin{bmatrix} f^{\dagger A}_{k,\sigma} & f^{\dagger B}_{k,\sigma} \end{bmatrix} \begin{bmatrix} -\mu_f & -t\chi^b \eta_k \\ -t\chi^b \eta^*_k & -\mu_f \end{bmatrix} \begin{bmatrix} f^A_{k,\sigma} \\ f^B_{k,\sigma} \end{bmatrix}, \tag{26}$$

$$H_h = \sum_{k} \begin{bmatrix} h^{\dagger A}_{k} & h^{\dagger B}_{k} \end{bmatrix} \begin{bmatrix} -\mu_h & -t\chi^f \eta_k \\ -t\chi^f \eta^*_k & -\mu_h \end{bmatrix} \begin{bmatrix} h^A_{k} \\ h^B_{k} \end{bmatrix}, \tag{27}$$

$$H_d = \sum_{k} \begin{bmatrix} d^{\dagger A}_{k} & d^{\dagger B}_{k} \end{bmatrix} \begin{bmatrix} -\mu_d & t\chi^f \eta_k \\ t\chi^f \eta^*_k & -\mu_d \end{bmatrix} \begin{bmatrix} d^A_{k} \\ d^B_{k} \end{bmatrix}, \tag{28}$$

$$E_0 = (\lambda - \mu_e)N + 6tN\chi^f\chi^b + \mu_e N_e, \tag{29}$$

where $\eta_k = \sum_{\vec{\delta}} e^{i\vec{k}.\vec{\delta}}$ and summation applies on the nearest neighbor vectors. μ_f, μ_h, μ_d are defined in previous section. Diagonalizing the above Hamiltonian gives the following energy eigen-values for fermions and holons and doublons:

$$E^f_{k,\sigma} = -\mu_f \pm t\chi^b |\eta_k|. \tag{30}$$

$$E^h_k = -\mu_h \pm t\chi^f |\eta_k|. \tag{31}$$

$$E^d_k = -\mu_d \pm t\chi^f |\eta_k|. \tag{32}$$

At the mean field level where the above auxiliary particles do not interact, the free energy at a given temperature T is given by:

$$F = -\frac{2}{\beta} \sum_{k,s=\pm} \ln(1+e^{-\beta E^f_{k,s}}) + \frac{1}{\beta} \sum_{k,s=\pm} \ln(1+e^{-\beta E^h_{k,s}}) + \frac{1}{\beta} \sum_{k,s=\pm} \ln(1+e^{-\beta E^d_{k,s}}) + (\lambda - \mu_e)N + 6tN\chi^f\chi^b \tag{33}$$

where $s = \pm$ stands for two branches of energy bands for each auxiliary particle. Minimizing free-energy leads to follow Self-consistency equations:

$$\frac{\partial F}{\partial \lambda} = 0 \quad \rightarrow \quad 2\sum_{k,s=\pm} f(E_{k,s}) + \sum_{k,s=\pm} n_B(E^h_{k,s}) + \sum_{k,s=\pm} n_B(E^d_{k,s}) - N = 0, \tag{34}$$

$$\frac{\partial F}{\partial \chi^f} = 0 \quad \rightarrow \quad \sum_{k,s=\pm} st|\eta_k| n_B(E^h_{k,s}) + \sum_{k,s=\pm} st|\eta_k| n_B(E^d_{k,s}) + 6tN\chi^b = 0, \tag{35}$$

$$\frac{\partial F}{\partial \chi^b} = 0 \quad \rightarrow \quad 2\sum_{k,s=\pm} st|\eta_k| f(E_{k,s}) + 6tN\chi^f = 0, \tag{36}$$

where $n_B(E)$ and $f(E)$ are Bose-Einstein and Fermi-Dirac distribution functions, respectively. If temperature is nonzero all of particles are distributed according to their corresponding distribution function among their energy levels and the self-consistent solution of the above equations determines various parameters. However in the limit of $T \rightarrow 0$, Bose particles may have a chance to undergo a Bose condensation that gathers them at the bottom of their energy band. Therefore in zero temperature if holons and doublons have zero energy they could be condensed in the bottom of their energy band that lies at $\vec{k} = 0$. Holons can condense when $\lambda = \mu_e - 3t\chi^f$. In such a case, the minimum energy of doublons becomes $U - 2\mu_e$. Consequently, depending on the value of μ_e holons or doublons or both of them could be condensed i.e. if $\mu_e = U/2$ we are in half-filling and have particle hole symmetry and both of holon and doublon condense, but for $\mu_e > U/2$ ($\mu_e < U/2$) only douboln
s (holons) condense. For example if system is doped by holes, following self-consistency equations at $T = 0$ becomes,

$$\lambda = \mu_e - 3t\chi^f, \qquad \chi^b = \frac{1}{2}\delta, \qquad \chi^f = \frac{1}{3N}\sum_k{}' |\eta_k|. \tag{37}$$

The prime in the above summation means that the sum is performed over the occupied fermionic states and δ is number of holons per site. By substituting χ^f we obtain $\lambda = \frac{t}{N}\sum'_k |\eta_k|$.

Away from half-filling where only one of bosons condenses, the boson conductivity will be a strong Dirac delta peak around the accessible energy ω_b from the condensed momentum. Therefore the boson conductivity will be zero away from ω_b. This makes the total conductivity of physical particles zero away from ω_b which is not consistent with the experiment of Ref. [11]. Hence as far as the doped graphene is concerned, the bosons will not be in the condensed phase as the light absorption in a wide energy range is non-zero.

5. Conclusions

For the intermediate values of U the presence of doublons are allowed and they are coupled to the fermions and holons by internal gauge field. In the absence of pairing parameters (i.e. in the normal state) we extended the Ioffe-Larkin composition rule to include the contributions of doublons in the physical conductivity. We also re-derived it using a general physical argument. We studied normal state of Hubbard model in zero temperature and showed in this case depending on electronic chemical potential, bosons or doublons or both of them can be condensed. We further concluded that in doped graphene, the bosons (either of holons or doublons) can not be condensed. Therefore one must consider non-zero temperatures in order to understand the optical absorption experiment of Ref. [11].

References

[1] Meng Z Y and Lang T C and Wessel S and Assaad F F and Muramatsu A 2010 *Nature* **464** 847-851

[2] Wehling T O and Sasioglu E and Friedrich C and Lichtenstein A I and Katsnelson M I and Blügel, S 2011 *Phys. Rev.* L **106** 236805
[3] Vaezi A and Wen X G 2010 *arXiv:* 1010.5744v1
[4] Nandkishore R and Levitov L and Chubakov A 2012 *Nature Phys* **8** 2
[5] Lee P A and Nagaosa N and Wen X G 2006 *Rev. Mod. Phys* **78** 17
[6] Nagaosa N and Lee P A 1990 *Phys. Rev.* L **64** 2450
[7] Gusynin V P and Sharapov S G 2006 *Phys. Rev.* B **73** 245411
[8] Ioffe L and Larkin A 1989 *Phys. Rev.* B **39** 8988
[9] Nagaosa N 1998 *Quantum Field Theory in Strongly Correlated Electronic Systems* (Berlin: Springer) p 150
[10] Lee P A and Nagaosa N 1992 *Phys. Rev. B* **46** 5621
[11] Li Z Q, Henriksen E A, Jiang Z, Hao Z, Martin M C, Kim P, Stormer H L and Basov D N 2008 *Nat. Phys.* **4** 532

Highly anisotropic physics in phosphorene

Motohiko Ezawa

Department of Applied Physics, University of Tokyo, Hongo 7-3-1, 113-8656, Japan

E-mail: ezawa@ap.t.u-tokyo.ac.jp

Abstract. Phosphorene, monolayer black phosphorus, is a highly anistropic material, where the band structure is Dirac like in one direction and Schrödinger like in the other direction. We present several analytic formulas to demonstrate the electronic properties. The conductance is highly anisotropic reflecting the anisotropy of the band structure. We also detemine the cyclotron motion in phosphorene and Landau level quantization by using the Bohr-Sommerfeld quantization. We show that the Landau-level energy behaves as $B^{2/3}$ as a function of external magnetic field, which is highly contrasted to the case of graphene where the Landau-level energy behaves as $\sqrt{B}$.

1. Introduction

Monolayer materials are one of the most active fields in condensed matter physics. Graphene, a monolayer carbon atom system, has various fascinating physics since its low-energy band structure is described by the Dirac theory[1]. The success in graphene excites a further search of monolayer materials. Silicene, germanene and stanene, which are monolayers of silicon, germanium and tin, respectively, are among the most remarkable materials since they are expected to be topological insulators in nature[2]. Transition-metal dichalcogenides also attract much attention in the context of the valleytronics[3], which uses the valley degree of freedom.

Recently, phosphorene, monolayer black phosphorus, has been experimentally manufactured by exfoliation method, which has activated a rapid growth of experimental and theoretical studies[4]. Phosphorene is a direct-gap semiconductor with the gap in the region of visible light, which will be suitable for future applications to light-emitting diodes and solar cells. Field-effect transistor made of phosphorene has already been realized experimentally[5]. It is intriguing that the band structure in the low-energy theory is highly anisotropic, where the one direction is Dirac like and the other direction is Schrödinger like[6, 7, 8]. It differentiates the other monolayer materials, where the basic physics is described by the Dirac theory. Recent experiments show highly angle-dependent conductance[9].

In this paper, we investigate the electronic properties of phosphorene which arise from the highly anisotropic energy dispersion. We derive several analytic formulas on the conductivity and the cyclotron motion. We have derived the Landau level energy, which behaves $B^{2/3}$ as a function of external magnetic field. We also show the results on the case of the gapless limit since the physical quantity shows an extremely simple scaling relation due to the scale free structure of the band structure.

2. Hamiltonian

Phosphorene is a direct-gap semiconductor, where the gap exists at the Γ point. The low-energy Hamiltonian for phosphorene is given by the Hamiltonian[6, 7, 8],

$$H_{2D} = \left(uk_y^2 + m\right)\tau_x + v_F k_x \tau_y, \quad (1)$$

where τ is the Pauli matrices and m is the gap acting as the Dirac mass. The Hamiltonian is linear in the k_x direction, while parabolic in the k_y direction, which produce highly anisotropic physics. This Hamiltonian emerges quite in general when two Dirac cones with the opposite chiralities merge into one cone. It is originally discussed in highly strained graphene[10, 11, 12], where the transfer integral is different over 2 times along the x and y directions. However this scenario is unrealistic since we cannot realize such a strong anisotropy in graphene by applying strain. Later the mimic of this energy is realized in optical lattice[13], where the transfer integral can be tuned arbitrarily. Phosphorene is the first materialization of the Hamiltonian, where the lattice structure is highly anisotropic along the x and y directions, and called the puckered structure. The energy spectrum reads

$$\varepsilon = \pm\sqrt{v_{\mathrm{F}}^2 k_x^2 + \left(u k_y^2 + m\right)^2}. \tag{2}$$

The energy dispersion is Dirac like in the k_y direction, but Schrödinger like in the k_y direction. This anisotropy leads to a highly anisotropic physics in phosphorene such as conductance.

The DOS is given by

$$\rho(\mu) = \frac{4}{v_{\mathrm{F}}\sqrt{u}}\mathrm{Re}\left[i\frac{|\mu|}{\sqrt{|\mu|+m}}K\left(\frac{2\mu}{|\mu|+m}\right)\right], \tag{3}$$

where μ is the chemical potential, $K(x)$ is the complete elliptic integrals of the first kind. In the massless limit, the DOS is given by

$$\lim_{m/\mu\to 0}\rho(\mu) = \frac{4}{v_{\mathrm{F}}\sqrt{u}}\mathrm{Re}\left[iK(2)\right]\sqrt{|\mu|}. \tag{4}$$

We show the DOS in Fig. 1 both in the massive and massless cases for comparison. It has a good agreement with the previous theoretical results based on the tight-binding Hamiltonian[14]. The DOS is exactly zero within the gap, while the DOS is finite at the band edge $|\mu| = m$. The DOS for the massive Hamiltonian approaches that of the massless Hamiltonian and behaves proportional to the square of the chemical potential for $|\mu| > m$.

The behavior is significantly different from the one in the massive Dirac system,

$$\rho(\mu) = \frac{4}{2\pi v^2}|\mu|\theta(|\mu| - m), \tag{5}$$

where the DOS is linear as a function of μ

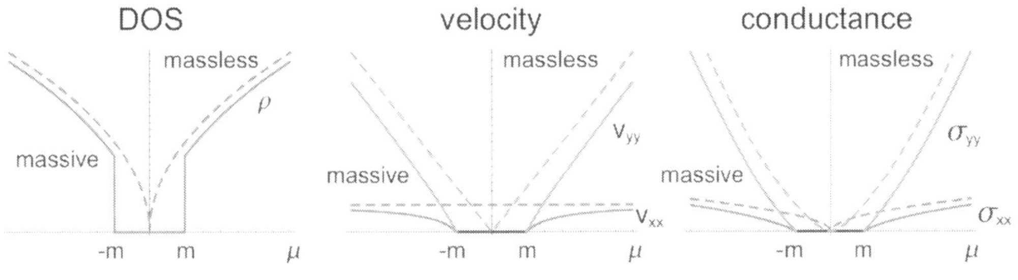

Figure 1. (a) DOS $\rho(\mu)$ as a function of the chemichal potential μ. (b) $\langle v_x^2\rangle$ and $\langle v_y^2\rangle$ as a function of the chemical potential μ. (c) σ_{xx} and σ_{yy} as a function of the chemical potential μ. Magenta (Cyan) curves represent the value along the x (y) direction. Solid (dotted) curves represent the massive (massless) case.

3. Velocity

We calculate the average of the Fermi surface velocity defined by

$$\langle v_x v_y \rangle = \frac{\int v_x v_y \delta(\varepsilon - \mu)\, dk}{\int \delta(\varepsilon - \mu)\, dk}, \tag{6}$$

where the velocity is given by differentiating the Hamiltonian with the corresponding momentum,

$$v_x = \frac{\partial \varepsilon}{\partial k_x} = \frac{v_{\mathrm{F}}^2 k_x}{\varepsilon}, \qquad v_y = \frac{\partial \varepsilon}{\partial k_y} = \frac{2uk_y\left(uk_y^2 + m\right)}{\varepsilon}. \tag{7}$$

Explicitly we have

$$\langle v_x^2 \rangle = \frac{2}{\rho(\mu)} \frac{4v_{\mathrm{F}}}{3|\mu|} \sqrt{\frac{|\mu| + m}{u}} \left[|\mu|\, K\left(\frac{m - |\mu|}{m + |\mu|}\right) - mE\left(\frac{m - |\mu|}{m + |\mu|}\right)\right], \tag{8}$$

where $E(x)$ is the complete elliptic integrals of the second kind, and

$$\begin{aligned} \langle v_y^2 \rangle &= \frac{2}{\rho(\mu)}[\frac{8\sqrt{u}\sqrt{m+|\mu|}}{15v_{\mathrm{F}}|\mu|}[(9\mu^2 - 2m^2)\, E\left(\frac{m - |\mu|}{m + |\mu|}\right) \\ &\quad + (2m - 9|\mu|)\sqrt{u}\sqrt{m + |\mu|}\mu K\left(\frac{m - |\mu|}{m + |\mu|}\right)]. \end{aligned} \tag{9}$$

Their massless limit reads

$$\lim_{m\to 0} \langle v_x^2 \rangle = \frac{2}{\rho(\mu)} \frac{4v_{\mathrm{F}}}{3} K(-1)\sqrt{|\mu|} \propto |\mu|^0, \tag{10}$$

and

$$\lim_{m\to 0} \langle v_y^2 \rangle = \frac{2}{\rho(\mu)} \frac{24}{5gv_{\mathrm{F}}} \left[E(-1) - K(-1)\right] |\mu|^{3/2} \propto |\mu|. \tag{11}$$

Thus, $\lim_{m\to 0} \langle v_x^2 \rangle$ is independent of the chemical potential, while $\lim_{m\to 0} \langle v_y^2 \rangle$ is proportional to the absolute value of the chemical potential. We show the average of the velocity in Fig. 1(b). The velocity is highly anisotropic. This is highly contrasted to the case of graphene, where the average velocity $\langle v_x^2 \rangle = \langle v_y^2 \rangle$ is constant and has no anisotropy.

4. DC Conductivity

Recent experiments have observed a strongly anisotropic conducting behavior[9]. Anisotropic transport behavior along different directions is a unique property of phosphorene. We now derive the analytic results on the conductance as a function of the chemical potential. The DC conductivity is obtained by solving the Boltzmann equation. The semiclassical conductivity tensor is given by using the average of the Fermi surface velocity

$$\sigma_{\alpha\beta} = e^2 \int dk \tau v_\alpha v_\beta \left(-\frac{\partial f}{\partial \varepsilon}\right) = e^2 \tau \rho(\mu) \langle v_\alpha v_\beta \rangle, \tag{12}$$

where τ is the life time due to the scattering. By substituting the mean velocity, we have

$$\sigma_{xx} = 2e^2\tau \frac{4v_{\mathrm{F}}}{3|\mu|} \sqrt{\frac{|\mu| + m}{u}} \left[\mu K\left(\frac{m - |\mu|}{m + |\mu|}\right) - mE\left(\frac{m - |\mu|}{m + |\mu|}\right)\right], \tag{13}$$

and

$$\sigma_{yy} = 2e^2\tau\frac{8\sqrt{u}\sqrt{m+|\mu|}}{15v_{\mathrm{F}}|\mu|}[(9\mu^2-2m^2)E\left(\frac{m-|\mu|}{m+|\mu|}\right) + (2m-9|\mu|)\sqrt{u}\sqrt{m+|\mu|}\,|\mu|\,K\left(\frac{m-|\mu|}{m+|\mu|}\right). \tag{14}$$

We show the conductance in Fig. 1(c). The conductance is highly anisotropic reflecting the anisotropy of the band structure. This is also highly contrasted with the case of graphene, where the conductance is isotropic and proportional to the chemical potential $\sigma \propto |\mu|$.

The Hall conductance σ_{xy} is zero since $\langle v_x v_y\rangle = 0$. We calculate the conductance along the general direction θ

$$\sigma_{\theta\theta} = e^2\tau\rho(\mu)\left\langle(v_x\cos\theta + v_y\sin\theta)^2\right\rangle = \sigma_{xx}\cos^2\theta + \sigma_{yy}\sin^2\theta. \tag{15}$$

The angle dependence of the conductance has a good agreement with the experimental results[9].

5. Cyclotron motion

To describe qualitatively the effect of weak magnetic field, we start with a simple semiclassical argument. Energy levels are quantized according to the Bohr-Sommerfeld quantization condition.

The equation of motion of an electron in magnetic field $\mathbf{B}$ is given by the Lorentz equation,

$$\frac{d\mathbf{k}}{dt} = -\frac{e}{c}\mathbf{v}\times\mathbf{B}. \tag{16}$$

Since the magnetic field does not affect the velocity, an electron performs a cyclotron motion along an equi-energy curve with ε given by (2), which is a quartic curve

$$v_{\mathrm{F}}^2k_x^2 + \left(uk_y^2+m\right)^2 = \varepsilon^2. \tag{17}$$

We show the cyclotron orbits in Fig. 2(b). The period is given as follows. Substituting the form of the velocity Eq. (7) into Eq. (16), we obtain

$$\frac{dk_x}{dt} = -2u\omega_0 k_y\left(m+uk_y^2\right), \qquad \frac{dk_y}{dt} = \omega_0 v_{\mathrm{F}}^2 k_x, \tag{18}$$

with $\omega_0 = eB/c\varepsilon$. Eliminating k_x in the above equation, we obtain

$$\frac{d^2k_y}{dt^2} = -2uv_{\mathrm{F}}^2\omega_0^2 k_y\left(m+uk_y^2\right). \tag{19}$$

Hence the period is given by

$$T = 2\int_{-k_{y,\max}}^{k_{y,\max}} dk_y \frac{dk_y}{\sqrt{-uv_{\mathrm{F}}^2\omega_0^2k_y^2\left(2m+uk_y^2\right)+C}}, \tag{20}$$

where the constant C is determined from the condition that $\dot{k}_y = 0$ when $k_y = k_{y,\max}$ as

$$C = uv_{\mathrm{F}}^2\omega_0^2k_{y,\max}^2\left(2m+uk_{y,\max}^2\right). \tag{21}$$

with

$$k_{y,\max} = \sqrt{\frac{|\varepsilon|-m}{u}} \tag{22}$$

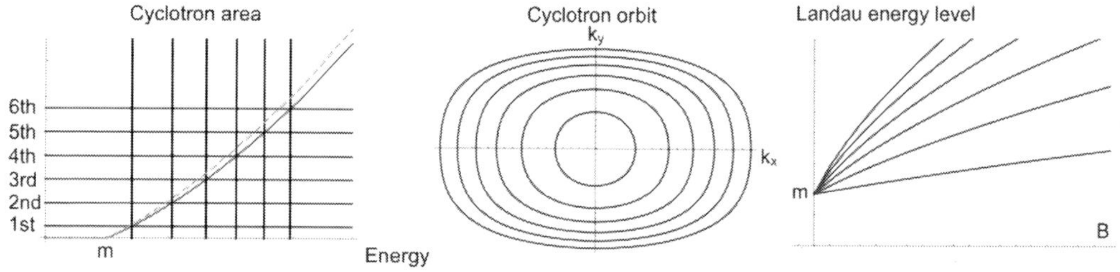

Figure 2. (a) Area of cyclotron motion as a function of the chemichal potential μ. The magenta curve represents the exact area Eq. (25), while the cyan curve represents the approximation given by Eq. (29). (b) Contours of cyclotron motion with the equal area. (c) Fan diagram representing the Landau level energy as a function of the magnetic field B.

for $|E| > m$. It is explicitly given as

$$T(\varepsilon) = \frac{4}{\sqrt{uv_{\mathrm{F}}^2\omega_0^2(m+|\varepsilon|)}} K\left(\frac{m-|\varepsilon|}{m+|\varepsilon|}\right). \tag{23}$$

In the massless limit, it reads

$$\lim_{m\to 0} T(\varepsilon) = \frac{2K(-1)}{\sqrt{uv_{\mathrm{F}}^2\omega_0^2}}\sqrt{\frac{1}{|\varepsilon|}}. \tag{24}$$

The cyclotron frequency is given by inverting the cyclotron time period, $\Omega_c = 2\pi/T$.

The area of the cyclotron orbit is given by

$$\begin{aligned} S(\varepsilon) &= \int_{-k_{y,\max}}^{k_{y,\max}} \frac{\sqrt{E^2-\left(uk_y^2+m\right)^2}}{v_{\mathrm{F}}} dk_y \\ &= \frac{4}{3v_{\mathrm{F}}}\sqrt{\frac{|\varepsilon|+m}{u}}\left[\mu K\left(\frac{m-|\varepsilon|}{m+|\varepsilon|}\right) - mE\left(\frac{m-|\varepsilon|}{m+|\varepsilon|}\right)\right]. \end{aligned} \tag{25}$$

We show the area $S(\varepsilon)$ in Fig. 2(a). The area of the cyclotron orbit will be observed by the de Haas van Alphen effect or the Shubnikov-de Haas effect[15]. The circumference of the cyclotron orbit is given by

$$\frac{dS}{d\varepsilon} = \frac{2|\varepsilon|}{v_{\mathrm{F}}\sqrt{u(m+|\varepsilon|)}} K\left(\frac{m-|\varepsilon|}{m+|\varepsilon|}\right). \tag{26}$$

The cyclotron mass is defined by[16]

$$m_c = \frac{1}{2\pi}\frac{dS}{d\varepsilon} \tag{27}$$

which can be observed by the cyclotron resonance.

The Landau level is derived by the Bohr-Sommerfeld quantization along one period of the cyclotron motion,

$$S(\varepsilon_n) = 2\pi(n+1/2)eB. \tag{28}$$

The energy of the Landau level is obtained by solving $S(\varepsilon_n)$ for ε_n. This is in general not possible. However, Eq. (25) is well approximated by

$$S(\varepsilon) \simeq \frac{4}{3v_{\mathrm{F}}\sqrt{u}} K(-1)\left(|\varepsilon|^{3/2} - |m|^{3/2}\right). \tag{29}$$

By inverting this approximation formula, we obtain the Landau level energy

$$|\varepsilon_n| = \left(\frac{3v_F\pi\sqrt{u}}{2}(n+1/2)eB + |m|^{3/2}\right)^{2/3}. \tag{30}$$

We show the Landau level energy as a function of B in Fig. 2(c). The $2/3$ dependence is highly contrasted to the case of the QHE in massive Dirac system, where

$$|\varepsilon_n| = \sqrt{\omega_c^2(n+1) + |m|^2}, \tag{31}$$

with ω_c being the cyclotron frequency in graphene. The difference will be checked experimentally in phosphorene.

We consider the massless limit ($m \to 0$). The area of the cyclotron orbit reads

$$\lim_{m\to 0} S(\varepsilon) = \frac{4}{3v_F\sqrt{u}} K(-1)|\varepsilon|^{3/2}, \tag{32}$$

and hence the cyclotron mass is

$$m_c = \frac{1}{2\pi}\lim_{m\to 0}\frac{dS}{d\varepsilon} = \frac{1}{v_F\pi\sqrt{u}}K(-1)\sqrt{|\varepsilon|}. \tag{33}$$

We may solve Eq. (28) to obtain

$$|\varepsilon_n| = \left(\frac{3v_F\pi\sqrt{u}}{2}(n+1/2)eB\right)^{2/3}, \tag{34}$$

which agrees with Eq. (30) with $m = 0$[10].

6. Discussion

We have derived several analytic formula of conductivity and quantum Hall effects in phosphorene. They are highly anisotropic reflecting the fact that the band structure is anisotropic. Especially $2/3$ dependence in the Landau-level energy is different from that in the case of graphene, which will be detected experimentally.

References

[1] Katsnelson M I *Graphene: Carbon in Two Dimensions* (Cambridge Univ. Press, Cambridge, 2012).
[2] Liu C-C, Feng W and Yao Y 2011 *Phys. Rev. Lett.* **107** 076802
[3] Xu X, Yao W, Xiao D and Heinz T F 2014 *Nat. Phys.* **10** 343
[4] Reich E G 2014 *Nature* **506** 19
[5] Li L, Yu Y, Ye G J, Ge Q, Ou X, Wu H, Feng D, Chen X H and Zhang Y 2014 *Nature Nanotech.* **9** 372
[6] Rodin A S, Carvalho A and Castro Neto A H 2014 *Phys. Rev. Lett.* **112** 176801
[7] Low T, Rodin A S, Carvalho A, Jiang Y, Wang H, Xia F and Castro Neto A H 2014 *Phys. Rev.* **B90** 075434
[8] Ezawa M 2014 *New J. Phys.* **16** 065015
[9] Liu H, Neal A T, Zhu Z, Xu X, Tomanek D and Ye P D 2014 *ACS Nano* **8** 4033
[10] Dietl P, Piechon F and Montambaux G 2008 *Phys. Rev. Lett.* **100** 236405
[11] Banerjee S and Pickett W E, *Phys. Rev.* **B86** 075124
[12] Montambaux G, Piechon F, Fuchs J-N and Goerbig M O 2009 *Eur. Phys. J. B* **73** 509
[13] Tarruell L, Greif D, Uehlinger T, Jotzu G and Esslinger T 2012 *Nature* **483** 302
[14] Yuan S, Rudenko A N and Katsnelson M I 2014 Transport and Optical Properties of Single- and Bilayer Black Phosphorus with Defects *Preprint* cond-mat/arXiv:1411.1681
[15] Li L, Ye G J, Tran V, Fei R, Chen G, Wang H, Wang J, Watanabe K, Taniguchi T, Yang L, Chen X H and Zhang Y 2014 Quantum Oscillations in Black Phosphorus Two-dimensional Electron Gas *Preprint* cond-mat/arXiv:1411.6572
[16] Falkovsky L A 2013 *JETP Letters* **97** 429

Journal of Physics: Conference Series **603** (2015) 012007 doi:10.1088/1742-6596/603/1/012007

Chirality effect on superconductivity

Takao Morinari

Graduate School of Human and Environmental Studies, Kyoto University, Kyoto 606-8501, Japan

E-mail: morinari.takao.5s@kyoto-u.ac.jp

Abstract. We consider electron systems characterized by chirality, such as Dirac fermion systems and iron-based superconductors. We investigate the chirality effect on superconducting states in such a system. We show that chirality effect leads to a nodal structure in the superconducting gap function. The node creation mechanism depends on the wave vector $\mathbf{q}$ of the pairing interaction and vorticity that characterizes chirality of electrons. The node creation effect due to chirality is significant for the case of Dirac fermions with $\mathbf{q} = (\pi, 0)$ and for the case of iron-based superconductors with $\mathbf{q} = (\pi, \pi)$.

1. Introduction

Massless Dirac fermions in condensed matter systems have attracted much attention since the discovery of the anomalous integer quantum Hall effect in graphene [1, 2]. In such a Dirac fermion system, the conduction and valence bands touch at a single point. This band structure is called the Dirac cone from its shape and the point of band contact is called the Dirac point. In graphene, the linear energy dispersion exists at the corners of the first Brillouin zone, so that the electrons at low energies are well described by the Dirac equation, with the speed of light being replaced by the Fermi velocity [3]. Physically Dirac fermions and conventional electrons in metals are quite different. For instance, there is qualitative difference in the energy spectrum under magnetic field. For the case of Dirac fermions, the Landau level energies are not equally spaced and the energies depend on the square root of the magnetic field. In particular, there is the zero energy Landau level. The presence of this zero energy Landau level leads to unusual half-integer quantum Hall effect [1, 2].

Another important physical properties of massless Dirac fermions is chirality. There are two types of Dirac fermions: right-handed and left-handed. The Dirac fermion with the wave vector $\mathbf{k}$ is either right-handed, denoted by pseudospin $\uparrow$, or left-handed, denoted by pseudospin $\downarrow$. In the Brillouin zone the pseudospins create a vortex or an anti-vortex configuration at Dirac points. It is well known that the backward scattering is suppressed in Dirac fermion systems because of the chirality effect [4]. In the backward scattering the Dirac fermion state $|\mathbf{k}, \sigma\rangle$ with σ being the pseudospin is scattered into the Dirac fermion state $|-\mathbf{k}, \sigma'\rangle$. Because of the nature of chirality, there is a matrix element for such a scattering process if $\sigma' = \bar{\sigma}$. Namely if the pseudospin is flipped, then there is a matrix element. However, those two energy states have different energies so that there is no matrix elements. The suppression of the backward scattering is understood from the Berry phase argument as well [4]. In addition, there is no Anderson localization. In conventional metals pure two-dimensional systems are insulators because of the Anderson localization effect. However, reflecting the absence of the Anderson localization, graphene, which is a pure two-dimensional system, is a good conductor.

Electrons in iron-based superconductors are also characterized by chirality. In the antiferromagnetic phase of iron-based superconductors, the Dirac fermion energy spectrum appears near the Fermi energy [5, 6, 7]. It was pointed out that hybridizations between the Fe 3d orbitals and the pnictide ion 4p orbitals give rise to the band degeneracy characterized by a nontrivial topology [5]. Contrary to a conventional spin density wave, there are gapless nodal points along the Fermi surface. The topology here is characterized by vorticity quantum number, which is associated with the phase winding defined through a two component spinor wave function. Although there is no Dirac fermion energy spectrum near the Fermi energy in the absence of the spin-density wave, the electrons in the iron-based superconductors have chirality with vorticity two.

In this paper, we consider superconducting states of electrons characterized by chirality. We study Dirac fermions and electrons with vorticity two. We show that there is a node creation mechanism that is associated with the chirality effect. The nodal structure of the superconducting gap function depends on vorticity. In Sec. 2 we briefly review chirality. In Sec. 3 we study chirality effect on superconducting states of Dirac fermion systems. In Sec. 4 we study chirality effect on superconducting states of iron-based superconductors. We summarize the result in Sec. 5

2. Chirality

In this section we briefly review chirality of Dirac fermions and chirality of electrons in iron-based superconductors.

We first consider Dirac fermion systems. In Dirac fermion systems, there are even number of Dirac points. If there is a Dirac point $\mathbf{k}_D$ in the Brillouin zone, there is another Dirac point at $-\mathbf{k}_D$. If we focus on a single Dirac point, and taking that point as the origin of $\mathbf{k}$ space, the Dirac fermion energy spectrum is described by the following 2×2 Hamiltonian:

$$\mathcal{H} = \hbar v \begin{pmatrix} 0 & k_x - ik_y \\ k_x + ik_y & 0 \end{pmatrix}. \tag{1}$$

Here v is the Fermi velocity. If we rewrite the Hamiltonian as

$$\mathcal{H} = \hbar v \mathbf{k} \cdot \boldsymbol{\sigma}, \tag{2}$$

the right hand side has a form of the Zeeman energy. The psuedospin is denoted by $\boldsymbol{\sigma}$ and $\mathbf{k}$ plays a role of magnetic field. Since $\boldsymbol{\sigma}$ is one-half, if we take the direction of $\mathbf{k}$ as the quantization axis for the pseudospin $\boldsymbol{\sigma}$, the eignestate is either psuedospin $\uparrow$ or $\downarrow$. The state $|\mathbf{k}, \uparrow\rangle$ has the eigen-energy $\varepsilon_k = \hbar v\sqrt{k_x^2 + k_y^2}$ and the state $|\mathbf{k}, \downarrow\rangle$ has the eigen-energy $-\varepsilon_k$.

One can define the vector field

$$\mathbf{n}_\mathbf{k} = \langle \mathbf{k}, \uparrow | \boldsymbol{\sigma} | \mathbf{k}, \uparrow \rangle. \tag{3}$$

For the case of the Hamiltonian above, the vector field $\mathbf{n}_\mathbf{k}$ exhibits radial configuration of pseudospins. The configuration becomes a vortex if we rotate each pseudospin by $\pi/2$ in counterclockwise direction. The Dirac points are represented by such a vortex in the Brillouin zone.

Now let us turn to iron-based superconductors. Iron-based superconductors are a multi-Fermi surface system. All the Fe $3d$ orbitals constitute the Fermi surfaces. There are hole and electron Fermi surfaces with almost equal sizes. Among the Fe $3d$ orbitals d_{xz} and d_{yz} orbitals play an important role. We can take the space of d_{xz} and d_{yz} orbitals as the pseudospin space.

Using the wave function at each point of the Fermi surfaces, we define the pseudospin. Figure 1 shows the pseudospin configuration in the Brillouin zone. As the model Hamiltonian for iron-based superconductors we take a five-band model [8]. We find pseudospins exhibits vortex configurations on each Fermi surface. Contrary to Dirac fermion case, vorticity is not one but two. In fact, the pseudospins on the outer hole Fermi surface at the Γ point rotate twice in clockwise direction when one goes around the Γ point. Similar rotations occur on other Fermi surfaces.

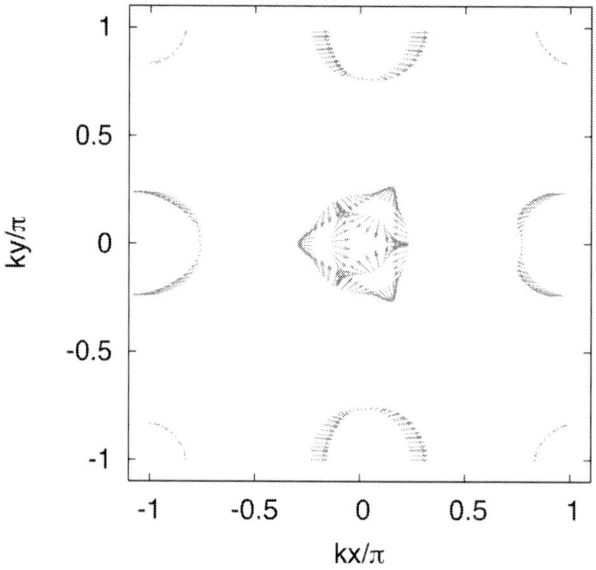

Figure 1. The psuedospin configuration on the Fermi surfaces of iron-based superconductors. The psuedospins on the outer hole Fermi surface at the Γ point rotate twice in clockwise direction when one goes around the Γ point. The psuedospins also rotate twice in either clockwise direction or counterclockwise direction when one goes around $(\pi, 0)$, (π, π), etc.

There are several possibilities of the pseudospin configuration as shown in Fig. 2 which are distinguished by vorticity. Figure 2(a) shows no vorticity case. The Dirac fermion case corresponds to Fig. 2(b). The pseudospins in iron-based superconductors exhibit a vortex configuration as shown in Fig. 2(c). In this case vorticity is two. In the following sections we consider superconducting states of electrons with vorticity.

3. Chirality effect on superconducting states of Dirac fermions

To study superconducting states of electrons characterized by vorticity, we introduce a pairing interaction by hand. We consider the following pairing interaction:

$$V_{\mathbf{k},\mathbf{k}'} = \frac{V_0}{(\mathbf{k} - \mathbf{k}' - \mathbf{q})^2 + \lambda^2}. \tag{4}$$

The scattering wave vector is denoted by $\mathbf{q}$. In the following analysis we consider the $\mathbf{q} = (\pi, 0)$ case and the $\mathbf{q} = (\pi, \pi)$ case. The parameter λ is the interaction range in the Brillouin zone.

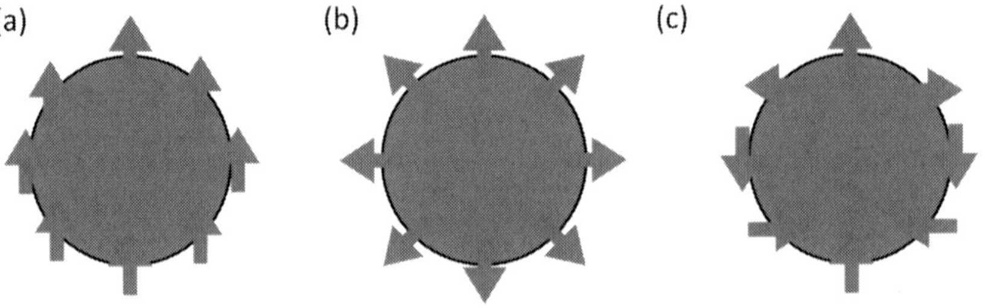

Figure 2. The pseudospin configuration for (a) vorticity zero, (b) vorticity one, and (c) vorticity two. The psudospins are denoted by arrows. The circles represent the Fermi surface.

As the pairing interaction we consider repulsive interactions only. It turns out that there is no significan effect associated with chirality for the case of attractive interactions.

For the analysis of the superconducting state, we solve the linearized BCS gap equation,

$$\Delta_{\mathbf{k}} = \int d\mathbf{k}'_{\parallel} \frac{V_{\mathbf{k},\mathbf{k}'}}{|\mathbf{v}_{\mathbf{k}'}|} \Delta_{\mathbf{k}'}. \tag{5}$$

Here $\Delta_{\mathbf{k}}$ is the gap function. The wave vector $\mathbf{k}$ and $\mathbf{k}'$ are taken those on the Fermi surface. The integral is taken over along the Fermi surface.

Figure 3 shows the result of $\mathbf{q} = (\pi, 0)$ and Fig. 4 shows the result of $\mathbf{q} = (\pi, \pi)$. The Dirac fermion case is Fig. 3(b) and Fig. 4(b) and the normal electron case is Fig. 3(a) and Fig. 4(a). We find that there is a clear qualitative difference for the case of $\mathbf{q} = (\pi, 0)$ between the Dirac fermion case and the normal electron case. The nodal structure is different between the two cases. Because of suppression of the pairing matrix elements due to the chirality effect there is additional nodal structure for the case of Dirac fermions. Contrary, there is no qualitative difference for the case of $\mathbf{q} = (\pi, \pi)$. The gap function has $d_{x^2-y^2}$-wave symmetry but this is just a consequence of the pairing interaction. For the interaction with $\mathbf{q} = (\pi, \pi)$ the chirality effect is not strong but weak. Therefore, there is no significant modification of the gap function.

4. Chirality effect on iron-based superconductors

Now we consider chirality effect on iron-based superconductors. In order to describe the energy band dispersion, we take a five band model with the energy dispersion perpendicular to the Fe-layers [9]. We take the k_x and k_y axises for the two dimensional Fermi surfaces associated with the Fe-layer and the k_z axis for the perpendicular direction. Numerical calculations of the pseudospins, which is defined in Sec. 2, show that the chirality somewhat depends on k_z as shown in Fig. 5. Although there is quantitative difference in the psuedospin configuration for different k_z, the important fact is that vorticity is conserved.

For the analysis of the superconducting gap function, we solve the linearized BCS gap equation as in Sec.3. The gap equation is,

$$\Delta_{\mathbf{k}\alpha} = \sum_{\beta} \int d\mathbf{k}'_{\parallel} \frac{V_{\mathbf{k}\alpha,\mathbf{k}'\beta}}{|\mathbf{v}_{\mathbf{k}'\beta}|} \Delta_{\mathbf{k}'\beta}. \tag{6}$$

Here α and β are indices for the Fermi surfaces.

Figure 6 and Fig. 7 show the result of $\mathbf{q} = (\pi, 0)$ and $\mathbf{q} = (\pi, \pi)$ at $k_z = 0$. The case of iron-based superconductors, which is characterized by vorticity two, is shown in Fig. 6(b) and

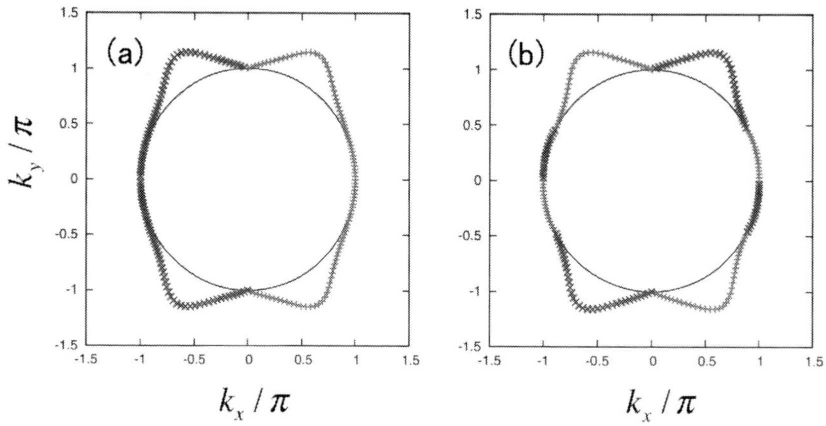

Figure 3. The superconducting gap function for (a) electrons with no vorticity and (b) Dirac fermions. For the pairing interaction (4) we take $\mathbf{q} = (\pi, 0)$ and $\lambda = 0.2$. The thin solid line is the Fermi surface. The thick red and blue lines show the gap function $\Delta_{\mathbf{k}}$. The red lines are for $\Delta_{\mathbf{k}} > 0$ and the blue lines are for $\Delta_{\mathbf{k}} < 0$. The distance in the radial direction between the thick lines and the thin line represents the magnitude of $|\Delta_{\mathbf{k}}|$ in an arbitrary scale.

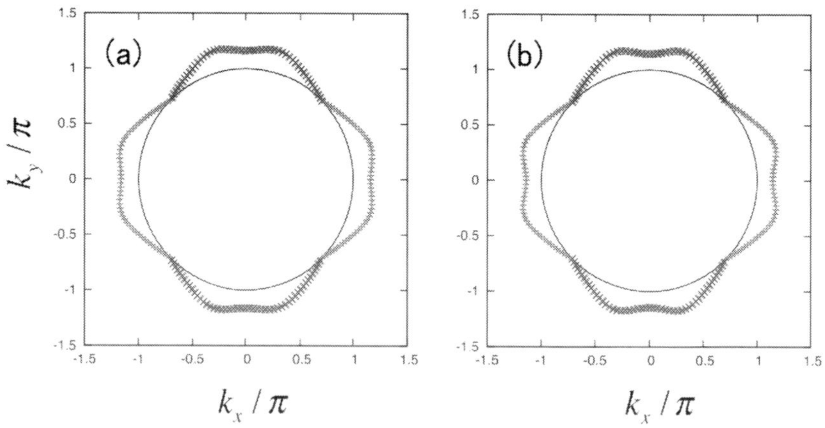

Figure 4. The superconducting gap function for (a) electrons with no vorticity and (b) Dirac fermions. For the pairing interaction (4) we take $\mathbf{q} = (\pi, \pi)$ and $\lambda = 0.2$. The definition of the thin line and the thick lines are the same as that in Fig. 3.

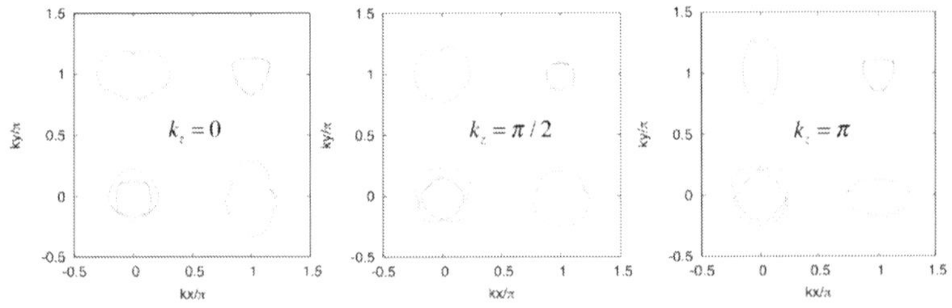

Figure 5. The pseudospin configuration on the Fermi surfaces of iron-based superconductors for different k_z. The change of the pseudospin configuration mainly comes from the change of the shape of the Fermi surfaces. Vorticity itself does not depend on k_z.

Fig. 7(b). The normal electron case is Fig. 6(a) and Fig. 7(a). Contrary to the Dirac fermion case we do not find qualitative difference for the case of $\mathbf{q} = (\pi, 0)$ between the vorticity two case and the normal electron case. Since vorticity is two, there is no suppression of the pairing matrix elements. The chiralities of $\mathbf{k}$ and $-\mathbf{k}$ on the same Fermi surface are equivalent, and so there is no suppression of the matrix elements.

Contrary to the case of $\mathbf{q} = (\pi, 0)$, there is clear qualitative difference for the case of $\mathbf{q} = (\pi, \pi)$. For the case of vorticity zero there are the nodal structures on the electron Fermi surface around $(\pi, 0)$ and $(0, \pi)$. These nodal structure disappear for the case of vorticity two. On the other hand, there are no nodal structure on hole Fermi surfaces around $(0, 0)$ and (π, π) for the case of vorticity zero. However, nodal structures appear on those hole Fermi surfaces for the case of vorticity two. These qualitative difference is associated with the chirality effect.

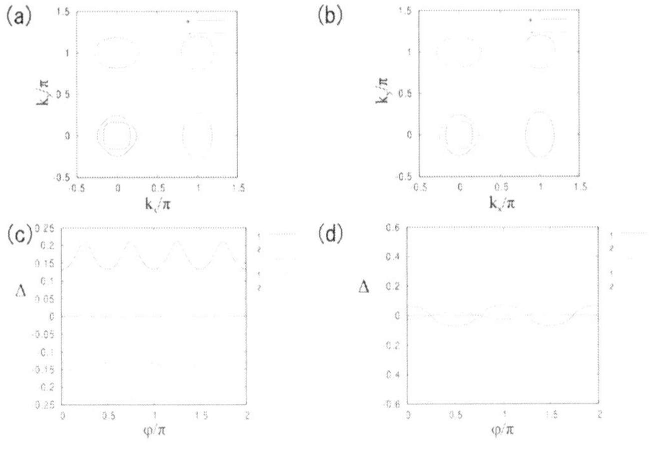

Figure 6. The superconducting gap function at $k_z = 0$ with $\mathbf{q} = (\pi, 0)$ for vorticity zero, (a) and (c) and for vorticity two, (b) and (d). The average electron number at Fe sites is 5.9, that is, a hole doped system. In (a) and (b), the sign of the gap functions on each Fermi surface is shown. The amplitude of the gap function on each Fermi surface is shown in (c) and (d). The inner and outer hole Fermi surfaces at Γ point are denoted by α_1 and α_2, respectively. The hole Fermi surface at (π, π) is denoted by γ and the electron Fermi surfaces at $(\pi, 0)$ and $(0, \pi)$ are denoted by β_1 and β_2, respectively.

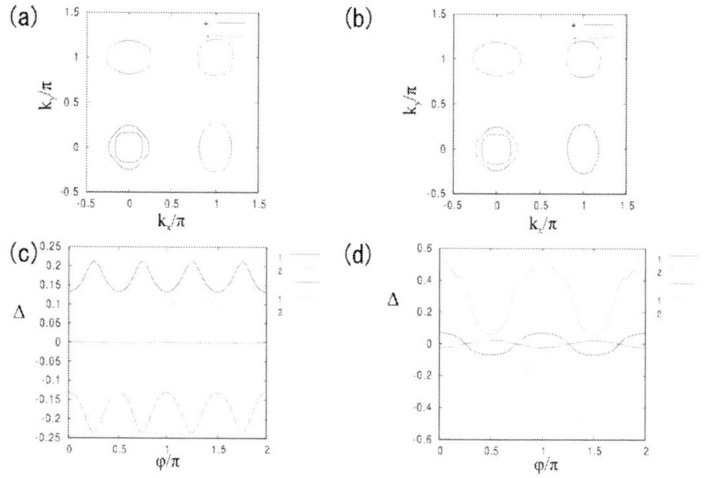

Figure 7. The superconducting gap function at $k_z = 0$ with $\mathbf{q} = (\pi, \pi)$ for vorticity zero, (a) and (c) and for vorticity two, (b) and (d). The average electron number at Fe sites is 5.9. In (a) and (b), the sign of the gap functions at each Fermi surface is shown. The amplitude of the gap function on each Fermi surface is shown in (c) and (d).

The gap function of the three dimensional Fermi surfaces is shown in Fig. 8. The difference between vorticity zero and vorticity two is clearly seen for the case of $\mathbf{q} = (\pi, \pi)$ ((c) and (d)). For the case of $\mathbf{q} = (\pi, 0)$ ((a) and (b)) there is some quantitative difference but no qualitative difference.

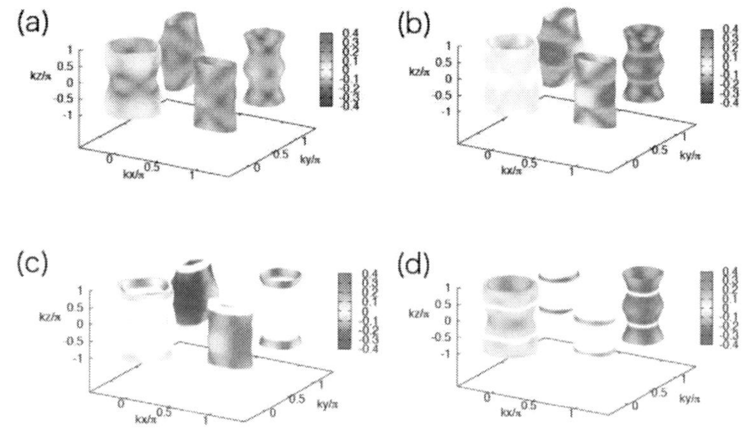

Figure 8. The gap function of the three dimensional Fermi surfaces. for vorticity zero, (a) and (c) and for vorticity two, (b) and (d). The average electron number at Fe sites is 5.9. The wave vector $\mathbf{q}$ of the pairing interaction is $\mathbf{q} = (\pi, 0)$ for (a) and (b) and $\mathbf{q} = (\pi, \pi)$ for (c) and (d).

5. Summary

To summarize, we have studied chirality effect on superconducting states of electrons characterized by chirality. Because of the chirality effect the superconducting gap function has a nodal structure. This node creation mechanism depends on the pairing interaction wave vector $\mathbf{q}$ and vorticity of the electrons. For the case of vorticity one, the node creation mechanism plays an important role for $\mathbf{q} = (\pi, 0)$. For the case of vorticity two, which corresponds to the case of iron-based superconductors, the node creation mechanism plays an important role for $\mathbf{q} = (\pi, \pi)$.

Although this node creation mechanism itself is interesting, the electrons with vorticity would be useful to detect the pairing symmetry of a superconducting state by a proximity effect. We will consider this problem in a future publication.

Acknowledgments

I would like to thank T. Tohyama and E. Kaneshita for helpful discussions. This work was supported by a Grant-in-Aid for Scientific Research (A) (No. 24244053), (C) (No. 24540370), and (B) (No. 25287089), from the Ministry of Education, Culture, Sports, Science, and Technology, Japan.

References

[1] Novoselov K S, Geim A K, Morozov S V, Jiang D, Katsnelson M I, Grigorieva I V, Dubonos S V and Firsov A A 2005 *Nature* **438** 197
[2] Zhang Y, Tan Y W, Stormer H L and Kim P 2005 *Nature* **438** 201
[3] Geim A K and Novoselov K S 2007 *Nat. Mater.* **6** 183–191
[4] Ando T, Nakanishi T and Saito R 1998 *J. Phys. Soc. Jpn.* **67** 2857
[5] Ran Y, Wang F, Zhai H, Vishwanath A and Lee D H 2009 *Phys. Rev. B* **79** 014505
[6] Richard P, Nakayama K, Sato T, Neupane M, Xu Y M, Bowen J H, Chen G F, Luo J L, Wang N L, Dai X, Fang Z, Ding H and Takahashi T 2010 *Phys. Rev. Lett.* **104** 137001
[7] Morinari T, Kaneshita E and Tohyama T 2010 *Phys. Rev. Lett.* **105** 037203
[8] Kuroki K, Onari S, Arita R, Usui H, Tanaka Y, Kontani H and Aoki H 2008 *Phys. Rev. Lett.* **101** 087004
[9] Graser S, Kemper A F, Maier T A, Cheng H P, Hirschfeld P J and Scalapino D J 2010 *Phys. Rev. B* **81** 214503

Journal of Physics: Conference Series **603** (2015) 012008 doi:10.1088/1742-6596/603/1/012008

Spin texture of the surface state of three-dimensional Dirac material Ca_3PbO

Toshikaze Kariyado

Division of Physics, University of Tsukuba, Tsukuba, Ibaraki, Japan

E-mail: `kariyado@rhodia.ph.tsukuba.ac.jp`

Abstract. The bulk and surface electronic structures of a candidate three-dimensional Dirac material Ca_3PbO and its family are discussed especially focusing on the spin texture on the surface states. We first explain the basic features of the bulk band structure of Ca_3PbO, such as emergence of Dirac fermions near the Fermi energy, and compare it with the other known three-dimensional Dirac semimetals. Then, the surface bands and spin-texture on them are investigated in detail. It is shown that the surface bands exhibit strong momentum–spin locking, which may be useful in some application for spin manipulation, induced by a combination of the inversion symmetry breaking at the surface and the strong spin-orbit coupling of Pb atoms. The surface band structure and the spin-textures are sensitive to the surface types.

1. Introduction

Recently, materials having three-dimensional linear dispersion near the Fermi energy attract much attention [1, 2, 3, 4, 5, 6] as a natural extension of graphene, typical two-dimensional material with linear dispersion near the Fermi energy. Such 3D materials are dubbed as 3D Dirac/Weyl semimetals depending on the degeneracy at the center of the linear dispersion, a Weyl node or a Dirac point. Low-energy physics in the Weyl semimetal is governed by a two-component Weyl equation, and there is two-fold degeneracy at the Weyl node. On the other hand a four-component Dirac equation dominates the low-energy physics in the Dirac semimetal, and we observe a four-fold degeneracy at the Dirac point. Weyl nodes and Dirac points are singularity of the band structure in momentum space. In general, such a singularity in the band structure leads to characteristic surface/edge states [7, 8, 9], for instance, as the edge mode in graphene ribbons. Surface/edge band structures depend crucially on how the Weyl nodes or the Dirac points are projected on the surface Brillouin zone [1].

In this paper, surface states of Ca_3PbO series are studied in detail, especially focusing on the spin texture on the surface bands. Ca_3PbO and its family, such as Ca_3SnO, Sr_3PbO, and Ba_3PbO, were first synthesized back in 1980 [10], and they are theoretically proposed as candidate materials having the 3D Dirac fermion with a tiny mass gap exactly at the Fermi energy [11]. Furthermore, it is also theoretically suggested that Ca_3PbO series is a potential topological crystalline insulator [12]. Then, further study on this series of material is highly demanded. In the following, we first give a brief review on the electronic structure of Ca_3PbO series. There, we explain the bulk band structure obtained by means of the first-principles calculation, and the surface states obtained using a low-energy effective model. In the review part, we also give some up-to-date arguments on 3D Dirac materials, for instance, comparison between Ca_3PbO series and other Dirac semimetals, Cd_3As_2 [4] and Na_3Bi [3], which are recognized as Dirac materials very recently. The review part is followed by new results on the spin texture of the surface bands of Ca_3PbO.

2. Electronic structure of Ca_3PbO and its family

Ca_3PbO and its family are cubic inverse perovskite materials, in which the metallic atoms, alkaline-earth metals in this case, octahedrally surround the oxygen atom. Experimentally, they are synthesized in 1980, and their lattice constants are reported at that time [10]. Because of the simple crystal structure, only a single parameter, the lattice constant, thoroughly determines the crystal structure.

The electronic structure of Ca_3PbO series was investigated by means of the first-principles calculation with the spin-orbit coupling using the experimental crystal parameters [11]. Close to the Fermi energy, there are almost completely filled Pb-6p orbital originated bands and almost empty Ca-3d bands. However, it is not a simple semiconductor with the closed Pb-6p shell. Instead, if we carefully examine the orbital character of each band, the top of Pb-6p band is above the bottom of Ca-3d bands. Basically, owing to the strong hybridization between Pb-6p and Ca-3d orbitals, Pb-6p and Ca-3d bands strongly repel each other and the gap is formed at the Fermi energy. However, restriction by the crystalline symmetry does not allow strong repulsion between Pb-6p and Ca-3d bands *on the k_x, k_y, and k_z axis in the Brillouin zone*, and the band crossing remains on these axis. Near the band crossing point, i.e., Dirac point, at which the Pb-6p and Ca-3d bands are degenerate, the band gap grows linearly as a function of the distance from the Dirac point in all three directions. This indicates that this material can be regarded as a 3D Dirac material. In short, the orbital overlapping and the protection of the band crossing by the crystalline symmetry play crucial roles in generating Dirac fermions. Strictly speaking, there is a tiny mass gap at the Dirac point induced by the *indirect* spin-orbit coupling [13]. Here, the *indirect* spin-orbit coupling means that it should involve orbitals fairly away from the Fermi energy, such as the Pb-5d orbitals. This indirect nature limits the size of the mass gap, and the smallness of the mass gap is guaranteed. Since the mass gap is very small, the linear dispersion, which is crucial in observing interesting phenomena associated with the Dirac fermions, is clearly identified in the band structure.

Here, we make a comparison between Ca_3PbO series and other Dirac semimetals, Cd_3As_2 [4] and Na_3Bi [3]. In Cd_3As_2, Cd-3d orbital originated bands and As-4p bands overlap near the Fermi energy. Similarly, in Na_3Bi, Bi-6p bands and Na-3s bands overlap in the vicinity of the Fermi energy. For both of Cd_3As_2 and Na_3Bi, the origin of the Dirac point is ascribed to the overlapping of orbitals with different characters and the crystalline symmetry protecting the band crossing, just as in the case of Ca_3PbO. However, it should be noted that p- and s-orbitals are involved in Cd_3As_2 or Na_3Bi while p- and d-orbitals are involved in Ca_3PbO. The difference between s- and d-orbitals is important in some sense, because the d-orbital contribution is essential for inducing a small but finite mass term in Ca_3PbO [13].

The surface states of Ca_3PbO was also theoretically studied in brief using a low-energy effective tight-binding model [13]. It was shown that there appear characteristic surface states, and the surface band structure is sensitive to the surface direction and termination. Because of the strong spin-orbit coupling served by Pb atoms, the spin texture on the surface bands is an interesting topic to study. However, the spin texture on the surface bands was not studied in detail in the previous work, and the latter half of this paper is devoted to this issue. Revealing the spin texture is also important in terms of application, such as the possible spin manipulation.

Before moving on to the next issue, we point out the other aspect of Ca_3PbO. So far, Ca_3PbO is regraded as a Dirac material because the gap is very small. However, strictly speaking, it is an insulator with a small but finite mass gap as we have noted. Very recently, it is proposed that Ca_3PbO is not a mere ordinary insulator, but is a topological crystalline insulator protected by the reflection symmetry [12]. Then, we can give a topological origin to the surface states discussed above, at least for the surface respecting the reflection symmetry.

3. Spin-texture on the surface states

Let us move on to the spin texture on the surface states of Ca_3PbO. For simplicity and clarity, the surface states and the spin texture on them are calculated using the low-energy effective tight-binding model introduced in Ref. [13]. The model is a 12 band (= 2 spins × 6 orbitals) model, where the six orbitals come from three Pb-6p orbitals and three Ca-3d orbitals. (See Ref. [13] for details.) There are 15 Ca-3d orbitals in a unit cell since three Ca atoms are contained in a unit cell, but only three of them are necessary to reproduce the low-energy band structure. The spin-orbit coupling is taken account of as the atomic $\boldsymbol{L}\cdot\boldsymbol{S}$ coupling on the Pb-6p orbitals, and the $\boldsymbol{L}\cdot\boldsymbol{S}$ coupling on the Ca-3d orbitals are neglected. The surface band structure is calculated using the system with the finite thickness slab geometry. We consider several kinds of surfaces with different surface directions and different terminations. For simplicity, we only treat terminations having perfectly flat surfaces. Further, we focus on the cases where the top and bottom surfaces of the slab are of the same sort of termination.

Let us begin with 001 surfaces. For the 001 surfaces, there exist two possible terminations leading to the flat surfaces, which are illustrated as Figs. 1(b) and 1(c). We call the surfaces in Fig. 1(b) and Fig. 1(c) as 001A and 001B respectively. The surface band structures for these surfaces are plotted in Figs. 1(d) and 1(e). The region filled with bands is a bulk continuum, which is contributed by the bulk part of the system. For both of 001A and 001B surfaces, we observe in-gap surface bands apart from the bulk continuum. Since introduction of 001A or 001B surfaces preserves the reflection symmetry that makes Ca_3PbO a topological crystalline insulator, these surface bands have a topological origin [12]. For both of the 001A and 001B surfaces, two surface bands are identified on the $\bar{\Gamma}$-$\bar{X}$ line, while only one surface band is identified on the $\bar{\Gamma}$-$\bar{M}$ line. [See Fig. 1(a) for names of the special points in the surface Brillouin zone.] That is, one of the two surface bands on the $\bar{\Gamma}$-$\bar{X}$ line is absorbed into the bulk continuum as we move from the $\bar{\Gamma}$-$\bar{X}$ line to the $\bar{\Gamma}$-$\bar{M}$ line. The existence of the surface states itself looks robust for the 001 surfaces, but the surface band structure is sensitive to the surface termination. For instance, the slope of the surface bands are opposite between the 001A and 001B surfaces.

The obtained surface bands are doubly degenerate, but this degeneracy is originated from the equivalence of the top and bottom surfaces of the slab. When each surface is examined separately, the surface bands are nondegenerate, i.e., they have no spin degeneracy. This is because there is strong spin-orbit coupling in this system, and the inversion symmetry is broken at the surface. In order to see this point more in detail, we calculate the band and momentum resolved spin moment on the atoms in the outermost layer of the slab. For the 001A surface, there are both of the Pb and Ca atoms at the surface [Fig. 1(b)], and both atoms contribute to the spin moment. On the other hand, for the 001B surface, only the Ca atoms appear at the surface [Fig. 1(c)], and the spin moment contains the contribution from Ca atoms only. (Recall that the O atoms are not included in the effective tight-binding model.)

The calculated spin textures are shown in Figs. 1(f)–1(i). Since we found at most two surface bands, the spin textures are calculated for two bands for each of 001A and 001B surface. Namely, Figs. 1(f) and 1(h) are for the lower surface bands, and Figs. 1(g) and 1(i) are for the upper ones. However, note that some part of the selected bands is merged into the bulk continuum as we have discussed. For instance, only a lower (upper) band among the two selected bands is an in-gap state on the $\bar{\Gamma}$-$\bar{M}$ line for 001A (001B) surface. In Figs. 1(f)–1(i), the spin moment parallel to the surface is expressed by arrows, and the one perpendicular to the surface is represented by the color map. For all the surface bands, the spin textures such that the spin moment winds around the $\bar{\Gamma}$-point are observed. More specifically, the spin moment directs along the isoenergy lines of the surface bands. This momentum-spin locking is caused by a induced Rashba term. Although

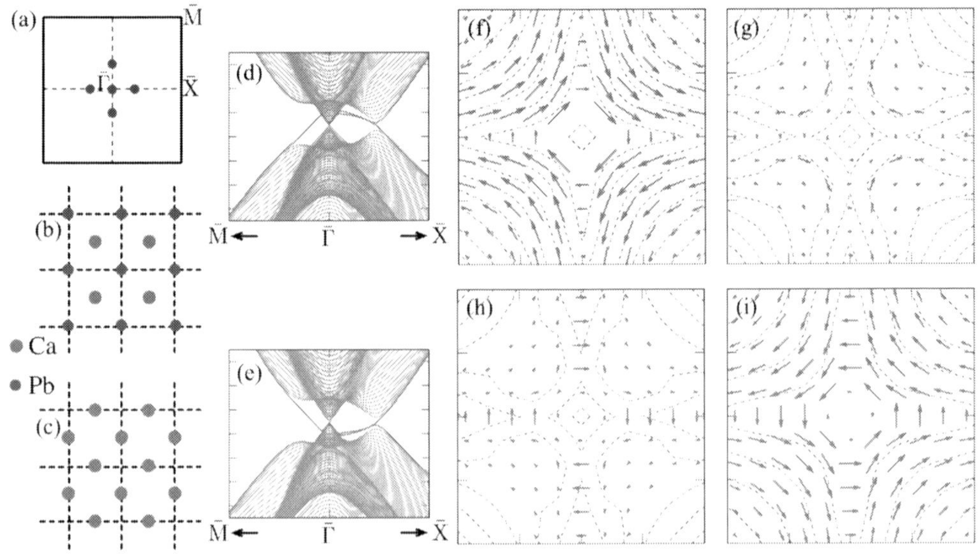

Figure 1. (a) Two-dimensional surface Brillouin zone for the 001 surface. Blue dots represent the projected Dirac points. (b) and (c) Schematic pictures for the 001A and 001B surfaces. O atoms, which are not included in the effective tight-binding model, are not displayed. (d) and (e) Surface band structure on the high symmetry line in the surface Brillouin zone. (See also Ref. [13].) (f) and (g) Spin texture on the surface bands for the 001A surface. (h) and (i) Spin texture on the surface bands for the 001B surface.

we have not added a Rashba term to the Hamiltonian, it is naturally induced by a combination of the inversion symmetry breaking and the bulk atomic $\boldsymbol{L} \cdot \boldsymbol{S}$ coupling. For the 001B surface, only Ca sites, where the atomic $\boldsymbol{L} \cdot \boldsymbol{S}$ coupling are neglected, appear at the surface and contribute to the surface spin moment, but the hybridization between Pb and Ca orbitals induce Rashba effect on Ca sites. Comparing Figs. 1(f)–1(i), we notice that the directions of winding of the spin moment are opposite between the lower and the upper surface bands. Another feature of the spin texture for the 001 surfaces is that there is no spin moment perpendicular to the surface. Figures 1(f)–1(i) also indicate that the spin textures are more clearly seen on the lower (upper) bands for the 001A (001B) surface than the other. This is natural because the upper (lower) bands for the 001A (001B) surface is merged into the bulk continuum. That is, first of all, the momentum-spin locking is caused by inversion symmetry breaking at the surface, and the effect of the inversion symmetry breaking at the surface is smeared and becomes less important if the band gains the bulk character as it is merged to the bulk continuum.

Next, we look at the 111 surface. For the 111 surfaces, there exist two possible flat terminations, but one of them contains only O atoms, which is not included in the low-energy effective tight binding model, at the outermost layer of the surface. Thus, we consider only one type of surface. (See Ref. [13].) As we can see from Fig. 2(b), there appear four surface bands apart from the bulk continuum. Two of the four surface bands, uppermost and lowermost ones form roughly conical band structure and are nearly isotropic in the surface Brillouin zone. On the other hand, middle two surface bands have strong anisotropy. That is, on the $\bar{\Gamma}$-$\bar{M}$ line, they are approximately flat, while on the $\bar{\Gamma}$-$\bar{X}$ line, they are dispersive. [See Fig. 2(a) for the names of the special points in the surface Brillouin zone.] This anisotropy is related to the projected Dirac points located on the $\bar{\Gamma}$-$\bar{M}$ line. The spin textures on these four surface bands are summarized

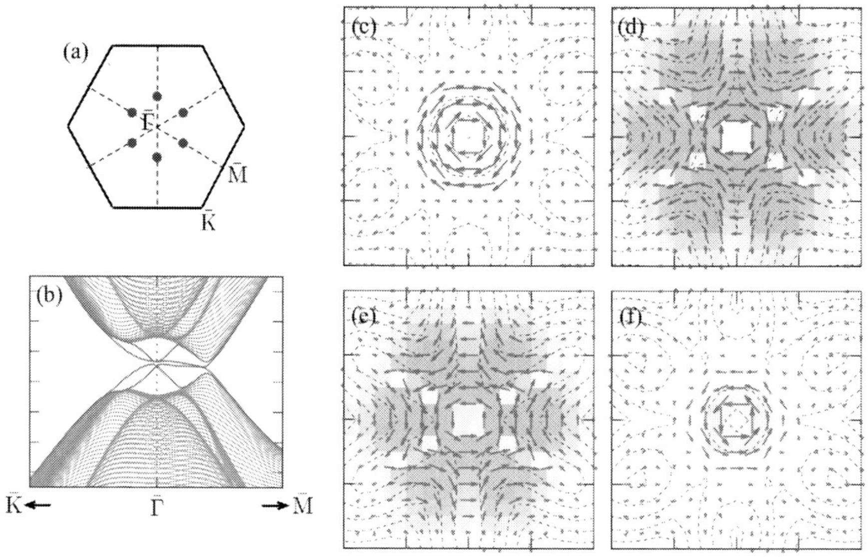

Figure 2. (a) Two-dimensional surface Brillouin zone for the 111 surface. Blue dots represent the projected Dirac points. (b) Surface band structure on the high symmetry line in the surface Brillouin zone. (See also Ref. [13].) (c)–(f) Spin texture on the surface bands for the 111 surface.

in Figs. 2(c)–2(f). Just as in the case of the 001 surface, we can see a winding structure of the spin moment, which is caused by the induced Rashba term. In contrast to the 001 surfaces, the spin moment normal to the surface is sizable especially for the middle two surface bands. Similar to the surface states of Bi_2Te_3, a representative time reversal invariant Z_2 topological insulator, the hexagonal warping of the surface bands induces the spin moment perpendicular to the surface [14]. Since the uppermost and lowermost surface bands have weak anisotropy, the spin moment perpendicular to the surface is negligible for these bands.

4. Summary

To summarize, we have reviewed the electronic structure of Ca_3PbO series, and investigated the spin texture on the surface bands. It is shown that the surface bands exhibit strong momentum–spin locking induced by a combination of the inversion symmetry breaking at the surface and the strong spin-orbit coupling given by Pb atoms. The surface band structure and the spin textures on them sensitively depend on the surface direction and also on the types of the termination at the surface. An interesting future work is to consider the possible application of these surface bands with the unique spin texture. Furthermore, it is also important to think about what kinds of termination are available in experiments.

Acknowledgments

I thank Prof. M. Ogata and Prof. H. Takagi for useful comments and discussions.

References

[1] Wan X, Turner A M, Vishwanath A and Savrasov S Y 2011 *Phys. Rev. B* **83** 205101
[2] Burkov A A and Balents L 2011 *Phys. Rev. Lett.* **107** 127205
[3] Wang Z, Sun Y, Chen X Q, Franchini C, Xu G, Weng H, Dai X and Fang Z 2012 *Phys. Rev. B* **85** 195320
[4] Wang Z, Weng H, Wu Q, Dai X and Fang Z 2013 *Phys. Rev. B* **88** 125427

[5] Liu Z K, Zhou B, Zhang Y, Wang Z J, Weng H M, Prabhakaran D, Mo S K, Shen Z X, Fang Z, Dai X, Hussain Z and Chen Y L 2014 *Science* **343** 864
[6] Neupane M, Xu S Y, Sankar R, Alidoust N, Bian G, Liu C, Belopolski I, Chang T R, Jeng H T, Lin H, Bansil A, Chou F and Hasan M Z 2014 *Nat. Commun.* **5** 3786
[7] Ryu S and Hatsugai Y 2002 *Phys. Rev. Lett.* **89** 077002
[8] Delplace P, Ullmo D and Montambaux G 2011 *Phys. Rev. B* **84** 195452
[9] Kariyado T and Hatsugai Y 2013 *Phys. Rev. B* **88** 245126
[10] Widera A and Schäfer H 1980 *Mater. Res. Bull.* **15** 1805
[11] Kariyado T and Ogata M 2011 *J. Phys. Soc. Jpn.* **80** 083704
[12] Hsieh T H, Liu J and Fu L 2014 *Phys. Rev. B* **90** 081112
[13] Kariyado T and Ogata M 2012 *J. Phys. Soc. Jpn.* **81** 064701
[14] Fu L 2009 *Phys. Rev. Lett.* **103** 266801

Longitudinal conductivity of a three-dimensional Dirac electron gas in magnetic field

Igor Proskurin[1,2], **Masao Ogata**[1], **and Yoshikazu Suzumura**[3]

[1] Department of Physics, The University of Tokyo, Bunkyo-ku, Tokyo 113-0033, Japan
[2] Institute of Natural Sciences, Ural Federal University, Ekaterinburg 620083, Russia
[3] Department of Physics, Nagoya University, Nagoya 464-8602, Japan

E-mail: iprosk@hosi.phys.s.u-tokyo.ac.jp

Abstract. We discuss a conductivity of a three-dimensional gas of Dirac electrons in the direction of magnetic field in high magnetic fields. Analytical expressions for the zero temperature conductivity are calculated in the linear response theory using the basis of relativistic Landau levels. The impurity scattering is treated in the self-consistent Born approximation which gives Landau level broadening increasing linearly in magnetic field. We demonstrate that in the special case of zero-gap Dirac semimetal this leads to the magnetic field and temperature independent conductivity in high magnetic field limit.

1. Introduction

Over the last year there has been a considerable progress in experimental investigations of transport properties in three-dimensional (3D) Dirac electron systems [1, 2, 3, 4]. A prominent real material where the existence of a zero gap 3D Dirac electron phase was predicted in the first principle calculations [5] and then confirmed experimentally [6] is Cd_3As_2. The electrical resistivity measurements in this material demonstrated that Cd_3As_2 had a giant positive magnetoresistance in the direction perpendicular to the magnetic field which is strongly sample dependent. In particular, the low-mobility samples showed linear magnetoresistance, while in the high-mobility samples the resistivity exhibited nearly quadratic magnetic-field dependence in the quantized magnetic field region where Shubnikov–de Haas oscillations were found [1, 2]. Although the theory of quantum linear transverse magnetoresistance was developed by Abrikosov [7], its origin in Cd_3As_2 [1, 2, 3], TlBiSSe [4] and other 3D Dirac semimetals is not completely understood at the present time. While the transverse magnetoresistance demonstrated a giant dependence on the magnetic field, the longitudinal resistivity in Cd_3As_2 in the same situation became nearly temperature independent in magnetic fields larger than 2 T and showed pronounced Shubnikov–de Haas oscillations.

In the present paper we provide a theoretical discussion of the longitudinal conductivity in 3D Dirac electron system. The case of transverse and Hall conductivities will be considered in the subsequent publication. From the theoretical point of view, the magnetoconductivity of the zero-gap 3D Weyl and Dirac semimetals was considered recently by Gorbar et al. [8]. In particular case of the longitudinal conductivity, the authors showed that in high magnetic fields Weyl and Dirac semimetals possessed the same conductivity, which contained "topological contribution" from the $n = 0$ Landau level which is independent of the temperature and the chemical potential and proportional to B/Γ_0 where B is a magnetic field and Γ_0 is a broadening of the zero Landau level .

The present paper is devoted to the longitudinal conductivity of Dirac materials with a finite gap Δ in the energy spectrum. For the zero-gap case we confirm the result obtained in Ref. [8]. However, in the present paper we calculate the Landau level broadening Γ in the self-consistent Born approximation (SCBA) which gives a magnetic field dependence of Γ. Similar to the two-dimensional case, where in SCBA $\Gamma \sim \sqrt{B}$ in the quantum limit [9], we found that in the 3D case Γ increases linearly in magnetic field. In the small-gap case the magnetic field dependence of Γ leads to the magnetic field (and temperature) independent conductivity in high magnetic fields.

The paper is organised as follows. In section 2 we develop a formalism. Some useful analytical expressions for the conductivity are also derived in Appendix A. In section 3 we discuss the magnetic-field and chemical potential dependences of the Landau level broadening and the conductivity. Section 4 is reserved for the summary.

2. Formalism

2.1. Energy spectrum and wave functions

We consider a relativistic tree-dimensional electron gas in external magnetic field with the Hamiltonian

$$\hat{H} = \begin{pmatrix} \Delta & v\boldsymbol{\sigma}\cdot(\boldsymbol{p}+e\boldsymbol{A}) \\ v\boldsymbol{\sigma}\cdot(\boldsymbol{p}+e\boldsymbol{A}) & -\Delta \end{pmatrix}, \tag{1}$$

where v is the Fermi velocity, Δ is a gap in an energy spectrum, $\boldsymbol{\sigma} = (\sigma_x, \sigma_y, \sigma_z)$ denotes the vector of the Pauli matrices, $\boldsymbol{p} = -i\hbar\boldsymbol{\nabla}$ is a momentum operator with $\hbar = h/2\pi$ being the Plank's constant, an electron charge is $-e$ ($e > 0$), and $\boldsymbol{A}$ is a magnetic vector potential. The vector potential is chosen in the Landau gauge $\boldsymbol{A} = (-By, 0, 0)$ which implies that the magnetic field $\boldsymbol{B}$ is applied along the z-axis.

In the Landau gauge the solution of an eigenvalue problem for the Hamiltonian (1) leads to relativistic Landau levels

$$E_n^{(\eta)} = \eta\sqrt{\epsilon_c^2 n + \epsilon_z^2 + \Delta^2}, \quad n = 0, 1, 2, \dots \tag{2}$$

where $\eta = \pm 1$ is a band index, $\epsilon_c = \sqrt{2eB\hbar v^2}$ is a cyclotron energy, and $\epsilon_z = vp_z = v\hbar k_z$. We imply that the system has a translation invariance along x- and z-directions where periodical boundary conditions are imposed, and k_x and k_z denote the corresponding wave numbers. The eigenfunctions have the following form [10, 11]

$$\psi_{ns}^{(\eta)}(\boldsymbol{r}) = \frac{e^{ik_x x + ik_z z}}{\sqrt{\ell L_x L_z}} \begin{pmatrix} \alpha_n^{(\eta)} A_n^{(s)} \phi_{n-1}(\xi) \\ -s\alpha_n^{(\eta)} A_n^{(-s)} \phi_n(\xi) \\ \eta s \alpha_n^{(-\eta)} A_n^{(s)} \phi_{n-1}(\xi) \\ -\eta \alpha_n^{(-\eta)} A_n^{(-s)} \phi_n(\xi) \end{pmatrix}, \tag{3}$$

where

$$\alpha_n^{(\eta)} = \left[\frac{1}{2}\left(1 + \eta\frac{\Delta}{E_n}\right)\right]^{\frac{1}{2}}. \tag{4}$$

and

$$A_n^{(s)} = \left[\frac{1}{2}\left(1 + \frac{s\,\epsilon_z}{\sqrt{E_n^2 - \Delta^2}}\right)\right]^{\frac{1}{2}}. \tag{5}$$

Hereafter, E_n denotes $\left|E_n^{(\eta)}\right|$. In equation (3), $\ell = \sqrt{\hbar/(eB)}$ is magnetic length, L_j ($j = x$, y, z) is the size of the system in the j-direction, $\xi = y/\ell + \ell k_x$, and ϕ_n is a wave function of a harmonic oscillator

$$\phi_n(\xi) = \frac{e^{-\frac{\xi^2}{2}}}{2^{\frac{n}{2}}\pi^{\frac{1}{4}}\sqrt{n!}} H_n(\xi), \tag{6}$$

with H_n being the nth Hermite polynomial. Each energy level is multiply degenerated with respect to k_x with the degeneracy factor $L_x L_y/(2\pi\ell^2)$, and each level with $n > 0$ is doubly degenerated with respect to $s = \pm 1$. The zero mode ($n = 0$ eigenstate) is obtained from equations (3), (4), and (5) if we set $s = -\epsilon_z/|\epsilon_z|$, $A_0^{(s)} = 0$, and $A_0^{(-s)} = 1$. The explicit form of the zero mode is

$$\psi_0^{(\eta)}(\boldsymbol{r}) = \frac{e^{ik_x x + ik_z z}}{\sqrt{\ell L_x L_z}}\phi_0(\xi)\begin{pmatrix} 0 \\ \alpha_0^{(\eta)} \\ 0 \\ -\frac{\epsilon_z}{|\epsilon_z|}\eta\alpha_0^{(-\eta)} \end{pmatrix}, \quad E_0^{(\eta)} = \eta\sqrt{\epsilon_z^2 + \Delta^2}. \tag{7}$$

Therefore, for $n > 0$ each quantum state is labelled by n, k_x, k_z, η, s, while for $n = 0$ state the quantum numbers are k_x, k_z, and η. Note that the variables defined in equations (1), (4), and (5) have k_z-dependence which is not indicated for brevity.

2.2. Longitudinal electrical conductivity

The diagonal part of the electrical conductivity of an electron gas in a magnetic field can be expressed in the following form [12, 13]

$$\sigma_{jj}(T,\mu) = -\frac{e^2\hbar}{\pi}\int_{-\infty}^{\infty} d\varepsilon \frac{\partial f(\varepsilon)}{\partial \varepsilon}\,\mathrm{Tr}\left[\hat{v}_j\,\mathrm{Im}\,\hat{G}(\varepsilon)\,\hat{v}_j\,\mathrm{Im}\,\hat{G}(\varepsilon)\right], \quad j = x, y, z, \tag{8}$$

where $f(\varepsilon) = \{1 + \exp[(\epsilon - \mu)/T]\}^{-1}$ is the Fermi–Dirac distribution function with chemical potential μ and temperature T, the Boltzmann constant $k_B = 1$, and $\mathrm{Im}\,\hat{G}(\varepsilon) = \frac{i}{2}[\hat{G}^{(+)}(\varepsilon) - \hat{G}^{(-)}(\varepsilon)]$ where

$$\hat{G}^{(\pm)}(\varepsilon) = \left(\varepsilon - \hat{\mathcal{H}} \pm i\delta\right), \quad \delta \to 0^+, \tag{9}$$

with $\hat{\mathcal{H}}$ being the total Hamiltonian which includes the scattering over impurities. An explicit form of $\hat{\mathcal{H}}$ will be specified later, however, hereafter we imply that the effect of impurities is included into the electrons self-energy $\hat{\Sigma}(\epsilon)$ since vertex corrections vanish for the conductivity of an electron gas in a magnetic field with short-ranged scatterers [12]. The velocity operator in this case is defined as $\hat{\boldsymbol{v}} = \frac{i}{\hbar}[\hat{H}, \boldsymbol{r}]$ which for Dirac electrons has the following explicit form

$$\hat{\boldsymbol{v}} = v\begin{pmatrix} 0 & \boldsymbol{\sigma} \\ \boldsymbol{\sigma} & 0 \end{pmatrix}. \tag{10}$$

In the present paper we discuss the conductivity along the direction of magnetic field, i. e. $\sigma_{zz}(T,\mu)$, which will be referred to as $\sigma_{\parallel}(T,\mu)$. If the trace in equation (8) is calculated with respect to the wave function (2), $\sigma_{\parallel}$ at $T = 0$ becomes

$$\sigma_{\parallel}(0,\mu) = \frac{e^2\hbar}{2\pi^2\ell^2}\int_{-\infty}^{\infty}\frac{dk_z}{2\pi}\sum_{nn'}\sum_{\eta\eta'}\sum_{ss'}\left|\langle n'\eta' s'|\hat{v}_z|n\eta s\rangle\right|^2 \mathcal{A}_{n'}^{(\eta')}(\mu)\,\mathcal{A}_n^{(\eta)}(\mu), \tag{11}$$

where we have already taken into account that matrix elements of $\hat{\boldsymbol{v}}$ are diagonal in k_z, and have no k_x dependence. The summation over k_z is changed to the integration and after the the limits of the integration are expanded to the infinity. The Green's functions in the r.h.s. are determined as

$$\mathcal{A}_n^{(\eta)}(\mu) = \frac{\Gamma_n^{(\eta)}(\mu)}{\left(\mu - E_n^{(\eta)} - \mathcal{R}_n^{(\eta)}(\mu)\right)^2 + \Gamma_n^{(\eta)2}(\mu)} \tag{12}$$

where $\mathcal{R}_n^{(\eta)}$ and $\Gamma_n^{(\eta)}$ refer to the real and imaginary part of the self-energy $\Sigma_n^{(\eta)}$ respectively. For $\hat{v}_z$, an explicit form of the matrix elements is

$$\langle n'\eta's'|\,\hat{v}_z\,|n\eta s\rangle = v\,\delta_{nn'}\left(s\eta\alpha_n^{(-\eta)}\alpha_{n'}^{(\eta')} + s'\eta'\alpha_{n'}^{(-\eta')}\alpha_n^{(\eta)}\right)\left[A_{n'}^{(s')}A_n^{(s)} - s's A_{n'}^{(-s')}A_n^{(-s)}\right]. \tag{13}$$

As far as there is no dependence on s or s' in equation (12), we can make a summation over s and s' in equation (11), taking into account that for the $n = 0$ we should make a substitution $A_0^{(s')} = A_0^{(s)} = 0$, $A_0^{(-s')} = A_0^{(-s)} = 1$, and remove the summation over s and s'. After algebraically cumbersome, but straight forward calculations, we obtain

$$\sum_{ss'}\left|\langle n'\eta's'|\hat{v}_z|n\eta s\rangle\right|^2 = v^2\alpha_n\left(\frac{1}{2} - \frac{1}{2}\eta\eta' + \eta\eta'\frac{\epsilon_z^2}{E_n^2}\right), \quad \alpha_n = 2 - \delta_{n0}. \tag{14}$$

With the help of equation (14), we arrive to the following expression for the zero temperature conductivity

$$\sigma_{\|}(0,\mu) = \frac{e^2v^2}{4\pi^3\ell^2}\int_{-\infty}^{\infty} dp_z \sum_{n=0}^{\infty}\sum_{\eta}\alpha_n\left[\frac{\epsilon_z^2}{E_n^2}\mathcal{A}_n^{(\eta)2} + \left(1 - \frac{\epsilon_z^2}{E_n^2}\right)\mathcal{A}_n^{(\eta)}\mathcal{A}_n^{(-\eta)}\right]. \tag{15}$$

As we discuss below, in the case when the self-energy in equation (12) depends only on μ, the summation over n or the integration over p_z in equation (15) can be performed analytically which produces expressions suitable for numerical calculations. The details can be found in Appendix A.

2.3. *Self-consistent Born approximation for impurity scattering*

We consider the total Hamiltonian in equation (9) in the following form

$$\hat{\mathcal{H}} = \hat{H} + u\sum_l \delta\left(\boldsymbol{r} - \boldsymbol{R}_l\right), \tag{16}$$

where we imply that the point like impurities are distributed randomly at the positions $\boldsymbol{R}_l$ with the concentration n_s and single-impurity potential u. As was demonstrated by Bastin et al. [12], for point-like impurities the self-energy in the self-consistent Born approximation is determined from the equation

$$\Sigma(\varepsilon) = \frac{n_s u^2}{V}\,\mathrm{Tr}\,\hat{G}(\epsilon), \tag{17}$$

where V is a volume of the system. Since $\hat{G}$ is diagonal in the Landau level basis in equation (2), the trace is calculated immediately, and an expanded form of equation (17) reads

$$\Sigma(\varepsilon) = \frac{n_s u^2}{2\pi\ell^2}\int_{-\infty}^{\infty}\frac{dk_z}{2\pi}\sum_{n=0}^{\infty}\sum_{\eta}\frac{\alpha_n}{\varepsilon - E_n^{(\eta)} - \Sigma(\varepsilon)}, \quad \alpha_n = 2 - \delta_{n0}, \tag{18}$$

The self-energy, therefore, depends only on ε. In order to have better understanding of equation (18), we consider the following particular cases.

2.3.1. Clean limit approximation. In the clean limit approximation we imply that the imaginary part of the self energy is negligible $\Gamma(\varepsilon) \to 0^+$, and replace r.h.s. of equation (18) by a δ-function. In this case, the self-consistency is lost and we are in situation of the first Born approximation. Integrating over k_z we obtain

$$\Gamma(\varepsilon) = \frac{n_s u^2}{vh\ell^2} \sum_{n=0}^{\infty} \alpha_n \frac{|\varepsilon|\,\theta(|\varepsilon| - \Delta)}{\sqrt{\varepsilon^2 - \Delta^2 - \epsilon_c^2\, n}}. \tag{19}$$

In the equation (19), $\Gamma(\varepsilon)$ as a function of ε has a characteristic saw-tooth profile, however, in the clean limit the r.h.s. contains divergences at $\varepsilon \approx E_n$ which are cured in a self-consistent case.

2.3.2. High magnetic field approximation and ζ-regularization. In order to obtain an expression for the self-energy suitable for the analyses in high magnetic fields, we make a summation over η in equation (18) and integrate over k_z which gives a self-consistent equation

$$\Sigma = -\frac{n_s u^2}{vh\ell^2} \sum_{n=0}^{\infty} \alpha_n \frac{\varepsilon - \Sigma}{\sqrt{\epsilon_c^2\, n + \Delta^2 - (\varepsilon - \Sigma)^2}}. \tag{20}$$

This expression is not restricted to the high magnetic field case, however, the r.h.s contains a divergent series which behaves as $n^{-\frac{1}{2}}$, and proper regularization is needed. In the case of quantizing magnetic fields, when energy levels are well separated, we can keep in the summation only the term with $\varepsilon \sim E_n^{(\eta)}$, and the divergence becomes irrelevant.

In the general case, to treat the divergent series of this type, one can try ζ-regularization, which is frequently used in the high-energy physics [14]. In the ζ-regularization technique, the series in the r.h.s. of equation (20) is expressed via the Hurwitz ζ-function $\zeta(s,a) = \sum_{n=0}^{\infty}(n+a)^{-s}$ which can be continued analytically for $s < 1$. By using analytical continuation, the uniquely defined finite value is assigned to the divergent series. With the help of this technique, we get the following answer

$$\Sigma = \frac{n_s u^2 \epsilon_c^2}{4\pi v^3 \hbar^3} (\varepsilon - \Sigma) \left[\frac{1}{\sqrt{\Delta^2 - (\varepsilon - \Sigma)^2}} - \frac{2}{\epsilon_c} \zeta\left(\frac{1}{2}, \frac{\Delta^2 - (\varepsilon - \Sigma)^2}{\epsilon_c^2}\right) \right], \tag{21}$$

Although equation (21) can be useful for practical calculations, the physical meaning of this regularization in the present case is obscure for the authors.

In order to have some insight on the meaning of ζ-regularization, it is instructive to consider asymptotic behaviour of equation (21) at $\epsilon_c \to \infty$. The Hurwitz ζ-function has the following asymptotic series [14]

$$\zeta(z+1, a) = \frac{1}{z} a^{-z} + \frac{1}{2} a^{-z-1} + \frac{1}{z} \sum_{k=2}^{\infty} \frac{B_k}{k!} \frac{\Gamma(z+k)}{\Gamma(z)} a^{-z-k}, \tag{22}$$

where B_k are Bernoulli numbers and $\Gamma(z)$ is the Euler's gamma-function. Using this expansion, equation (21) in the $\epsilon_c \to \infty$ limit reads

$$\Sigma = \frac{n_s u^2}{\pi v^3 \hbar^3} (\varepsilon - \Sigma) \sqrt{\Delta^2 - (\varepsilon - \Sigma)^2}. \tag{23}$$

This expression has to be compared with the self-energy calculated at zero magnetic field directly from the Hamiltonian in equation (1).

Journal of Physics: Conference Series **603** (2015) 012009 doi:10.1088/1742-6596/603/1/012009

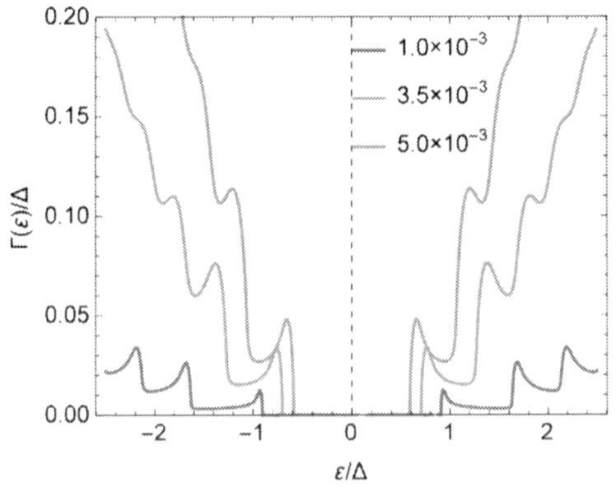

Figure 1. (color online) $\Gamma(\varepsilon)$ as a function of ε for $\varkappa = 0.001$, 0.0035, and 0.005 with $\epsilon_c/\Delta = 1.5$ calculated from equation (20) with 150 Landau levels.

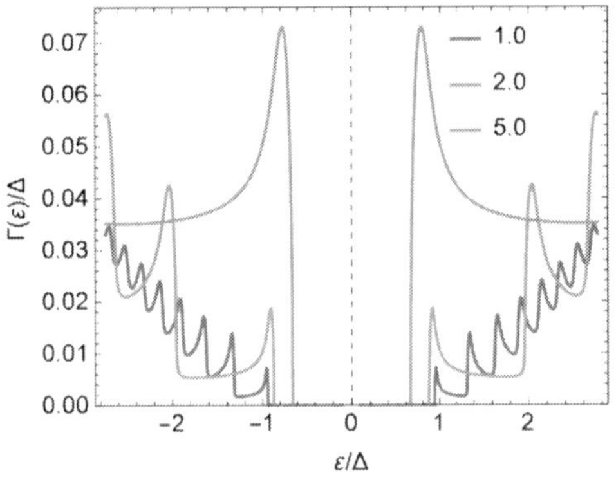

Figure 2. (color online) $\Gamma(\varepsilon)$ as a function of ε with fixed $\varkappa = 0.001$ for $\epsilon_c/\Delta = 1$, 2, and 5 calculated from equation (20) with 150 Landau levels.

2.3.3. Zero magnetic field case. In the zero magnetic field the energy spectrum of the Dirac equation consists of two bands $E_{\boldsymbol{p}}^{(\eta)} = \eta\sqrt{\Delta^2 + v^2p^2}$, each band is doubly degenerated with respect to the two possible projection of electron spin on the direction of $\boldsymbol{p}$. The self-consistent equation (17) in this case gives

$$\Sigma = \frac{2n_s u^2}{\hbar^3}\int_{\mathrm{BZ}} \frac{d\boldsymbol{p}}{(2\pi)^3}\sum_{\eta}\frac{1}{\varepsilon - E_{\boldsymbol{p}}^{(\eta)} - \Sigma}, \tag{24}$$

where the integration is over the Brillouin zone. In the clean limit approximation, where $\Gamma(\varepsilon) \to 0^+$, we obtain the following result

$$\Gamma(\varepsilon) = \frac{n_s u^2}{\pi\hbar^3 v^3}|\varepsilon|\sqrt{\varepsilon^2 - \Delta^2}\,\theta(|\varepsilon| - \Delta). \tag{25}$$

In the general case, if we consider free electrons, the r.h.s in equation (24) diverges linearly

$$\Sigma = -2\,(\varepsilon - \Sigma)\,\frac{n_s u^2}{\pi^2\hbar^3}\int_0^\infty dp\frac{p^2}{v^2p^2 + \Delta^2 - (\varepsilon - \Sigma)^2} \tag{26}$$

Using the cut-off parameter $\frac{\pi}{2}\Lambda$ for the integral, we obtain the self-consistent equation

$$\Sigma = -\frac{n_s u^2}{\pi\hbar^3 v^3}\,(\varepsilon - \Sigma)\left[\Lambda - \sqrt{\Delta^2 - (\varepsilon - \Sigma)^2}\right]. \tag{27}$$

which differs from equation (23) by a presence of cut-off parameter. For a practical usage of equation (21), we can add manually the term $-4\Lambda/\epsilon_c^2$ to the brackets in the r.h.s of (21) to have consistence with equation (27).

3. Results and discussion

3.1. Landau level broadening in magnetic field

We start the discussion with analysing the behaviour of the Landau level broadening in the magnetic field. In figure 1, $\Gamma(\varepsilon)$ is shown as a function of ε/Δ for several values of dimensionless parameter $\varkappa = \Delta/\epsilon_s$ where $\epsilon_s = 4\pi v^3\hbar^3/(n_s u^2)$ is a characteristic energy of the scattering. At small $\varkappa$, Γ as a function of ε has a characteristic for a three-dimensional system saw-tooth profile in agreement with the clean limit approximation in equation (19) where in the self-consistent approximation the divergences at $\epsilon = E_n$ become regularized. With increasing $\varkappa$, the Landau levels become broader and the gap between $E_0^{(-)}$ and $E_0^{(+)}$ energy levels reduces leading, finally, to the disappearance of the Landau quantization and the closing of the gap for large impurity concentration.

Energy dependence of the $\Gamma(\varepsilon)$ for several ϵ_c is demonstrated in figure 2. With increasing ϵ_c both level separation and level broadening increase. In order to understand how $\Gamma(\varepsilon)$ depends on the magnetic field with fixed ε, let us consider equation (20) in the clean limit $\epsilon_c/\epsilon_s \ll 1$ when Landau levels are well separated. In high magnetic fields we can keep only the term with $n = 0$ in the r.h.s.

$$\Sigma_0 = -\frac{n_s u^2 \epsilon_c^2}{4\pi v^3 \hbar^3} \frac{\varepsilon - \Sigma_0}{\sqrt{\Delta^2 - (\varepsilon - \Sigma_0)^2}}. \tag{28}$$

In the case of negligible Δ, we can neglect Δ^2 with respect to $(\varepsilon - \Sigma_0)^2$, which gives $\Gamma_0(\varepsilon)$ independent from ε for $|\varepsilon| < \epsilon_c$ and proportional to the magnetic field

$$\Gamma_0(\varepsilon) = \frac{\epsilon_c^2}{\epsilon_s} = \frac{n_s u^2 e B}{2\pi \hbar^2 v}. \tag{29}$$

Figure 3 illustrates $\Gamma(\varepsilon)$ in the zero-gap case for several values of ϵ_c/ϵ_s which shows flat $\Gamma(\varepsilon)$ in the vicinity of $\varepsilon = 0$. From equation (29) we find that with increasing ϵ_c, the Landau level broadening increases faster than the level separation which leads to the violation of the clean limit approximation when $\epsilon_c \approx \epsilon_s$ where the magnetic field dependence of Γ_0 changes.

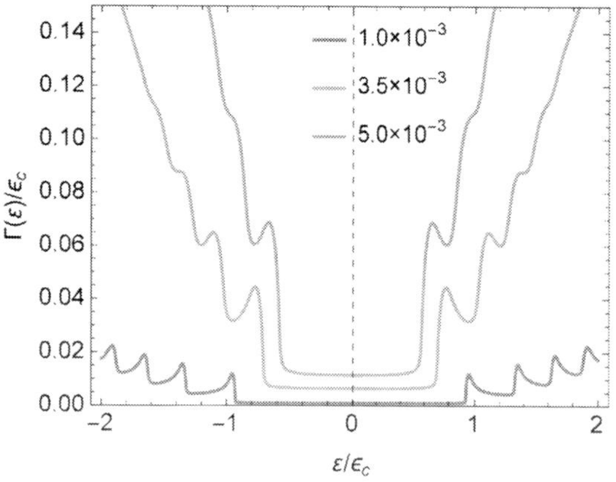

Figure 3. (color online) $\Gamma(\varepsilon)$ as a function of ε/ϵ_c for several ϵ_c/ϵ_s with $\Delta = 0$. In the vicinity of $\varepsilon = 0$, $\Gamma(\varepsilon)$ becomes independent of ε and given by equation (29).

3.2. Conductivity as a function of chemical potential

First, we would like to comment on the difference between the constant Γ approximation, which is frequently used to analyse transport properties, and the self-consistent Born approximation applied in the present paper. Figure 4 shows conductivity as a function of μ/ϵ_c for $\Delta/\epsilon_c = 0$, 0.25, and 1 calculated with constant $\Gamma/\epsilon_c = 0.01$ from equation (A.7) using numerical summation over 150 Landau levels. An important point is that $\sigma_\parallel$ with $\Gamma =$ const increases monotonously with increasing $|\mu|$. In the zero gap case $\sigma_\parallel$ becomes flat in the segment $|\mu| < \epsilon_c$ where it is given by equation (A.8). This result coincide with a longitudinal conductivity for a Weyl semimetal

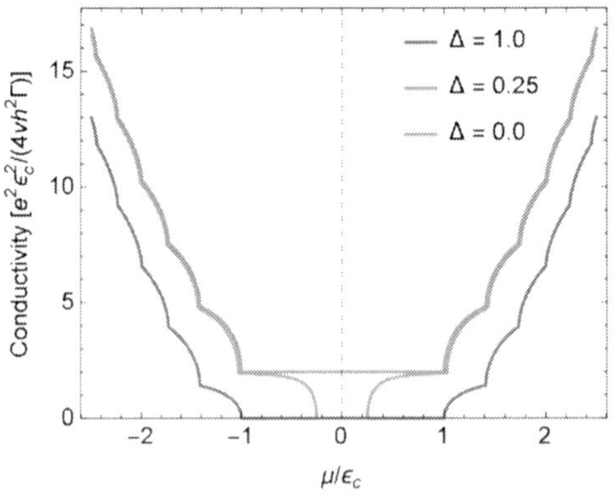

Figure 4. (color online) $\sigma_{\parallel}(0,\mu)$ as a function of μ for several Δ in the constant-Γ approximation with $\Gamma/\epsilon_c = 0.01$.

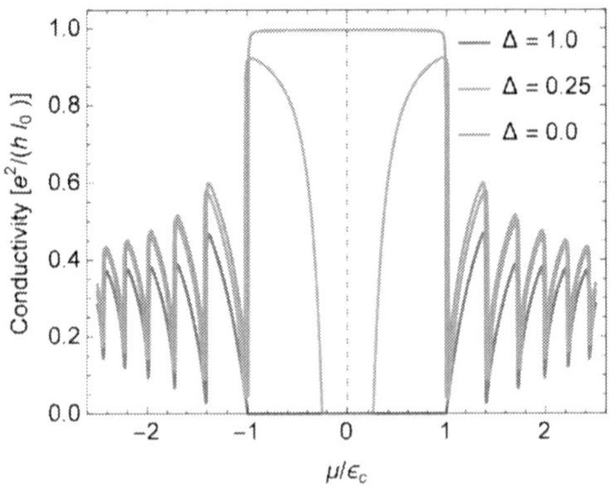

Figure 5. (color online) $\sigma_{\parallel}(0,\mu)$ as a function of μ for several Δ/ϵ_c in the self-consistent Born approximation.

obtained by Gorbar et al. [8]. The conductivity at $\mu = 0$ in this case increases linearly in magnetic field and independent of temperature for $\epsilon_c > k_B T$.

However, if we use self-consistent Born approximation for $\Gamma(\varepsilon)$ the result changes considerably. Figure 5 shows the $\sigma_{\parallel}(0,\mu)$ with the self-energy calculated from equation (19) for $\epsilon_c/\epsilon_s = 10^{-4}$. In the self-consistent Born approximation $\sigma_{\parallel}$ as a function of μ shows pronounced oscillations due to oscillating behaviour of $\Gamma(\epsilon)$ (see figure 2). In particular, in the zero gap case $\Gamma(\epsilon) \sim B$ and conductivity for $|\mu| < \epsilon_c$ becomes independent of magnetic field and chemical potential

$$\sigma_0 = \frac{e^2}{h}\frac{1}{\ell_0}, \tag{30}$$

where $\ell_0 = 2n_s u^2/(\hbar^2 v^2)$ is some characteristic length scale which is proportional to the impurity concentration.

Finally, we show the conductivity as a function of a chemical potential for several ϵ_c with finite gap and $\varkappa = 10^{-4}$. Similar, to the zero-gap case, the conductivity for $\mu \approx \Delta$ is independent of the magnetic field for fixed μ. However, in real materials with finite doping the chemical potential depends on magnetic field since electron concentration is fixed and the conductivity with finite gap becomes magnetic field and temperature dependent.

4. Summary

In summary, we have considered the longitudinal conductivity of a 3D gas of Dirac fermions in the magnetic field in the framework of the Kubo linear response theory. The scattering on the impurities was treated using the self-consistent Born approximation. The application of the self-consistent Born approximation results in the linear magnetic field dependence of the Landau level broadening in high magnetic fields and clean system. Using this level broadening we calculated the chemical potential dependence of the conductivity. In the zero gap case our result for the conductivity is in agreement with the result obtained previously by Gorbar at al. for the case of a 3D Weyl semimetal [8]. However, in our case linear magnetic-field dependence of the level broadening leads to the magnetic field independent conductivity at $\mu = 0$ in high magnetic fields where the contribution from the $n = 0$ Landau level becomes dominant.

Finally, we comment on the application of our results to the recent experiments. The longitudinal transport measurements in zero-gap 3D Dirac electron systems demonstrated that the magnetoresistance in the direction of the magnetic field was negligible with respect to the

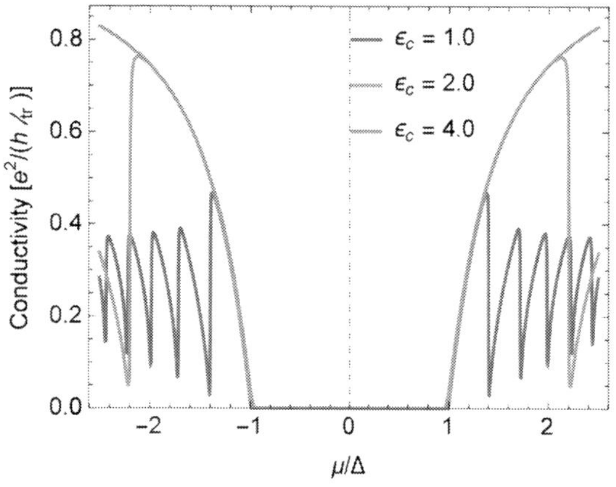

Figure 6. (color online) Conductivity as a function of μ for several ϵ_c with $\varkappa = 10^{-4}$ calculated from equations (A.7) and (20) with 150 energy levels.

giant transverse magnetoresistance [1, 2, 3, 4]. This is in agreement with our result that the zero-gap conductivity becomes independent of the magnetic field in quantum limit. However, in Cd_3As_2 negative temperature independent magnetoresistance was reported in the magnetic fields greater than 2 T for the low mobility sample [1]. This might indicate that in Cd_3As_2 magnetic-field dependence of the level broadening is weaker than linear.

Acknowledgments
This work was supported by Grants-in-Aid for Scientific Research (A) (No. 24244053) and (C) (No. 26400355) from the Ministry of Education, Culture, Sports, Science and Technology of Japan. One of the authors (I. P.) acknowledges the Russian Foundation for Basic Research, Grant No. 14-02-92104.

Appendix A. Analytical expressions for $\sigma_{\parallel}(0,\mu)$

Here we discuss several analytical expressions for $\sigma_{\parallel}(0,\mu)$ in equation (15) which can be useful for numerical calculations. We suppose that self-energy in equation (12) depends only on μ. The real part of the self-energy is absorbed in the chemical potential for brevity. Let us start with the summation over η in the expression

$$\sum_{\eta}\left[\frac{\epsilon_z^2}{E_n^2}\mathcal{A}_n^{(\eta)2}+\left(1-\frac{\epsilon_z^2}{E_n^2}\right)\mathcal{A}_n^{(\eta)}\mathcal{A}_n^{(-\eta)}\right]. \tag{A.1}$$

At first, we note that

$$\mathcal{A}_n^{(\eta)}\mathcal{A}_n^{(-\eta)}=\frac{i\Gamma}{4\mu}\left(\frac{1}{E_n^2-\mu_-^2}-\frac{1}{E_n^2-\mu_+^2}\right), \tag{A.2}$$

where the shorthand notation $\mu_\pm=\mu\pm i\Gamma$ is introduced. Next, we regroup the terms in equation (A.1) and make the η summation in the expression

$$\sum_{\eta}\left(\mathcal{A}_n^{(\eta)2}-\mathcal{A}_n^{(\eta)}\mathcal{A}_n^{(-\eta)}\right)=-E_n^2\left(\frac{1}{E_n^2-\mu_-^2}-\frac{1}{E_n^2-\mu_+^2}\right)^2, \tag{A.3}$$

where the following identity was used

$$\mathcal{A}_n^{(\eta)}-\mathcal{A}_n^{(-\eta)}=-i\eta E_n\left(\frac{1}{E_n^2-\mu_-^2}-\frac{1}{E_n^2-\mu_+^2}\right). \tag{A.4}$$

Combining together equations (A.2) and (A.3), we arrive to the following representation of expression (A.1)

$$\sum_\eta \left[\frac{\epsilon_z^2}{E_n^2}\mathcal{A}_n^{(\eta)2} + \left(1-\frac{\epsilon_z^2}{E_n^2}\right)\mathcal{A}_n^{(\eta)}\mathcal{A}_n^{(-\eta)}\right] = \frac{2\left(\epsilon_z^2+\Gamma^2\right)}{\left(E_n^2-\mu_-^2\right)\left(E_n^2-\mu_+^2\right)} - \frac{\epsilon_z^2}{\left(E_n^2-\mu_-^2\right)^2} - \frac{\epsilon_z^2}{\left(E_n^2-\mu_+^2\right)^2}. \tag{A.5}$$

With the help of equation (A.5), the conductivity in equation (15) can be rewritten in the following form

$$\sigma_\parallel(0,\mu) = \frac{e^2\epsilon_c^2}{4\pi^3\hbar^2 v}\int_0^\infty d\epsilon \sum_{n=0}^\infty \alpha_n \left[\frac{2\left(\epsilon^2+\Gamma^2\right)}{\left(\epsilon_c^2 n+\epsilon^2+\Delta_+^2\right)\left(\epsilon_c^2 n+\epsilon^2+\Delta_-^2\right)} - \frac{\epsilon^2}{\left(\epsilon_c^2 n+\epsilon^2+\Delta_+^2\right)^2} - \frac{\epsilon^2}{\left(\epsilon_c^2 n+\epsilon^2+\Delta_-^2\right)^2}\right]. \tag{A.6}$$

where $\Delta_\pm^2 = \Delta^2 - \mu_\pm^2$. In equation (A.6) either summation over n or integration over ϵ can be performed analytically. In the following, we consider these two cases separately.

High magnetic fields. In the case of high magnetic fields, it is reasonable to make integration over ϵ since the summation over Landau levels can be usually restricted to several dozens of terms. The integration is performed straightforwardly

$$\sigma_\parallel(0,\mu) = \frac{e^2\epsilon_c^2}{4vh^2}\sum_{n=0}^\infty \alpha_n \frac{4\Gamma^2 - \left(\sqrt{\epsilon_c^2 n+\Delta_+^2} - \sqrt{\epsilon_c^2 n+\Delta_-^2}\right)^2}{\sqrt{\left(\epsilon_c^2 n+\Delta_+^2\right)\left(\epsilon_c^2 n+\Delta_-^2\right)}\left(\sqrt{\epsilon_c^2 n+\Delta_+^2}+\sqrt{\epsilon_c^2 n+\Delta_-^2}\right)}. \tag{A.7}$$

Note that the series oven is convergent. In the important practical case $n = 0$ and $\Delta = 0$ this expression gives simple answer [8]

$$\sigma_0 = \frac{e^2}{2h^2 v}\frac{\epsilon_c^2}{\Gamma}. \tag{A.8}$$

Another important example is the conductivity at $\mu = 0$. In this case $\Delta_+^2 = \Delta_-^2 = \Delta^2 + \Gamma^2$, and equation (A.7) gives

$$\sigma_\parallel(0,0) = \frac{e^2\Gamma^2}{vh^2\epsilon_c}\left[\zeta\left(\frac{3}{2}, \frac{\Delta^2+\Gamma^2}{\epsilon_c^2}\right) - \frac{1}{2}\frac{\epsilon_c^3}{\left(\Delta^2+\Gamma^2\right)^{3/2}}\right]. \tag{A.9}$$

In the $\epsilon \to 0$ limit, with the help of equation (22), this expression gives

$$\sigma_\parallel(0,0) = \frac{2e^2}{vh^2}\frac{\Gamma^2}{\sqrt{\Delta^2+\Gamma^2}}. \tag{A.10}$$

Moderate magnetic fields. In order to study a behaviour at low magnetic fields, it maybe useful to perform a summation in equation (A.6) via special functions. With the help of the definition of Hurwitz ζ-function and using the identity

$$\sum_{n=0}^\infty \frac{1}{(n+a)(n+b)} = \frac{1}{b-a}\left(\psi(b)-\psi(a)\right), \tag{A.11}$$

where $\psi(z) = \Gamma'(z)/\Gamma(z)$ is digamma function, we obtain $\sigma_{\|} = \sigma_{\rm I} + \sigma_{\rm II}$ where

$$\sigma_{\rm I} = \frac{e^2}{\pi^3\hbar^2 v}\int_0^\infty d\epsilon \left\{ \frac{(\epsilon^2+\Gamma^2)}{\Delta_+^2 - \Delta_-^2}\left[\psi\left(\frac{\epsilon^2+\Delta_+^2}{\epsilon_c^2}\right) - \psi\left(\frac{\epsilon^2+\Delta_-^2}{\epsilon_c^2}\right) + \frac{1}{2}\frac{\epsilon_c^2}{\epsilon^2+\Delta_+^2} - \frac{1}{2}\frac{\epsilon_c^2}{\epsilon^2+\Delta_-^2}\right]\right\}, \quad (A.12)$$

and

$$\sigma_{\rm II} = -\frac{e^2}{2\pi^3\hbar^2 v\epsilon_c^2}\int_0^\infty d\epsilon\epsilon^2 \left[\zeta\left(2, \frac{\epsilon^2+\Delta_+^2}{\epsilon_c^2}\right) + \zeta\left(2, \frac{\epsilon^2+\Delta_-^2}{\epsilon_c^2}\right) - \frac{1}{2}\frac{\epsilon_c^4}{(\epsilon^2+\Delta_+^2)^2} - \frac{1}{2}\frac{\epsilon_c^4}{(\epsilon^2+\Delta_-^2)^2}\right]. \quad (A.13)$$

Note that $\sigma_{\rm I}$ and $\sigma_{\rm II}$ are introduced for convenience and have no individual meaning since the integrals over ϵ are divergent. However, $\sigma_{\rm I} + \sigma_{\rm II}$ is convergent.

Zero magnetic field. The $\epsilon_c \to 0$ limit of $\sigma_{\rm I}$ can be studied using the expansion $\psi(z) = \log z - 1/(2z) - 1/(12z^2) + \ldots$, which gives

$$\sigma_{\rm I} = \frac{e^2}{\pi^3\hbar^2 v}\int_0^\infty d\epsilon\,(\epsilon^2+\Gamma^2)\frac{\log(\epsilon^2+\Delta_+^2) - \log(\epsilon^2+\Delta_+^2)}{\Delta_+^2 - \Delta_-^2}, \quad (A.14)$$

where

$$\frac{\log(\epsilon^2+\Delta_+^2) - \log(\epsilon^2+\Delta_+^2)}{\Delta_+^2 - \Delta_-^2} = \frac{1}{2|\mu|\Gamma}\begin{cases} \tan^{-1}\dfrac{2|\mu|\Gamma}{\epsilon^2+\Delta^2+\Gamma^2-\mu^2}, & \epsilon^2 > \mu^2 - \Delta^2 - \Gamma^2, \\ \pi + \tan^{-1}\dfrac{2|\mu|\Gamma}{\epsilon^2+\Delta^2+\Gamma^2-\mu^2}, & \epsilon^2 < \mu^2 - \Delta^2 - \Gamma^2. \end{cases} \quad (A.15)$$

Finally, we arrive to the following expression for $\sigma_{\rm I}$:

$$\sigma_{\rm I} = \frac{e^2}{\pi^3\hbar^2 v}\int_0^\infty d\epsilon\,\frac{\epsilon^2+\Gamma^2}{2\mu\Gamma}\tan^{-1}\frac{2\mu\Gamma}{\epsilon^2+\Delta^2+\Gamma^2-\mu^2}, \quad \text{if} \quad \mu < \sqrt{\Delta^2+\Gamma^2}, \quad (A.16)$$

and

$$\sigma_{\rm I} = \frac{e^2}{\pi^3\hbar^2 v}\int_0^{\sqrt{\mu^2-\Delta^2-\Gamma^2}} d\epsilon\,\frac{\epsilon^2+\Gamma^2}{2|\mu|\Gamma}\left(\pi + \tan^{-1}\frac{2|\mu|\Gamma}{\epsilon^2+\Delta^2+\Gamma^2-\mu^2}\right) + \frac{e^2}{\pi^3\hbar^2 v}\int_{\sqrt{\mu^2-\Delta^2-\Gamma^2}}^\infty d\epsilon\,\frac{\epsilon^2+\Gamma^2}{2\mu\Gamma}\tan^{-1}\frac{2\mu\Gamma}{\epsilon^2+\Delta^2+\Gamma^2-\mu^2}, \quad \text{if} \quad \mu > \sqrt{\Delta^2+\Gamma^2}. \quad (A.17)$$

The zero-field limit for $\sigma_{\rm II}$ is straightforward

$$\sigma_{\rm II} = -\frac{e^2}{2\pi^3\hbar^2 v}\int_0^\infty d\epsilon\,\epsilon^2\left(\frac{1}{\epsilon^2+\Delta_+^2} + \frac{1}{\epsilon^2+\Delta_-^2}\right). \quad (A.18)$$

As we mentioned above, although each of $\sigma_{\rm I}$ and $\sigma_{\rm II}$ diverges, the physical meaning has only the sum $\sigma_{\rm I} + \sigma_{\rm II}$ which is well defined, as it can be clearly seen in the $\mu = 0$ case where equations (A.16) and (A.18) give the same result as equation (A.10).

References

[1] Liang T, Gibson Q, Ali M N, Liu M, Cava R J and Ong N P 2014 *Nature Materials* 10.1038/nmat4143

[2] He L P, Hong X C, Dong J K, Pan J, Zhang Z, Zhang J and Li S Y 2014 *Phys. Rev. Lett.* **113**(24) 246402

[3] Feng J, Pang Y, Wu D, Wang Z, Weng H, Li J, Dai X, Fang Z, Shi Y and Lu L 2014 *ArXiv e-prints* (*Preprint* 1405.6611)

[4] Novak M, Sasaki S, Segawa K and Ando Y 2014 *ArXiv e-prints* (*Preprint* 1408.2183)

[5] Wang Z, Weng H, Wu Q, Dai X and Fang Z 2013 *Phys. Rev. B* **88**(12) 125427

[6] Borisenko S, Gibson Q, Evtushinsky D, Zabolotnyy V, Büchner B and Cava R J 2014 *Phys. Rev. Lett.* **113**(2) 027603

[7] Abrikosov A A 1998 *Phys. Rev. B* **58**(5) 2788–2794

[8] Gorbar E V, Miransky V A and Shovkovy I A 2014 *Phys. Rev. B* **89**(8) 085126

[9] Shon N H and Ando T 1998 *J. Phys. Soc. Jpn.* **67** 2421–2429

[10] Canuto V and Chiuderi C 1970 *Phys. Rev. D* **1**(8) 2219–2226

[11] Kaminker A and Yakovlev D 1981 *Theoret. and Math. Phys.* **49** 1012–1020

[12] Bastin A, Lewiner C, Betbeder-Matibet O and Nozieres P 1971 *J. Phys. Chem. Solids* **32** 1811 – 1824

[13] Středa P and Smrčka L 1975 *Phys. Status Solidi B* **70** 537–548

[14] Elizalde E, Odintsov S D, Romeo A, Bystrenko A A and Zerbini S 1994 *Zeta Regularization Techniques with Applications* (World Scientific/Singapore, New Jersey, London, Hong Kong)

Journal of Physics: Conference Series **603** (2015) 012010 doi:10.1088/1742-6596/603/1/012010

Plaquette chirality patterns for robust zero-gap states in α-type organic conductor

Frédéric Piéchon[1], Yoshikazu Suzumura[2] and Takao Morinari[3]

[1] Laboratoire de Physique des Solides, CNRS UMR 8502, Université Paris-Sud, F-91405 Orsay Cedex, France

[2] Department of Physics, Nagoya University, Chikusa-ku, Nagoya 464-8602, Japan

[3] Graduate School of Human and Environment Studies, Kyoto University, Kyoto 606-8501, Japan

E-mail: suzumura@s.phys.nagoya-u.ac.jp

Abstract. Dirac electrons with a zero-gap state (ZGS) in organic conductor α-(BEDT-TTF)$_2$I$_3$ result from a fine tuning of the seven nearest neighbors transfer integrals $(a_1, a_2, a_3, b_1, b_2, b_3, b_4)$ between the four molecules of the unit cell. In this work we show that for given moduli $|a_1|, \ldots |b_4|$, the possibility of having Dirac electron with a ZGS at 3/4 (or 1/4) filling strongly depends on the specific configurations of signs of the seven transfer integral. More precisely it is possible to classify the sign configurations into essentially four classes determined by $\chi_a = \text{sign}(a_2 a_3)$ and $\chi_b = \text{sign}(b_1 b_2 b_3 b_4)$. Using extended numerics, we show that for both weak and large inhomogeneity in the moduli, the class $(\chi_a, \chi_b) = (-, -)$ is the most favorable to find Dirac electrons with ZGS at 3/4 (or 1/4) filling. For the class $(\chi_a, \chi_b) = (+, +)$ in the opposite case, we never found any ZGS at either 1/4 or 3/4 filling. The last two classes given by $(\chi_a, \chi_b) = (+, -)$ and $(\chi_a, \chi_b) = (-, +)$ corresponding to an intermediate situation; they allow for ZGS at 3/4 (resp. 1/4) filling but are much less favorable than class $(\chi_a, \chi_b) = (-, -)$. As a matter of fact, all previous numerical studies of Dirac electrons and ZGS in α-(BEDT-TTF)$_2$I$_3$ correspond to class $(\chi_a, \chi_b) = (-, +)$.

1. Introduction

It is now well established that organic conductors α-(BEDT-TTF)$_2$I$_3$ [1], exhibit a zero gap-state (ZGS) characteristic of the presence of massless Dirac electrons close to the Fermi energy [2]. However, compared to graphene, the Dirac cones are strongly tilted resulting in a strong anisotropy of the Dirac electron velocity components. Furthermore, the fact that α-(BEDT-TTF)$_2$I$_3$ has a unit cell composed of four inequivalent molecules leads to some unique physical properties associated to this ZGS. In particular, as revealed by recent NMR measurements, the anisotropy of the Dirac particles appears strongly related to their local spectral weight on the four inequivalent molecules per unit cell; such that the smallest (largest) velocity Dirac particles have their largest weight on the electron rich (poor) molecule site [3, 4, 5].

From a theoretical point of view, the electronic properties of α-(BEDT-TTF)$_2$I$_3$ are usually described by a tight-binding model with seven nearest-neighbor transfer integrals $a_1, \cdots, b_4$ between the four molecules A, A', B, and C of the unit cell. As illustrated on figure Fig. 1, this model presents an inversion symmetry with inversion centers located on B,C and the middle point between A and A'.[1]. Thank's to this inversion symmetry, recent studies have used Fu-Kane topological argument in order to determine quantitatively the existence domain of Dirac

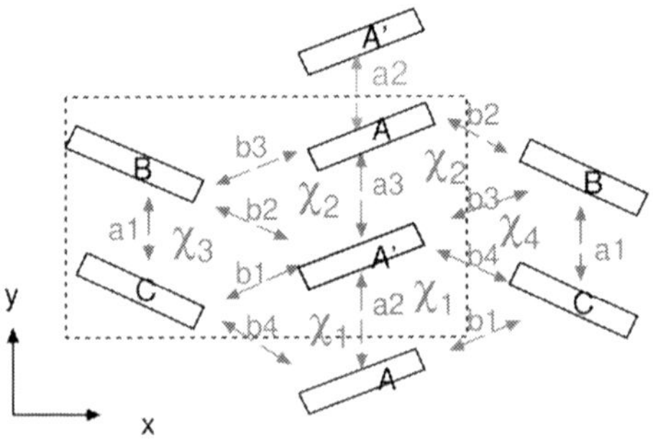

Figure 1. (Color online) Structure of α-(BEDT-TTF)$_2$I$_3$, with four molecules A, A', B, and C per unit cell (dotted square). Inversion centers are located on sites B, C, and the middle point between A and A'.[1]. The seven nearest-neighbor transfer integrals are shown by $a_1, \cdots, b_4$. The chirality of four distinct triangular plaquettes is shown by χ_1, χ_2, χ_3, and χ_4.

electrons from the knowledge of the energy and inversion parity eigenvalues at the four time reversal invariant momenta (TRIM) [6, 7, 8]. Despite their usefulness, these studies do not allow to distinguish ZGS systems in which Dirac electrons are at the Fermi energy (with a vanishing density of states at the Fermi energy) from metallic systems in which there is an indirect overlapp between valence and conduction band such that the Dirac electrons are below (or above) the Fermi energy [9, 10].

The purpose of the present work is to clarify the relative role played by the the inhomogeneity in the moduli and in the signs of the seven transfer integrals on the appearance and stability of Dirac electrons with ZGS. In fact as explained in more details below, for given moduli $|a_1|, ...|b_4|$, the possibility of having Dirac electron with a ZGS at 3/4 (or 1/4) filling strongly depends on the specific configurations of signs of the seven transfer integral. More precisely it is possible to classify the sign configurations into essentially four classes determined by $\chi_a = \text{sign}(a_2 a_3)$ and $\chi_b = \text{sign}(b_1 b_2 b_3 b_4)$. Using extensive numerics, we show that for weak, large and no inhomogeneity in the moduli, the class $(\chi_a, \chi_b) = (-, -)$ is most favorable to find Dirac electrons with ZGS at 3/4 (or 1/4) filling. The classes $(\chi_a, \chi_b) = (+, -)$ and $(\chi_a, \chi_b) = (-, +)$ constitute equivalent but less favorable situation for the occurrence of Dirac electrons with ZGS at 3/4 (or 1/4) filling. Finally, the last class is $(\chi_a, \chi_b) = (+, +)$ and for the range of inhomogeneity that was explored no ZGS was obtained for either 1/4 or 3/4 filling.

The paper is organized as follows. In §2 we present the model and the main quantities of interest. In §3 we first introduce the chirality patterns and their properties and then using properties established in Appendix A we explained the relation between the chirality patterns and the energy band structure and local density of states. In §4 we present our extensive numerical studies of the existence of ZGS at 3/4 filling for the three pertinent classes of chirality patterns. In §5 we summarized our main results.

2. Tight-binding model

The starting point is the 4×4 Bloch Hamiltonian matrix $H(\boldsymbol{k})$ corresponding to the real space tight-binding model shown in Fig. 1:

$$H(\boldsymbol{k}) = \begin{pmatrix} 0 & a & b & c \\ a^* & 0 & b^* e^{ik_x} & c^* e^{ik_x + ik_y} \\ b^* & b e^{-ik_x} & 0 & f \\ c^* & c e^{-ik_x - ik_y} & f^* & 0 \end{pmatrix}. \tag{1}$$

with $a = a_3 + a_2 e^{ik_y}$, $b = b_3 + b_2 e^{ik_x}$, $c = e^{ik_y}(b_4 + b_1 e^{ik_x})$ and $f = a_1(1 + e^{ik_y})$ (the lattice constant is taken as unity). The Boch basis $|j\boldsymbol{k}\rangle$ $(j = A, A', B, C)$ was chosen such that the Hamiltonian matrix is Bloch periodic $H(\boldsymbol{k}+\boldsymbol{G}) = H(\boldsymbol{k})$ with $\boldsymbol{G}$ a reciprocal lattice vector. In this Bloch basis, the energy eigenstate $|E_n(\boldsymbol{k})\rangle$ of the nth band is represented by a four components vector $u_n(\boldsymbol{k})^T \equiv (u_n^A, u_n^{A'}, u_n^B, u_n^C)$ such that $H(\boldsymbol{k})u_n(\boldsymbol{k}) = E_n(\boldsymbol{k})u_n(\boldsymbol{k})$ where by convention $E_1(\boldsymbol{k}) < E_2(\boldsymbol{k}) < E_3(\boldsymbol{k}) < E_4(\boldsymbol{k})$. The local density of states (LDOS) on each site is then obtained as $(j = A, A', B, C)$

$$D_j(E) = \int \frac{d\boldsymbol{k}}{4\pi^2} \sum_n |u_n^j(\boldsymbol{k})|^2 \delta(E - E_n(\boldsymbol{k})). \tag{2}$$

For 3/4 filling, the existence of a pair of Dirac points at $\pm\boldsymbol{k}_D$ implies $E_4(\boldsymbol{k}_D) = E_3(\boldsymbol{k}_D) = E_{\boldsymbol{k}_D}$ with $\boldsymbol{k}_D \neq \boldsymbol{G}/2$; the existence of a ZGS further necessitates that $E_4(\boldsymbol{k}) > E_{\boldsymbol{k}_D} > E_3(\boldsymbol{k})$ for any $\boldsymbol{k}$. In that situation the LDOS verify $D_j(E) \simeq \frac{|E-E_D|}{v_j^2}$ for E near E_D where v_j is the site dependent angular averaged Fermi velocity.

3. Classification in terms of chirality patterns

For fixed moduli $(|a_1|, \cdots |b_4|)$, the energy bands $E_n(\boldsymbol{k})$ and the eigenfunction components $u_n^j(\boldsymbol{k})$ have the freedom of the sign of the seven hopping parameters, as a result one may expected to obtain 2^7 distinct LDOS $D_j(E)$. As explained in Appendix A, the number of distinct LDOS $D_j(E)$ is entirely determined by the number of patterns of the four chiralities $\chi_i = \pm$ $(i = 1, 2, 3, 4)$ [11]

$$\chi_1 = \mathrm{sign}(a_2 b_1 b_4), \quad \chi_2 = \mathrm{sign}(a_3 b_2 b_3), \quad \chi_3 = \mathrm{sign}(a_1 b_1 b_2), \quad \chi_4 = \mathrm{sign}(a_1 b_3 b_4), \tag{3}$$

where each chirality χ_i encodes the sign of the product of the transfer integrals for each of the four distinct triangular plaquette that compose the unit cell. The are in total $2^4 = 16$ chirality patterns $(\chi_1, \chi_2, \chi_3, \chi_4)$ which are summarized in Table 1.

Interestingly it appears that the chirality patterns are not all *independent*, more precisely one can essentially define five classes of patterns with independent real space *topology*: $\mathscr{C}_1 = (A_1, A_2, A_7, A_8)$, $\mathscr{C}_2 = (A_3, A_4, A_5, A_6)$, $\mathscr{C}_3 = (S_3, S_4, S_5, S_6)$, $\mathscr{C}_{4a} = (S_1, S_8)$, $\mathscr{C}_{4b} = (S_2, S_7)$. As an illustration of their distinct *topology*, the chirality patterns associated to A_3, A_7, S_6, S_2, S_1 are represented on Fig. 2(a)...(e). Each class is characterized by the fact that it is globally invariant under the following three transformations: (i) chirality conjugaison C_χ: $(\chi_1, \chi_2, \chi_3, \chi_4) \to (-\chi_1, -\chi_2, -\chi_3, -\chi_4)$, (ii) mirror along y-axis chains M_y: $(\chi_1, \chi_2, \chi_3, \chi_4) \to (\chi_1, \chi_2, \chi_4, \chi_3)$, (iii) translation by half unit along y-axis T_y: $(\chi_1, \chi_2, \chi_3, \chi_1) \to (\chi_2, \chi_1, \chi_4, \chi_3)$. As example one easily verifies the identities $A_8 = T_y[A_7] = M_y[A_7]$, $A_1 = C_\chi T_y[A_7]$ and $A_2 = C_\chi[A_7]$. Note that for patterns S_i in classes $\mathscr{C}_{4a}, \mathscr{C}_{4b}$ one has $S_i = M_y[S_i] = T_y[S_i]$; in that sens it is more meaningful to define a single class $\mathscr{C}_4 = (S_1, S_8, S_2, S_7)$. In fact it then appears that the four classes $\mathscr{C}_i$ are equivalent to classify the 16 chirality patterns in terms of the two effective chiralities

$$\chi_a = \chi_1\chi_2\chi_3\chi_4 = \mathrm{sign}(\mathrm{a}_2\mathrm{a}_3), \qquad \chi_b = \chi_3\chi_4 = \mathrm{sign}(\mathrm{b}_1\mathrm{b}_2\mathrm{b}_3\mathrm{b}_4), \tag{4}$$

such that $\mathscr{C}_i \equiv (\chi_a, \chi_b)$ with $\mathscr{C}_1 = (-,-)$, $\mathscr{C}_2 = (-,+)$, $\mathscr{C}_2 = (+,-)$ and $\mathscr{C}_4 = (+,+)$.

We terminate this section by listing other properties relating chirality patterns and LDOS. P_1: Since there are only 2^4 patterns, it implies that there are 2^3 (out of 2^7) configurations of signs of the seven transfer integrals that leads to the same pattern $(\chi_1, \chi_2, \chi_3, \chi_4)$ and to the same LDOS $D_j(E)$ (see Appendix B). In other words, without loss of generality, it is always possible to fix arbitrarily the sign of three out of the seven transfer integrals and hereafter the convention $(b_1, b_2, b_3) > 0$ is adopted in all numerical computations.

Table 1. Chirality patterns $(\chi_1, \chi_2, \chi_3, \chi_4)$. The group A_i is associated to $\chi_1\chi_2\chi_3\chi_4 = -$ whereas the group S_i is associated to $\chi_1\chi_2\chi_3\chi_4 = +$.

	A_1	A_2	A_3	A_4	A_5	A_6	A_7	A_8	S_1	S_2	S_3	S_4	S_5	S_6	S_7	S_8
χ_1	-	-	-	-	+	+	+	+	-	-	-	-	+	+	+	+
χ_2	-	-	+	+	-	-	+	+	-	-	+	+	-	-	+	+
χ_3	-	+	-	+	-	+	-	+	-	+	-	+	-	+	-	+
χ_4	+	-	-	+	-	+	+	-	-	+	+	-	+	-	-	+

P_2: From the properties of the moments of the LDOS (Appendix A), it is straightforward to show that chirality conjugation C_χ is associated to $D_j(E) \to D_j(-E)$. More physically it means that if a pattern exhibits a ZGS at 1/4 (resp. 1/2, 3/4) filling then its chirality conjugate pattern exhibits a ZGS at 3/4 (resp. 1/2, 1/4) filling.
P_3: In the cases $(|a_2| = |a_3|, |b_1| = |b_3|, |b_2| = |b_4|)$, using eq.(A.6) it easily shown that for any pattern $(\chi_1, \chi_2, \chi_3, \chi_4)$, its M_y (or T_y) transformed pattern has the same DOS.

As a consequence of properties (P_2, P_3), for the cases $(|a_2| = |a_3|, |b_1| = |b_3|, |b_2| = |b_4|)$ it is thus sufficient to study the ZGS and LDOS properties of patterns A_3, A_7, S_6, S_2, S_1 to deduce the properties of all other patterns. Lastly, we note that all previous numerical studies of Dirac points and ZGS in α-(BEDT-TTF)$_2$I$_3$ corresponds to the pattern A_3.

4. Existence and stability of ZGS at 3/4 fillings

In the following, we concentrate our numerics on the study of ZGS at 3/4 filling for patterns A_7, A_3, S_6 of classes $\mathscr{C}_1, \mathscr{C}_2, \mathscr{C}_3$ respectively. We stress that patterns A_8, A_4, S_5 present very similar properties as A_7, A_3, S_6. We discard patterns of class $\mathscr{C}_4$ because for the range of parameters that we have studied we never found ZGS at either 1/4 or 3/4 filling for these patterns.

4.1. ZGS and LDOS properties for homogeneous modulus $|a_1| = |a_2| = ... = |b_4| = 1$

As a first illustration of the importance of the chirality pattern on the appearance of ZGS, Fig. 3 represents respectively the LDOS $D_j(E)$ $(j = A, B, C)$ for patterns A_3, A_7, S_6 in the case of homogeneous moduli $|a_1| = |a_2| = ... = |b_4| = 1$. On Fig. 3, for each pattern, the energies corresponding to 1/4 and 3/4 fillings are indicated by vertical arrows.

For pattern A_3, the A, B, C sites have strongly different LDOS; in addition at 1/4 and 3/4 fillings the system is metallic, moreover there is a strong 1D like VanHove singularity for $D_{A,B,C}(E)$ at 3/4 filling. For pattern A_7, the LDOS verify $D_B(E) = D_C(E) \neq D_A(E)$; in addition there is a *wide* ZGS at 3/4 filling and a metallic state at 1/4 filling with a strong VanHove singularity for $D_A(E)$. For pattern S_6, the LDOS verify $D_B(E) = D_C(-E) \neq D_A(E)$ moreover there is a *narrow* ZGS both for 1/4 and 3/4 fillings whereas there is a VanHove singularity at 1/2 filling.

From these results, one can already anticipate that pattern A_3 which shows a VanHove singulary at 3/4 filling may thus require some peculiar and probably strong anisotropy in the moduli in order to exhibit a ZGS state at this filling. In a similar manner, one can anticipate that the ZGS at 3/4 filling obtained for patterns A_7 and S_6 may have distinct robustness against the introduction of an anisotropy in the moduli. More precisely the ZGS of pattern A_7 which exhibits a linear LDOS over wide energy range is anticipated to be more robust than the ZGS of pattern S_6 which is relatively narrow in energy.

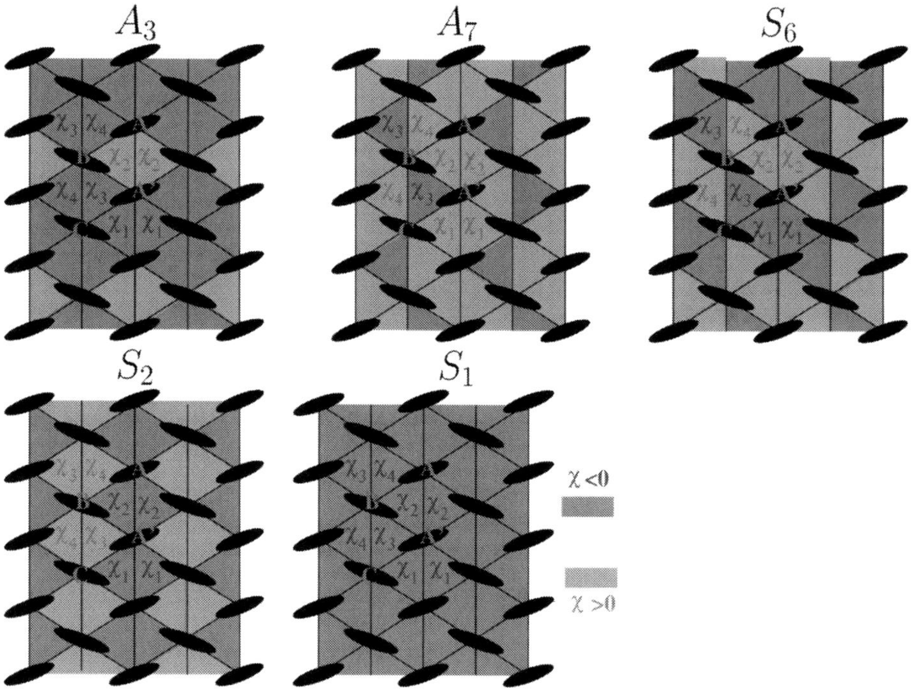

Figure 2. (Color online) Real space representation of the five distinct triangular plaquette chirality patterns A_7, A_3, S_6 , S_1 , S_2. Triangular plaquette with $\chi_i = +$ ($\chi_i = -$) are in red (blue). Chirality conjugaison C_χ amounts to switching the color of each plaquette, in that way one obtains the corresponding five patterns A_2, A_6, S_5 , S_8 , S_7. By applying M_a the five patterns A_7, A_3, S_6 , S_1 , S_2 transform into respectively A_8, A_3, S_5 , S_1 , S_2. Similarly, by applying T_a they transform into respectively A_8, A_4, S_3 , S_1 , S_2.

4.2. ZGS and LDOS properties for BEDT moduli at pressure $P = 6Kbars$

As a first illustration of the combined role of chirality pattern and inhomogeneity in the moduli, Fig 4 represents the LDOS $D_j(E)$ ($j = A, B, C$) for patterns A_3, A_7, S_6 in the case of BEDT moduli at pressure $P = 6$ Kbars ($|a_2|,|a_3|,|b_1|,|b_2|,|b_3|,|b_4|$) $=|a_1|(2.23, 0.39, 2.86, 3.46, 1.70, 0.58)$[12, 13], for which it is known that pattern A_3 exhibits a ZGS at 3/4 filling. The value of $|a_1|$ is taken such that the second moment of the DOS is the same as for homogeneous modulus.

At first sight, the main consequence of the strong inhomogeneity in the moduli is that the shape of the LDOS are now qualitatively much more similar for all patterns. In particular, now all patterns exhibit a ZGS at 3/4 filling, however for pattern A_3 the ZGS is very narrow in energy and with strong VanHove singularities very near the Dirac point indicating a strong tilt of the Dirac cones. Note that for pattern S_6 there is no more a ZGS at 1/4 filling where it is now metallic with a strong VanHove singularity. Overall, these results seem to indicate that chirality patterns might play a weaker role for strong inhomogeneity.

4.3. ZGS properties for inhomogeneous moduli

The combined roles of chirality pattern and inhomogeneity of moduli on the existence and stability of ZGS at 3/4 fillings are now examined more systematically for the three patterns A_3, A_7, S_6. To characterize the inhomogeneity of the moduli it is convenient to define the

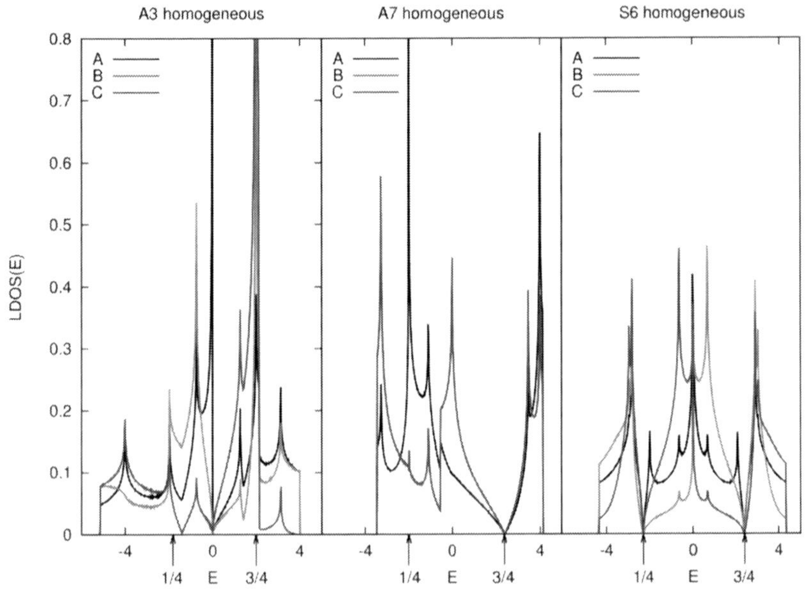

Figure 3. (Color online) Local densities of states for patterns A_3, A_7, S_6 in the case of homogeneous modulus $|a_1| = |a_2| = ... = |b_4| = 1$, $D_A(E) = D_{A'}(E)$ black line, $D_B(E)$ red line and $D_C(E)$ blue line.

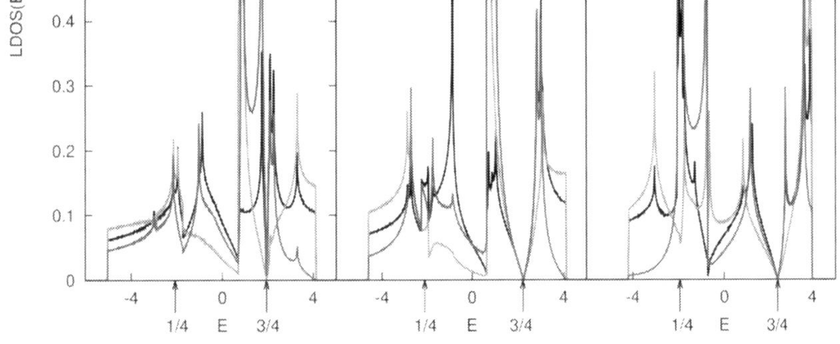

Figure 4. (Color online) Local densities of states for patterns A_3, A_7, S_6 in the case of BEDT moduli at pressure $P = 6$ Kbars $(|a_2|,|a_3|,|b_1|,|b_2|,|b_3|,|b_4|) = |a_1|(2.23, 0.39, 2.86, 3.46, 1.70, 0.58)$ [12, 13]. $D_A(E) = D_{A'}(E)$ black line, $D_B(E)$ red line and $D_C(E)$ blue line.

following six dimensionless parameters:

$$\begin{aligned} \Delta_a &= \frac{|a_2|-|a_3|}{|a_2|+|a_3|}, \quad \Delta_b = \frac{|b_1|-|b_3|}{|b_1|+|b_3|}, \quad \Delta_{b'} = \frac{|b_2|-|b_4|}{|b_2|+|b_4|}, \\ z_a &= \frac{|a_2|+|a_3|}{2|a_1|}, \quad z_b = \frac{|b_1|+|b_3|}{2|a_1|}, \quad z_{b'} = \frac{|b_2|+|b_4|}{2|a_1|}. \end{aligned} \tag{5}$$

The quantities $-1 \leq \Delta_{a,b,b'} \leq 1$ characterize the *dimerization* along the chains (AA') and the *trimerization* along the two effective chains (BACA'). The ratio $z_a, z_b, z_{b'}$ quantify essentially the relative anisotropy between the different chains as compared to chain (BC). Within this parametrization, the case of homogeneous moduli corresponds to $\Delta_{a,b,b'} = 0$ and $z_{a,b,b'} = 1$, whereas the case of BEDT moduli at $P = 6$ corresponds to $(z_a, z_b, z_{b'}, \Delta_a, \Delta_b, \Delta_{b'}) = (1.31, 2.27, 2.02, 0.70, 0.25, 0.71)$. In the following, for a given set of parameters $(z_{a,b,b'}, \Delta_{a,b,b'})$, the value of a_1 is determined so as to keep the second moment of the DOS equal to that of the homogeneous case; in that way the total bandwidth is roughly constant: $|a_1| = \sqrt{6/[1 + z_a^2(1+\Delta_a^2) + 2z_b^2(1+\Delta_b^2) + 2z_{b'}^2(1+\Delta_{b'}^2)]}$ With the convention $(b_1, b_2, b_3) > 0$, the seven hopping parameters read:

$$\begin{aligned} a_1 &= \chi_3|a_1|, \\ a_2 &= \chi_1\chi_3\chi_4 z_a(1+\Delta_a)|a_1|, \\ a_3 &= \chi_2 z_a(1-\Delta_a)|a_1|, \\ b_1 &= z_b(1+\Delta_b)|a_1|, \\ b_3 &= z_b(1-\Delta_b)|a_1|, \\ b_2 &= z_{b'}(1+\Delta_{b'})|a_1|, \\ b_4 &= \chi_3\chi_4 z_{b'}(1-\Delta_{b'})|a_1|, \end{aligned} \tag{6}$$

For a given chirality pattern and a given set of parameters $(z_{a,b,b'}, \Delta_{a,b,b'})$, to characterize the existence of a ZGS at 3/4 filling we proceed as follows. In a first step the energy eigenvalues $E_n(\mathbf{k})$ are computed over a regular mesh of 400×400 k-points in the BZ such that the maximum $E_n^{\max}(\mathbf{k}_n^{\max})$ and minimum $E_n^{\min}(\mathbf{k}_n^{\min})$ of each band are determined. In a second step a ZGS at filling 3/4 is obtained whenever the following three conditions are simultaneously verfied: (i): $\mathbf{k}_4^{\min}, \mathbf{k}_3^{\max} \neq \mathbf{G}/\mathbf{2}$; (ii): $|\mathbf{k}_4^{\min} - \mathbf{k}_3^{\max}| \ll \delta$; (iii): $0 < E_4^{\min}(\mathbf{k}_4^{\min}) - E_3^{\max}(\mathbf{k}_3^{\max}) \leq \delta \ll 1$ where in practice $\delta \sim (\pi/100)$ was taken. If either (i) or (ii) are not verified or if $E_4^{\min}(\mathbf{k}_4^{\min}) - E_3^{\max}(\mathbf{k}_3^{\max}) < 0$ the system is considered as metallic, whereas it is considered as gapped if $E_4^{\min}(\mathbf{k}_4^{\min}) - E_3^{\max}(\mathbf{k}_3^{\max}) > \delta$. Note that taking a smaller value for δ will a priori decrease the number of configuration which can be considered as metallic or ZGS.

For each of the three patterns (A_3, A_7, S_6), the numerical determination of possible ZGS at 3/4 fillings is done systematically for parameters $z_{a,b,b'} \in (0, 0.2, 0.4, ..., 4)$ and for six distinct dimerization configurations $(\Delta_a, \Delta_b, \Delta_{b'})$ as given in table 2; such that configuration I is equivalent to no dimerization, II to weak and VI to strong dimerization. For a given pattern and a given dimerization configuration the existence of ZGS is thus explored for $(21)^3 = 9261$ distinct configurations of parameters $z_{a,b,b'}$.

4.3.1. histrogram distribution of the number of gapped, ZGS and metallic phases The first statistic Fig.5 represents the histogram distribution of the number of gapped, ZGS and metallic configurations for patterns (A_3, A_7, S_6) at 3/4 filling and for the six distinct dimerization configurations (I,II,...VI).

The common and expected feature is that the number of gapped configurations (black bars) increases with the strength of the dimerization along the different chains; for all patterns it eventually becomes the dominant phase for strong dimerization (VI). Despite this common feature, we find it relatively surprising that for pattern S_6, the number of gapped configurations is always relatively large even for low or no dimerization. As far as the number of metallic

Table 2. Dimerization configurations $(\Delta_a, \Delta_b, \Delta_{b'})$

	I	II	III	IV	V	VI
Δ_a	0	0.25	0.25	0.5	0.75	0.75
Δ_b	0	0.25	0.75	0.5	0.25	0.75
$\Delta_{b'}$	0	0.25	0.25	0.5	0.75	0.75

configurations (blue bars) is concerned, its behaviour is very different for the three distinct patterns (A_3, A_7, S_6). For pattern A_3, it represents the dominant phase for small dimerization (I) and then it steadily decreases with increasing dimerization and it becomes the rarest phase for large dimerization (VI); this type behaviour is somehow as expected. For pattern A_7, whatever the dimerization, the number of metallic configurations is surprisingly always very small in proportion such that roughly speaking this pattern exhibits either a ZGS or a gapped phase. For pattern S_6, the number of metallic configurations slowly decreases when dimerization is increased and it is always below that of gapped and ZGS phases excepted for dimerization (III) where it represents the dominant phase. Overall, we find it quite remarkable that for most dimerization configurations and for the three patterns (A_3, A_7, S_6), the number of metallic configurations remains always smaller than the number of ZGS phases. In fact, regarding the number of ZGS configurations, the three patterns (A_3, A_7, S_6) show distinctive behaviours. For patterns (A_3,S_6) the number of ZGS configurations shows no clear tendency as a function of dimerization strength; for each dimerization configuration it constitutes around 1/3 of the 9261 anisotropy configurations, slightly higher for S_6 and slightly lower for A_3. More remarkably, for pattern A_7, even if the number of ZGS configurations decreases steadily when the dimerization strength is increased, it still constitutes the largely dominant phase over the entire studied range of dimerization. In summary, the main message to retain from Fig.5 is that over the range of anisotropy and dimerization that we have considered, the pattern A_7 is by far the most favorable for the appearance of stable ZGS at 3/4 filling whereas patterns A_3, S_6 are more or less similarly favorable.

4.3.2. Anisotropy versus dimerization dependency of ZGS properties The previous histograms mainly provide quantitative information of the role played by the dimerization parameters $(\Delta_a, \Delta_b, \Delta_{b'})$ on the appearance of ZGS at 3/4 filling, however they do not allow to understand the role played by the anisotropy parameters $z_{a,b,b'}$. In order to quantify the role played by the anisotropy between the different chains, for each dimerization configuration we have computed both the average value and the variance of the anisotropy for configurations $z_{a,b,b'}$ that lead to ZGS at 3/4 filling for patterns A_3, A_7, S_6. The results are summarized on Fig. 6. On this figure the grey region corresponds to the case of a uniform distribution of ZGS configurations over the explored interval of anisotropy : $(z_a, z_b, z_{b'}) \in [0, 4]$; we reminds that no anisotropy correspond to $z_a = z_b = z_{b'} = 1$. For pattern A_3, for no or weak dimerization there is a strong deviation from the uniform distribution, more precisely the more favorable anisotropy configurations correspond to large value of z_a and small value of $z_b, z_{b'}$. By increasing the dimerization the average values of $z_a, z_b, z_{b'}$ enter the region corresponding to a uniform distribution. For pattern S_6 the behaviour is somewhat symmetric to that of pattern A_3, for small dimerization small value of z_a and larger value of $z_b, z_{b'}$ are more favorable to obtain a ZGS at 3/4 filling. Finally for pattern A_7 the distribution of ZGS is almost independent on both the anisotropy and dimerization. The main message to retain from this Fig. 6, is that the larger the dimerization the less important is the role of the anisotropy; in other words the strength of the dimerization along the different chains

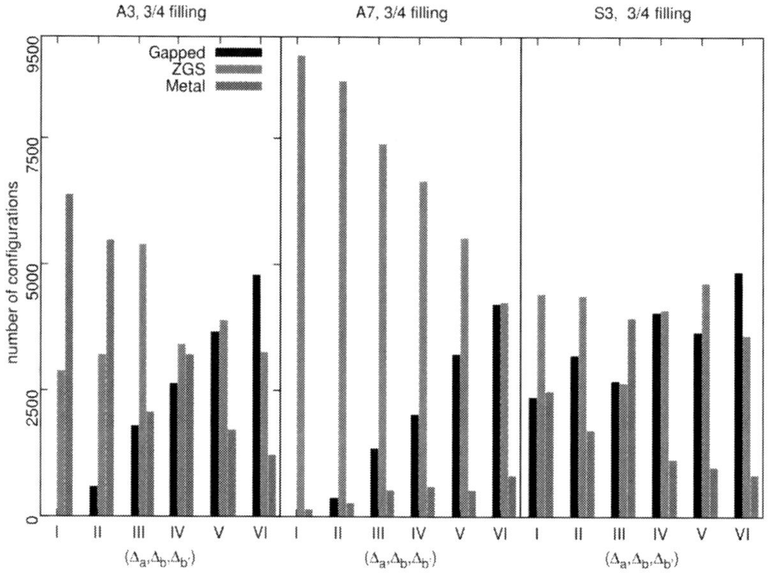

Figure 5. histogram distribution of the number of configurations that are gapped (black), ZGS (red) or metallic (blue) for patterns A_3, A_7, S_6 at 3/4 filling for the six distinct dimerization configurations (I,II,...VI).

seems a more pertinent parameter than the anisotropy between the chains.

4.3.3. k-space distribution of Dirac points associated to ZGS phases As a last characterization of the ZGS properties, the following Fig. 7, represents the evolution of the k-space distribution of the Dirac points associated to the ZGS at 3/4 for pattern A_3 as function of the dimerization. Figs. 8,9 represent this same quantity but for pattern A_7 and S_6 respectively. The most striking feature on these figures is that the case of no dimerization (I) is very different from all the other cases. In fact, in the absence of dimerization the distribution of Dirac points associated to ZGS is much less spreaded over the Brillouin zone, moreover it differs strongly from one pattern to the other. By increasing the dimerization, the k-space distribution of ZGS's Dirac points for patterns A_3 and A_7 become more and more similar, in fact they seem to only differ by number of Dirac points (intensity) as already seen in their corresponding histogram distribution. For sufficiently large dimerization (V,VI), the k-space distribution of ZGS's Dirac points for all patterns are very similar. All these features are further indications that the larger the dimerization the less important is the role played by the form of the chirality pattern.

Journal of Physics: Conference Series **603** (2015) 012010 doi:10.1088/1742-6596/603/1/012010

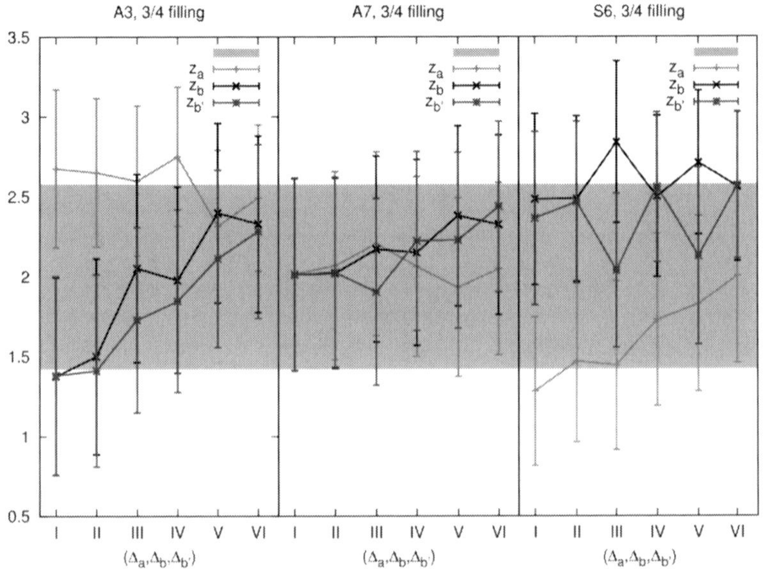

Figure 6. (color online) Averaged anisotropy parameter values z_a (red), z_b (black) $z_{b'}$ (blue) versus dimerization configurations (I,II,...VI) for the ZGS configurations found for patterns A_3, A_7, S_3 at 3/4 filling. The errorbars encode the meansquare fluctuations. For comparison, the grey region corresponds to the case of a uniform distribution of ZGS configurations over the interval $(z_a, z_b, z_{b'}) \in [0, 4]$.

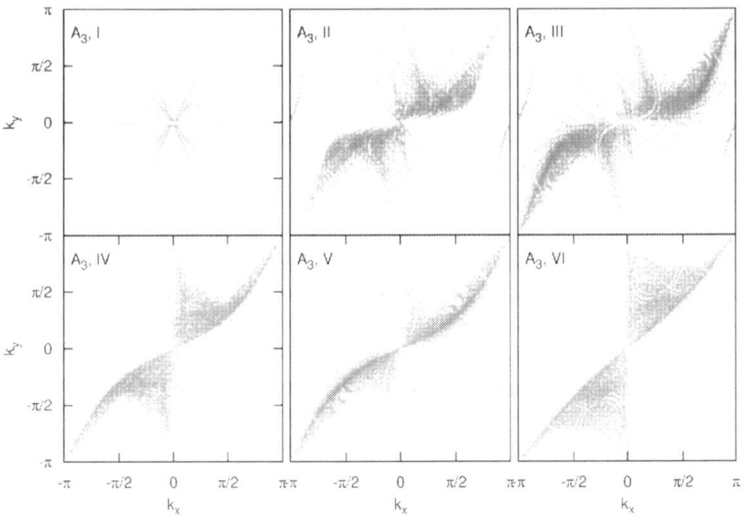

Figure 7. (color online) k-space distribution of the Dirac points associated to the ZGS configurations obtained for patterns A_3 at 3/4 filling for the six distinct dimerization configurations (I,II,...VI).

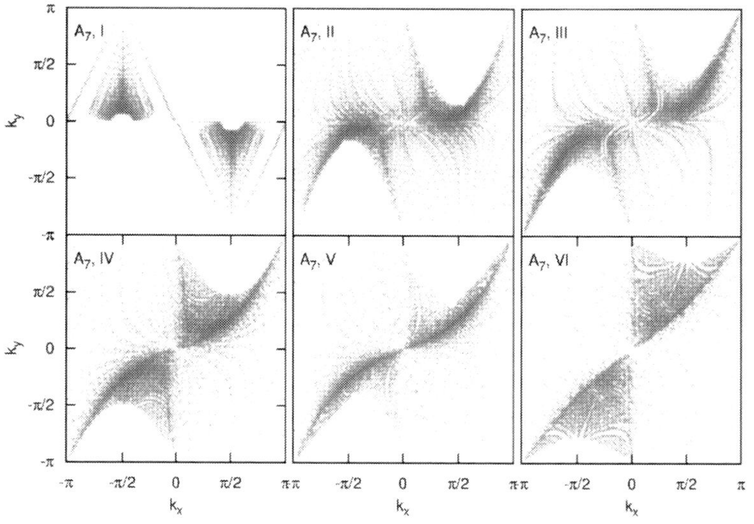

Figure 8. k-space distribution of the Dirac points associated to the ZGS configurations obtained for patterns A_7 at 3/4 filling for the six distinct dimerization configurations (I,II,...VI).

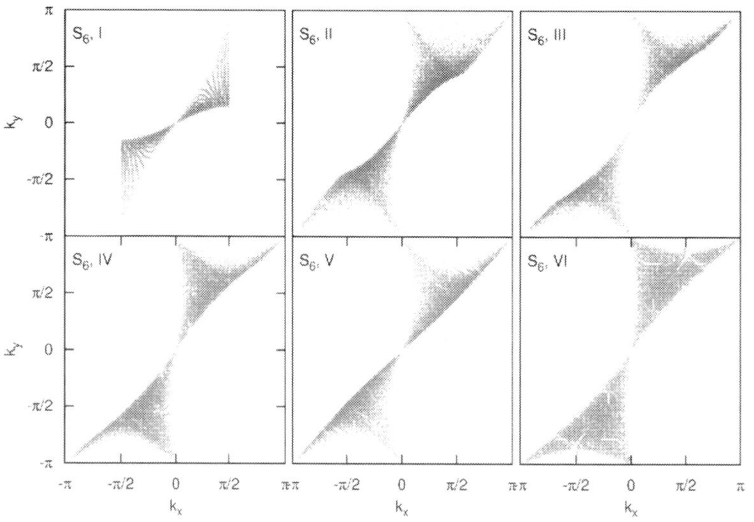

Figure 9. k-space distribution of the Dirac points associated to the ZGS configurations obtained for patterns S_6 at 3/4 filling for the six distinct dimerization configurations (I,II,...VI).

5. Summary

In summary, we have examined the existence of ZGS at 3/4 filling for a tight binding model with seven nearest neighbors transfer integrals $(a_1, a_2, a_3, b_1, b_2, b_3, b_4)$ between the four molecules of the unit cell, as in α-type organic conductor. We have systematically studied the combined role of the sign and moduli of the transfer integrals on the appearance of ZGS at 3/4 filling. We have shown that for given moduli, the are only 16 independent band structures $A_1, \cdots, S_8$. Each band structure is characterized by a chirality pattern $(\chi_1, \chi_2, \chi_3, \chi_4)$, where each chirality encodes the sign of the product of three transfer integrals that compose one of the four distinct triangular plaquette of the unit cell $(\chi_1 = \text{sign}(a_2b_1b_4), \chi_2 = \text{sign}(a_3b_2b_3), \chi_3 = \text{sign}(a_1b_1b_2), \chi_4 = \text{sign}(a_1b_3b_4))$. In term of this classification, it appears that all previous works on α-(BEDT-TTF)$_2$I$_3$ corresponds to pattern A_3 with $(\chi_1, \chi_2, \chi_3, \chi_4) \equiv (-, +, -, -)$.

As far as the properties of the LDOS and ZGS at 3/4 (resp. 1/4) filling are concerned, the 16 patterns can be further classified into only four different classes determined by two effective chiralities $\chi_a = \text{sign}(a_1a_2)$ and $\chi_b = \text{sign}(b_1b_2b_3b_4)$. Using extensive numerics we have established that for both large, or no inhomogeneity in the moduli $|a_1|, ...|b_4|$, the patterns A_7, A_8 (resp. A_1, A_2) of class $(\chi_a, \chi_b) = (-, -)$ are the most favorable patterns for the existence of ZGS at 3/4 (resp. 1/4) filling. At the opposite, for patterns (S_1, S_2, S_7, S_8) of class $(\chi_a, \chi_b) = (+,+)$ we never found any ZGS at either 1/4 or 3/4 filling. Finally, patterns (S_5, S_6) (resp. S_3, S_4) and A_5, A_6 (resp. A_3, A_4) of class $(\chi_a, \chi_b) = (+, -)$ and $(\chi_a, \chi_b) = (-, +)$ present an intermediate situation; they allow for ZGS at 3/4 (resp. 1/4) filling but are much less favorable that patterns of class $(\chi_a, \chi_b) = (-, -)$. We note however that the role played by the chirality pattern is dependent on strength of the dimerization in the different effective chains formed by the transfer integrals. More precisely for weak or no dimerization the form of the chirality pattern strongly determines the probability to find a ZGS at 1/4 or 3/4 filling whereas for strong dimerization the possibility of having a ZGS become more or less equivalent for the three classes $(-, -), (-, +)$ and $(+, -)$. From that perspective we note that the usual transfer integral parameters taken to modelize α-(BEDT-TTF)$_2$I$_3$ are in the strong dimerization regime.

On a larger perspective, we note that for bipartite lattice (like square lattice) when all plaquette chiralities are negative it corresponds to the so called uniform π-flux phase for which it is well known that there is a ZGS at 1/2 filling. Our results suggest that on a non-bipartite lattice like the triangular lattice, it is necessary to have a fractional density p/q of π-flux plaquettes to obtain a ZGS at some filling r/q; the difficulty in such a situation is that there appears many non equivalent chirality patterns.

Acknowledgments
One of the authors (Y.S.) thanks G. Montambaux and K. Awaga for useful discussion. This work was supported by a Grant-in-Aid for Scientific Research (A) (No. 24244053) and (C) (No. 26400355) from the Ministry of Education, Culture, Sports, Science, and Technology, Japan.

Appendix A. Local density of states and chirality patterns

Consider the α-BEDT tight binding model with only nearest-neighbor hopping on the triangular lattice

$$H = \sum_i \sum_{<j,i>} [t_{ij}|i\rangle\langle j| + \text{h.c.}] , \tag{A.1}$$

with $t_{i,j} \in (a_1, ..., b_4)$. The one particle local Greens function on a site j is $G_{jj}(E) = \langle j|(E - H)^{-1}|j\rangle$. It can be expanded as

$$G_{jj}(E) = \frac{1}{E} \sum_\ell \frac{\langle j|H^\ell|j\rangle}{E^\ell}. \tag{A.2}$$

Journal of Physics: Conference Series **603** (2015) 012010 doi:10.1088/1742-6596/603/1/012010

The local density of states is obtained from the identity $D_j(E) = \frac{-1}{\pi}\Im m\{G_{jj}(E+i0)\}$, which allows to write [14, 15]

$$\mathscr{M}_j^\ell = \int_{-\infty}^{\infty} \mathrm{d}E\ E^\ell D_j(E) = \langle j|H^\ell|j\rangle = \sum_{<i_1,j>}\sum_{<i_2,i_1>}\cdots\sum_{<i_{\ell-1},j>} t_{ji_1}t_{i_1i_2}...t_{\ell-1j}. \tag{A.3}$$

This last identity means that the ℓth moment $\mathscr{M}_j^\ell$ of the local density of states on site j is equal to sum over all closed path of ℓ steps that start from j and return to j. The key property is that each path contributes with a weight proportionnal to the product of moduli of all the hoppings along the path and with a sign given by the product of the chiralities of all the triangular plaquettes encircled by the path. For the tight binding model of α-BEDT, the moment for each site up to $\ell = 3$ are easily obtained as:

$$\begin{aligned}
\mathscr{M}_A^2 &= a_2^2 + a_3^2 + b_1^2 + b_2^2 + b_3^2 + b_4^2,\\
\mathscr{M}_B^2 &= 2(a_1^2 + b_2^2 + b_3^2),\\
\mathscr{M}_C^2 &= 2(a_1^2 + b_1^2 + b_4^2),\\
\mathscr{M}_A^3 &= 2(\chi_2|a_3b_2b_3| + \chi_1|a_2b_1b_4|) + \chi_3|a_1b_1b_2| + \chi_4|a_1b_3b_4|,\\
\mathscr{M}_B^3 &= 2(\chi_2|a_3b_2b_3| + \chi_3|a_1b_1b_2| + \chi_4|a_1b_3b_4|),\\
\mathscr{M}_C^3 &= 2(\chi_1|a_2b_1b_4| + \chi_3|a_1b_1b_2| + \chi_4|a_1b_3b_4|).
\end{aligned} \tag{A.4}$$

More generally, odd and even moment have the generic form:

$$\begin{aligned}
\mathscr{M}_j^{2\ell+1} &= \textstyle\sum_\alpha \chi_\alpha\ x_{\alpha,j}^\ell + \sum_{\alpha\neq\beta\neq\gamma}\chi_\alpha\chi_\beta\chi_\gamma\ x_{\alpha\beta\gamma,j}^\ell,\\
\mathscr{M}_j^{2\ell} &= y_j^\ell + \textstyle\sum_{\alpha\neq\beta}\chi_\alpha\chi_\beta\ y_{\alpha\beta,j}^\ell + \chi_1\chi_2\chi_3\chi_4\ y_{1234,j}^\ell
\end{aligned} \tag{A.5}$$

where $(x_{\alpha,j}^\ell, x_{\alpha\beta\gamma,j}^\ell, y_j^\ell, y_{\alpha\beta,j}^\ell, y_{1234,j}^\ell) > 0$ depend only on the moduli $(|a_1|, ..., |b_4|)$. For $|a_2| = |a_3|$, $|b_1| = |b_3|$ and $|b_2| = |b_4|$, one easily obtains that the global moments $\mathscr{M}^\ell = 2\mathscr{M}_A^\ell + \mathscr{M}_B^\ell + \mathscr{M}_C^\ell$ verify

$$\begin{aligned}
\mathscr{M}^{2\ell+1} &= (\chi_1+\chi_2)(x_1^\ell + \chi_3\chi_4 x_{134}^\ell) + (\chi_3+\chi_4)(x_3^\ell + \chi_1\chi_2 x_{123}^\ell),\\
\mathscr{M}^{2\ell} &= y^\ell + (\chi_1+\chi_2)(\chi_3+\chi_4)y_{13}^\ell + \chi_1\chi_2 y_{12}^\ell + \chi_3\chi_4 y_{34}^\ell + \chi_1\chi_2\chi_3\chi_4\ y_{1234}^\ell.
\end{aligned} \tag{A.6}$$

Appendix B. The 8 transfer integral configuration associated to pattern A_3

Table B1. The $8 = 2^3$ configurations of signs of the seven transfer integrals that lead to the chirality pattern A_3. All these configurations have the same LDOS $D_j(E)$. Strictly speaking for k-space dependent properties it is still necessary to distinguish two groups, i.e configurations (1,2,5,6) with ($b_1b_4 < 0$ and $b_2b_3 < 0$) and congigurations (3,4,7,8) with ($b_1b_4 > 0$ and $b_2b_3 > 0$). The energy bands $E_n(\mathbf{k})$ of the first group become equal to that of the second group after the substitution $k_x \to k_x + \pi$.

A_3	(1)	(2)	(3)	(4)	(5)	(6)	(7)	(8)
sign(a_1)	+	+	+	+	-	-	-	-
sign(a_2)	+	+	-	-	+	+	-	-
sign(a_3)	-	-	+	+	-	-	+	+
sign(b_1)	+	-	+	-	+	-	+	-
sign(b_2)	-	+	-	+	+	-	+	-
sign(b_3)	+	-	-	+	-	+	+	-
sign(b_4)	-	+	+	-	-	+	+	-

References

[1] Mori T, Kobayashi A, Sasaki T, Kobayashi H, Saito G, and Inokuchi H 1984 *Chem. Lett.* **13** 957
[2] Katayama S, Kobayashi A, and Suzumura Y 2006 *J. Phys. Soc. Jpn.* **75** 054705
[3] Tajima N and Kajita K 2009 *Sci. Technol. Adv. Mater.* **10** 024308
[4] Kobayashi A, Katayama S, and Suzumura Y 2009 *Sci. Technol. Adv. Mater.* **10** 024309
[5] Kajita K, Nishio Y, Tajima N, Suzumura Y, and Kobayashi A 2014 *J. Phys. Soc. Jpn.* **83** 072002
[6] Fu L and Kane C L 2007 *Phys. Rev.* B **76** 045302
[7] Piéchon F and Suzumura Y 2013 *J. Phys. Soc. Jpn.* **82** 033703
[8] Mori T 2013 *J. Phys. Soc. Jpn.* **82** 034714
[9] Mori T 2010 *J. Phys. Soc. Jpn.* **79** 014701
[10] Suzumura Y, Morinari T and Piéchon F 2013 *J. Phys. Soc. Jpn.* **82** 023708
[11] Piéchon F, Suzumura Y, and Morinari T 2014 *presented at JPS September Meeeting, 2014*
[12] Kondo R, Kagoshima S and Maesato M 2003 *Phys. Rev.* B **67** 134519
[13] Kondo R, Kagoshima S, Tajima N and Kato R 2009 *J. Phys. Soc. Jpn.* **78** 114714
[14] Nori F and Lin Y-L 1994 *Phys. Rev.* B **49** 4131
[15] Lin Y-L and Nori F 1996 *Phys. Rev.* B **53** 13374

Reflectance of Dirac electrons in organic conductor

Yoshikazu Suzumura[1], **Igor Proskurin**[2,3], **and Masao Ogata**[2]
[1]Department of Physics, Nagoya University, Nagoya 464-8602, Japan
[2]Department of Physics, University of Tokyo, Bunkyo-ku, Tokyo 113-0033, Japan
[3]Department of Theoretical Physics, Institute of Natural Sciences, Ural Federal University ,Mira 19, Ekaterinburg 620002, Russia

E-mail: suzumura@s.phys.nagoya-u.ac.jp

Abstract. We examine the reflectance R of Dirac electrons of organic conductor α-(BEDT-TTF)$_2$I$_3$ with a tilted cone which is characterized by a degree of the tilting $\eta(= 0.8)$ and the angle α measured from the applied electric field. The reflectance is calculated using the complex dielectric constant, $\varepsilon_1 + i\varepsilon_2$, which is evaluated from the in-plane dynamical conductivity with a constant damping Γ. The bulk reflectance on the two-dimensional plane is estimated by taking account of an interlayer distance of organic conductor α-(BEDT-TTF)$_2$I$_3$. We compare a difference of R between $\alpha = 0$ (i) and $\pi/2$ (ii), which correspond to the tilting being parallel and perpendicular to the electric field, respectively. The frequency ω dependence of R is examined for several magnitudes of the chemical potential μ to find a crossover from the zero doping to a finite doping. With increasing ω, R decreases from unity. For the small doping with $\mu < \Gamma$, a difference of R between (i) and (ii) is small, while, for the large doping $\Gamma < \mu$, a noticeable difference of R between (i) and (ii) exists in a certain region of ω around 2μ. The behavior that R of (ii) is larger than that of (i) is enhanced by the doping. We discuss the case of the clean limit and the role of both the intraband excitation and the interband excitation in the presence of Γ.

1. Introduction

The Dirac electrons in organic conductors have been studied extensively since the discovery in organic conductor α-(BEDT-TTF)$_2$I$_3$ [1] based on the X-ray Diffraction experiment [2]. There are several characteristics of Dirac electron in organic conductor, e.g., the presence of the tilting of Dirac cone and four sites per unit cell [3] compared with that of graphene [4].

Using an effective Hamiltonian which describes the electrons close to the Dirac point [5], the transport properties such as Nernst effect [6] and electric conductivity [7, 8] have been examined where the effect of tilting gives rise to a noticeable anisotropy of the conductivity for both the zero doping and finite but small doping in contrast to the isotropic Dirac cone [9]. The effect of tilting on the dynamical conductivity appears even for large frequency where the chemical potential is close to the onset of interband excitation [10].

In the present paper, we examine the reflectance based on the previous calculation of the in-plane dynamical conductivity [8]. By comparing with the isotropic case[11, 12], we show how the effect of tilting appears in the crossover from the intraband excitation to the interband excitation with increasing a frequency ω.

In section 2, the formulation is given where the reflectance R on the plane is estimated from the dielectric constant based on the dynamical conductivity with a frequency ω. In section

3, R as a function of ω is examined to show the crossover from the intraband excitation to the interband excitation which depends on the magnitude of the chemical potential, i.e., the degree of doping. The section 4 is devoted to summary and discussion on the relevance to the reflectance in organic conductors.

2. Formulation

We consider an effective Hamiltonian for the organic conductor expressed as

$$H = \mathbf{v}_0 \cdot \mathbf{p} + v_x p_x \sigma_x + v_y p_y \sigma_y \ , \tag{1}$$

where $\mathbf{p} = (p_x, p_y)$ is a two-dimensional momentum and $\sigma = (\sigma_x, \sigma_y)$ are Pauli matrices. The velocities $\mathbf{v} = (v_x, v_y)$ and $\mathbf{v}_0 = (v_{0x}, v_{0y})$ determine the anisotropy of the Dirac cone and the tilting of the cone axis, respectively. The eigenvalues of the Hamiltonian in Eq. (1) are given by

$$E_{\mathbf{p}}^{(\pm)} = \mathbf{v}_0 \cdot \mathbf{p} \pm \sqrt{v_x^2 p_x^2 + v_y^2 p_y^2} \ . \tag{2}$$

Defining α as the angle between the projection of the tilted Dirac cone axis on the x-y-plane and the x-axis, and using the parametrization $v_{0x} = \eta v_x \cos\alpha$, $v_{0y} = \eta v_y \sin\alpha$ with $\eta = \sqrt{(v_{0x}/v_x)^2 + (v_{0y}/v_y)^2}$, $v_x p_x = k\cos\theta_{\mathbf{p}}$, and $v_y p_y = k \sin\theta_{\mathbf{p}}$, we can rewrite the eigenvalues in Eq. (2) as

$$E_{\mathbf{p}}^{(\pm)} = \eta k \cos(\theta_{\mathbf{p}} - \alpha) \pm k \ , \tag{3}$$

with the corresponding eigenfunctions given by $\psi_{\mathbf{p}}^{(+)} = \left[-e^{-i\theta_{\mathbf{p}}}, 1\right]^T/\sqrt{2}$ and $\psi_{\mathbf{p}}^{(-)} = \left[1, e^{i\theta_{\mathbf{p}}}\right]^T/\sqrt{2}$. The parameter η ($0 \le \eta < 1$) represents degree of the tilting. The electron current density operators $\mathbf{j} = -e\partial H/\partial\mathbf{p}$ in the same parametrization can be expressed as

$$\begin{aligned} j_x &= -ev_x(\eta\cos\alpha + \sigma_x) \ , \\ j_y &= -ev_y(\eta \sin\alpha + \sigma_y) \ , \end{aligned} \tag{4}$$

where $-e$ denotes the charge of the electron.

Within the framework of the Kubo formula, the dynamical conductivity with a frequency ω can be written in the well-known form: [6, 7, 8, 13, 14]

$$\begin{aligned} \sigma_{\mu\nu}(\omega) &= \frac{i}{\omega}\int_{\infty}^{\infty} d\varepsilon f(\varepsilon) \\ &\times \ \mathrm{Tr}\left\{ j_\mu \left[G^{(+)}(\varepsilon + \hbar\omega) - G^{(+)}(\varepsilon)\right] j_\nu A(\varepsilon) - j_\mu A(\varepsilon) j_\nu \left[G^{(-)}(\varepsilon) - G^{(-)}(\varepsilon - \hbar\omega)\right]\right\} , \end{aligned} \tag{5}$$

where $f(\varepsilon)$ is the Fermi distribution function with a chemical potential μ and a temperature $T(\to 0)$. $A = \frac{i}{2\pi}\left[G^{(+)} - G^{(-)}\right]$ and Green's functions are defined by $G^{\pm}(\varepsilon) = (\varepsilon - H \pm i\Gamma)^{-1}$ with a damping constant Γ. We treat Γ as a constant parameter while the dynamical effect on Γ within the self-consistent Born approximation [15] is discussed in Ref. [8]. In Eq. (5) the conductivities per spin and per valley are considered, implying that the final result should be multiplied by 4.

The conductivity σ_{xx} is expressed as ($v_x = v_y$)

$$\sigma(\omega, \eta, \alpha) = \sigma_{\mathrm{intra}} + \sigma_{\mathrm{inter}} = \sigma_1(\omega) + i\,\sigma_2(\omega) \ , \tag{6}$$

where σ_1 and σ_2 denote the real part and the imaginary part. Quantities σ_{intra} and σ_{inter} denote the intraband and interband contributions coming from the excitation within the same energy band and the excitation from $E_{\mathbf{p}}^{(-)}$ to $E_{\mathbf{p}}^{(+)}$, respectively. The real part $\sigma_1(\omega)$ of $\sigma(\omega,\eta,\alpha)$ is calculated as $\sigma_1 = \mathrm{Re}\{\sigma_{\mathrm{intra}}^{(+)}\} + \mathrm{Re}\{\sigma_{\mathrm{intra}}^{(-)}\} + \mathrm{Re}\{\sigma_{\mathrm{inter}}\}$ where

$$\mathrm{Re}\{\sigma_{\mathrm{intra}}^{(\pm)}\} = \left\langle \int_0^\infty \mathrm{d}y\, y\, (\eta\cos\alpha \pm \cos\theta)^2 f(E^{(\pm)}, E^{(\pm)}) \right\rangle_\theta , \tag{7}$$

$$\mathrm{Re}\{\sigma_{\mathrm{inter}}\} = \left\langle \int_0^\infty \mathrm{d}y\, y\, (\sin\theta)^2 \left[f(E^{(+)}, E^{(-)}) + f(E^{(-)}, E^{(+)}) \right] \right\rangle_\theta , \tag{8}$$

where $\theta = \theta_{\mathbf{p}}$, and $\langle \cdots \rangle_\theta$ denotes $(2\pi)^{-1} \int_0^{2\pi} d\theta \cdots$. The quantity $f(a,b)$ is defined by

$$f(a,b) = \frac{e^2}{\hbar^2\omega} \left[\int_{-\infty}^{\mu} \mathrm{d}z \left\{ \frac{\Gamma}{(z-a)^2+\Gamma^2} \times \frac{\Gamma}{(z+\hbar\omega-b)^2+\Gamma^2} - (\omega \to -\omega) \right\} \right] . \tag{9}$$

The real part and imaginary parts of $\sigma(\omega,\eta,\alpha)$ satisfy the relations $\sigma_1(\omega) = \sigma_1(-\omega)$ and $\sigma_2(\omega) = -\sigma_2(-\omega)$, respectively. The imaginary part $\sigma_2(\omega)$ is calculated from the real part using the Kramers-Kronig relation,

$$\sigma_2(\omega) = -\frac{1}{\pi} \int_{-\omega_c}^{\omega_c} \mathrm{d}z\, \frac{\sigma_1(z)}{z-\omega} , \tag{10}$$

where the numerical integration is performed for $\omega_c = 30$. The limiting behavior of large μ gives the conductivity in the clean limit $\Gamma \to +0$ (Appendix A) [9, 10, 8] in which the real part of $\sigma_{\mathrm{intra}} + \sigma_{\mathrm{inter}}$ is expressed as

$$\mathrm{Re}\{\sigma_{\mathrm{intra}}\} = \frac{e^2}{\pi h}\pi^2\mu\delta(\hbar\omega) \left\{ \frac{1-\sqrt{1-\eta^2}}{\eta^2}\cos^2\alpha + \frac{1-\sqrt{1-\eta^2}}{\eta^2\sqrt{1-\eta^2}}\sin^2\alpha \right\} , \tag{11}$$

$$\mathrm{Re}\{\sigma_{\mathrm{inter}}\} = \frac{e^2}{\pi h}\frac{\pi^2}{8}F_c(\alpha) . \tag{12}$$

$F_c(\alpha) = 0$ for $\omega < 2\mu/(1+\eta)$, $F_c(\alpha) = 1$ for $2\mu/(1-\eta) < \omega$, and $F_c(\alpha) = \pi^{-1}(\varphi_0 - \sin\varphi_0\cos\varphi_0\cos(2\alpha)) < 1$ with $\cos\varphi_0 = (2\mu/\omega - 1)/\eta$ for $2\mu/(1+\eta) < \omega < 2\mu/(1-\eta)$. Note that $F_c(\alpha)$ with a fixed ω takes a maximum (minimum) at $\alpha = \pi/2$ and a minimum (maximum) at $\alpha = 0$ for $2\mu/(1+\eta) < \omega < 2\mu$ $(2\mu < \omega < 2\mu/(1-\eta))$.

In terms of Eq. (6), the dielectric constant is calculated from

$$\varepsilon(\omega,\eta,\alpha) = 1 + \mathrm{i}\frac{4\pi\tilde{\sigma}(\omega,\eta,\alpha)}{\omega} = \varepsilon_1(\omega,\eta,\alpha) + \mathrm{i}\varepsilon_2(\omega,\eta,\alpha) , \tag{13}$$

where $\tilde{\sigma}(\omega,\eta,\alpha) = C\sigma(\omega,\eta,\alpha)/d$ with an interlayer distance d and $C = 4$ for the freedom of spin and valley. Hereafter we introduce normalized quantities as

$$\frac{\hbar\omega}{\Gamma} \to \overline{\omega} , \quad \frac{\sigma}{e^2/\pi h} \to \overline{\sigma} , \quad \frac{\mu}{\Gamma} \to \overline{\mu} ,$$

where h denotes the Plank constant. Using the dynamical conductivity $\overline{\sigma}(\overline{\omega},\eta,\alpha)$ for both zero and finite doping[8], the real part and imaginary parts ε_1 and ε_2 in Eq. (13) are obtained as

$$\varepsilon_1(\overline{\omega},\eta,\alpha) = 1 - 4\pi F\overline{\sigma}_2(\overline{\omega},\eta,\alpha)/\overline{\omega} , \tag{14}$$

$$\varepsilon_2(\overline{\omega},\eta,\alpha) = 4\pi F\overline{\sigma}_1(\overline{\omega},\eta,\alpha)/\overline{\omega} , \tag{15}$$

where

$$F = \frac{2e^2}{\pi^2 d\Gamma} , \tag{16}$$

with d being an interlayer distance. The quantity F is estimated as $F \sim 660$ when $\Gamma/k_B \simeq 3$ K [6] and $d \simeq 1.7 \times 10^{-9}$ m [16]. Finally the reflectance R is calculated as

$$R(\bar{\omega}, \eta, \alpha) = \left| \frac{\sqrt{\varepsilon_1 + i\varepsilon_2} - 1}{\sqrt{\varepsilon_1 + i\varepsilon_2} + 1} \right|^2 . \tag{17}$$

Note that F corresponds to the plasma frequency where the reflectance decreases for $\bar{\omega} > F$ owing to ε_2.

The characteristics of Dirac electrons of organic conductors is examined by choosing $\eta = 0.8$ with two limiting angles $\alpha = \pi/2$ and $\alpha = 0$ which correspond to the tilting direction being perpendicular to the electric field and parallel to the electric field, respectively. The case of $\eta =$ 0 without tilting of the isotropic Dirac cone is shown for the comparison.

3. Reflectance

In Fig. 1, the conductivity as a function of $\bar{\omega}$ is shown for both the real part $\bar{\sigma}_1$ and the imaginary part $\bar{\sigma}_2$ with four cases of $\bar{\mu} = 0$ (a) , 2 (b), 4 (c) and 8 (d). The solid and dashed lines represent the case for $\eta = 0.8$ with $\alpha = \pi/2$ and $\alpha = 0$, while the dotted line represents that for $\eta = 0$. The asymptotic value of $\bar{\sigma}_1$ ($\bar{\sigma}_2$) for large $\bar{\omega}$ is given by $\pi^2/8$ (0). For $\bar{\mu} = 0$, $\bar{\sigma}_1$ takes a maximum while that of $\eta = 0$ increases monotonously. The imaginary part shows $\bar{\sigma}_2 > 0$ for $\alpha = \pi/2$ and both $\bar{\sigma}_2 < 0$ and $\bar{\sigma}_2 > 0$ for $\alpha = 0$ in contrast to that of $\eta = 0$ showing $\bar{\sigma}_2 < 0$ with a minimum. With increasing $\bar{\mu} = 2$, 4, and 8, $\bar{\sigma}_1(\alpha, \eta) (= \bar{\sigma}_1(\bar{\omega}, \eta, \alpha))$ at $\bar{\omega} = 0$ increases owing to the increase of the density of states at $\bar{\omega} = \bar{\mu}$. There is a characteristic frequency $\bar{\omega}$ showing a crossover where $\bar{\sigma}_1(\pi/2, 0.8) > \bar{\sigma}_1(0, 0.8) > \bar{\sigma}_1(0, 0)$ for small $\bar{\omega}$ and $\bar{\sigma}_1(0, 0) > \bar{\sigma}_1(0, 0.8) > \bar{\sigma}_1(\pi/2, 0.8)$ for the large $\bar{\omega}$. The behavior around the characteristic frequency suggests the noticeable effect of the interband excitation as shown in Eq. (12), i.e., in the clean limit, where the characteristic frequency is given by $\bar{\omega} = 2\bar{\mu}$ for $\bar{\mu} >> 1$. As for the imaginary part $\bar{\sigma}_2$ of $\eta = 0.8$, there is a maximum which increases with increasing $\bar{\mu}$ due to the increase of $\bar{\sigma}_1$ as seen from Eq. (10). For $\eta = 0$, a change from $\bar{\sigma}_2(0, 0) < 0$ into $\bar{\sigma}_2((0, 0) > 0$ occurs with increasing $\bar{\mu}$ where $\bar{\sigma}_2$ for the large $\bar{\mu}$ takes both maximum and minimum. The effect of the interband excitation becomes prominent for the large $\bar{\mu}$ (i.e., $\mu >> \Gamma$).

In Fig. 2, the $\bar{\omega}$ dependence of the dielectric constant $\varepsilon(\alpha, \eta) = \varepsilon(\bar{\omega}, \eta, \alpha)$ is shown corresponding to Fig. 1. It is found that $\varepsilon_2(\alpha, \eta) > 0$ from Eq. (15) and $\bar{\sigma}_1 > 0$. For $\bar{\mu} = 0$, the difference of these three lines of ε_2 reduces rapidly since $\bar{\sigma}_1$ is reduced to $\pi^2/8$ rapidly with increasing $\bar{\omega}$. The difference of three lines of ε_1 comes from $\bar{\sigma}_2$ as seen from Eq. (14) and Fig 1 (a). The introduction of μ gives the enhancement of the difference among these lines for both $\bar{\sigma}_1$ and $\bar{\sigma}_2$. For $\bar{\mu} = 2$, the behavior of $\bar{\sigma}_1$ gives $\varepsilon_2(\pi/2, 0.8) > \varepsilon_2(0, 0.8)$ for small $\bar{\omega}$ (not shown in Fig. 2) and $\varepsilon_2(0, 0.8) < \varepsilon_2(\pi/2, 0.8)$ for large $\bar{\omega}$. The maximum of ε_1 for $\alpha = 0$ is reduced owing to the increase of $\bar{\sigma}_2$. For $\bar{\mu} = 4$ and 8, ε_2 exhibits a crossover where $\varepsilon_2(\pi/2, 0.8) > \varepsilon_2(0, 0.8) > \varepsilon_2(0, 0)$ for small $\bar{\omega}$ and $\varepsilon_2(0, 0) > \varepsilon_2(0, 0.8) > \varepsilon_2(\pi/2, 0.8)$ for large $\bar{\omega}$. Such a relation is the same as that of $\bar{\sigma}_1$ in Fig. 1 as seen from Eqs. (15) and (12). The real part of $\varepsilon(\alpha, \eta)$ shows $\varepsilon_1(0, 0) > \varepsilon_1(0, 0.8) > \varepsilon_1(\pi/2, 0.8)$ in the most region of $\bar{\omega}$ where the negative value of ε_1 is enhanced by $\bar{\mu}$ suggesting a metallic behavior.

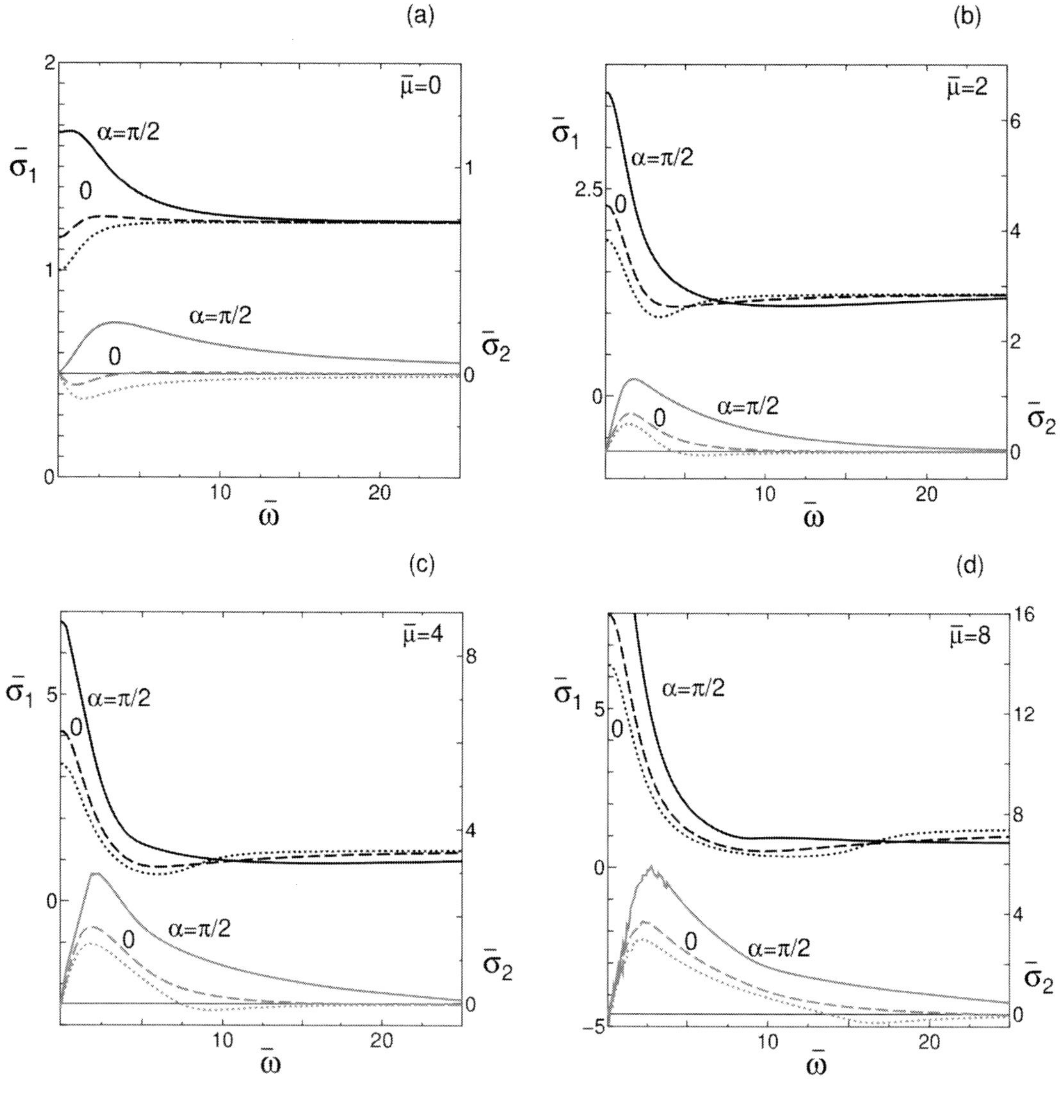

Figure 1. Frequency dependence of the conductivity $\overline{\sigma}(\alpha,\eta)(=\overline{\sigma}(\overline{\omega},\eta,\alpha))$ for the real part $\overline{\sigma}_1$ and imaginary part $\overline{\sigma}_2$ for $\overline{\mu}=0$ (a), 2 (b), 4 (c), and 8 (d). The horizontal and vertical axes are expressed by normalized quantities $\overline{\omega}(=\hbar\omega/\Gamma)$, and $\overline{\sigma}(=\sigma/(e^2/\pi h))$. The solid curve and dashed curve denote $\alpha=\pi/2$ and $\alpha=0$, respectively, with $\eta=0.8$. The dotted curve corresponds to $\overline{\sigma}_1$ and $\overline{\sigma}_2$ for η=0. The dashed line is always located between the solid line and the dotted line.

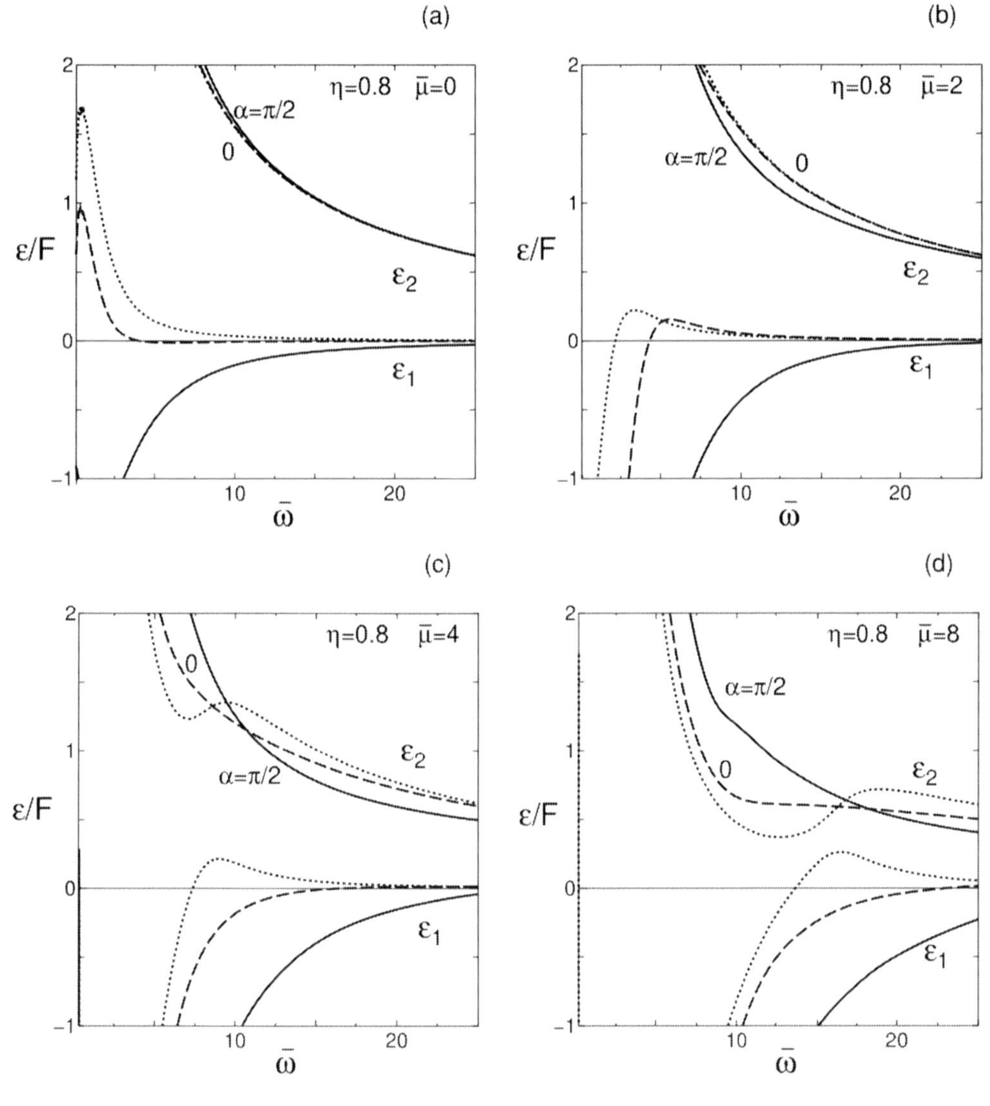

Figure 2. $\overline{\omega}$ dependence of the real part $\varepsilon_1(\overline{\omega}, \eta, \alpha)$ and the imaginary part $\varepsilon_1(\overline{\omega}, \eta, \alpha)$ of the dielectric constant for $F = 660$ with $\overline{\mu}$ =0 (a), 1 (b), 4 (c), and 8 (d). The horizontal axis denotes $\overline{\omega}(= \hbar\omega/\Gamma)$. The solid line and the dashed line denote $\alpha = \pi/2$ and 0 for η= 0.8, respectively. The dotted line corresponds to η= 0.

In Fig. 3, the $\overline{\omega}$ dependence of the reflectance $R(\alpha, \eta)(= R(\overline{\omega}, \eta, \alpha))$ corresponding to Fig. 1 is shown for $\overline{\mu}$ =0 (a), 2 (b), 4 (c), and 8 (d) where $F = 660$. The solid line and the dashed line denote $R(\pi/2, 0.8)$ and $R(0, 0.8)$ while the dotted line denotes $R(0, 0)$, i.e., without tilting. For $\overline{\mu} = 0$, the monotonous decrease of R is seen with a relation $R(\pi/2, 0.8) > R(0, 0.8) > R(0, 0)$. The fact that $R(\pi/2, 0.8)$ is slightly larger than $R(0, 0.8)$ and $R(0, 0)$ comes from $\varepsilon_1(\pi/2, 0.8) < 0$ suggesting the effect of $\overline{\sigma}_2$ being larger than that of $\overline{\sigma}_1$. For $\overline{\mu} = 2$, $R(\pi/2, 0.8)$ is enhanced owing to the presence of the Fermi surface. There is a crossover from $R(0, 0.8) > R(0, 0)$ to $R(0, 0) > R(0, 0.8)$ around $\overline{\omega} \sim 4$ with increasing $\overline{\omega}$. The effect of μ is enhanced for $\overline{\mu} = 4$

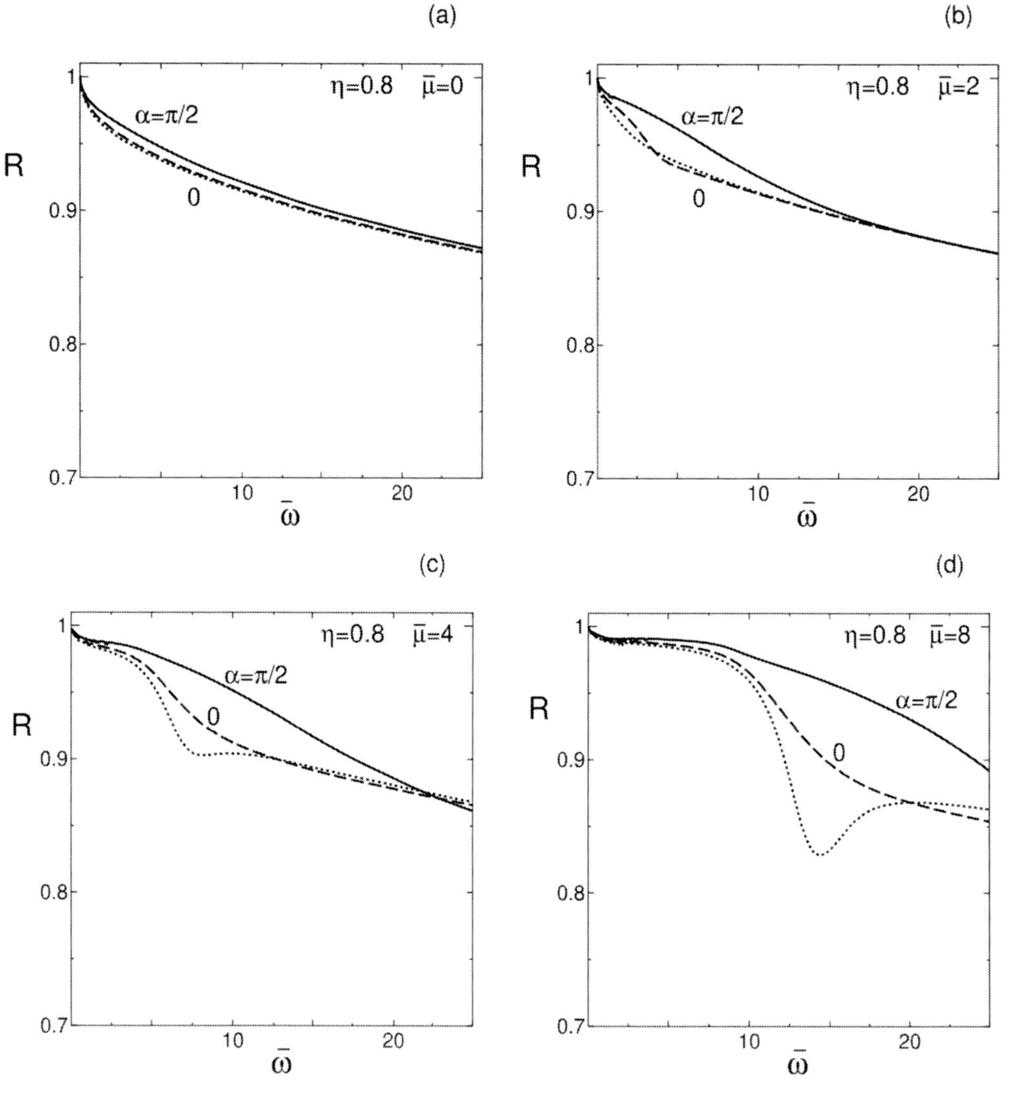

Figure 3. $\overline{\omega}$ dependence of reflectance $R(\overline{\omega}, \eta, \alpha)$ for $F = 660$ with $\overline{\mu} = 0$ (a), 2 (b), 4 (c), and 8 (d). The horizontal axis denotes $\overline{\omega}(= \hbar\omega/\Gamma)$. The solid line and the dashed line denote $\alpha = \pi/2$ and 0 for $\eta = 0.8$, respectively. The dotted line corresponds to $\eta = 0$.

where the increase of $\overline{\omega}$ gives the variation of such as $R(\pi/2, 0.8) > R(0, 0.8) > R(0, 0)$, $R(\pi/2, 0.8) > R(0, 0) > R(0, 0.8)$, and $R(0, 0) > R(0, 0.8) > R(\pi/2, 0.8)$. When $\overline{\mu}$ is large, a metallic behavior is found in the region where $R(\pi/2, 0.8)$ as a function of $\overline{\omega}$ becomes convex upward. A shoulder is seen in $R(0, 0)$. For $\overline{\mu} = 8$, the difference among these three lines becomes large and the behavior in Fig. 3(c) is enhanced, i.e., the noticeable crossover from $R(\pi/2, 0.8) > R(0, 0.8) > R(0, 0)$ to $R(0, 0) > R(0, 0.8) > R(\pi/2, 0.8)$. The dip exists in $R(0, 0)$ as already found in the isotropic case [12]. This behavior comes from a crossover from intraband excitation to the interband excitation, i.e., the former (the latter) comes from $\overline{\omega} < 2\overline{\mu}$ ($\overline{\omega} > 2\overline{\mu}$). The dip of the dotted line ($\eta = 0$) disappears for small $\overline{\mu}$ owing to the strong effect of Γ.

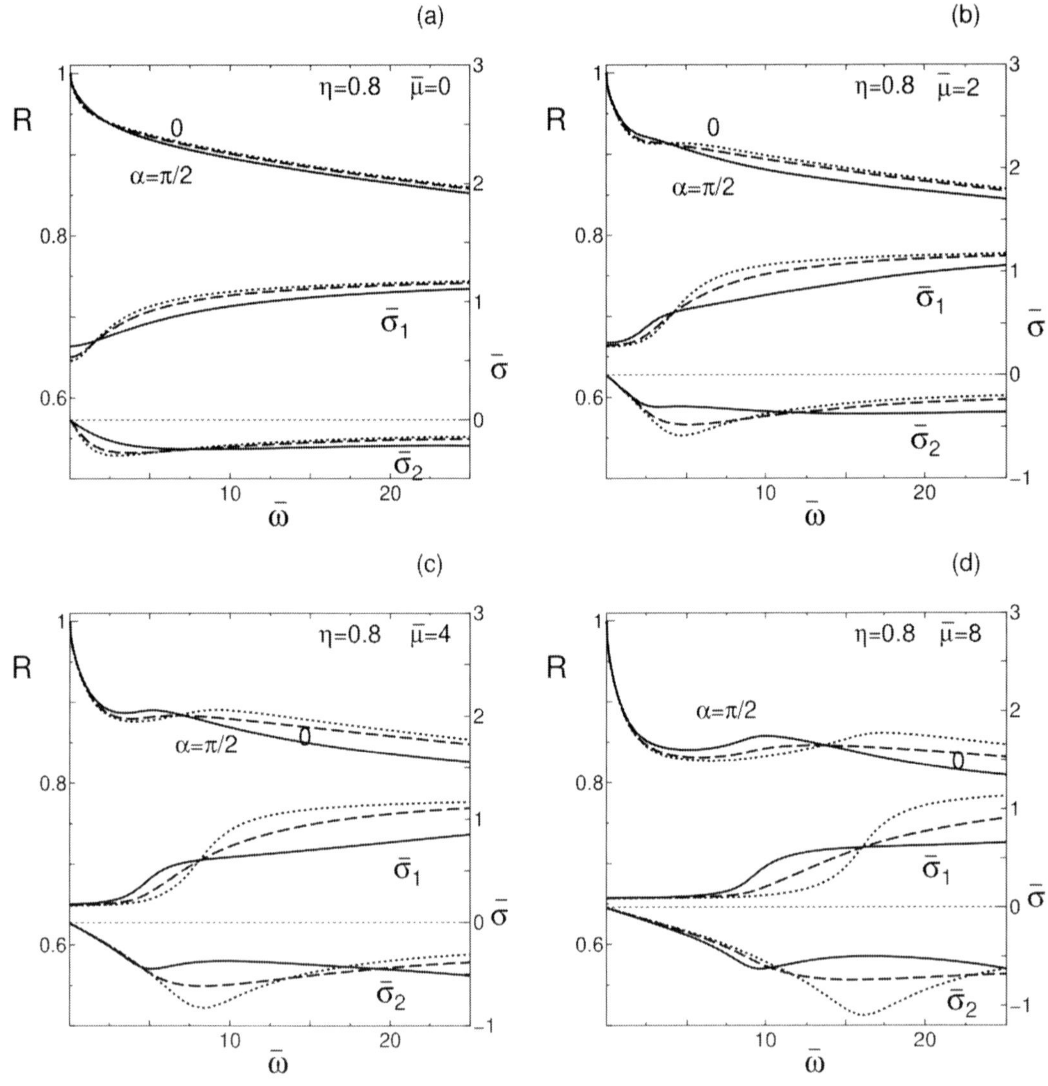

Figure 4. Inter band component of reflectance $R(\overline{\omega},\eta,\alpha)$ with the real part $\varepsilon_1(\overline{\omega},\eta,\alpha)$ and the imaginary part $\varepsilon_2(\overline{\omega},\eta,\alpha)$ of the dielectric constant with $\overline{\mu}$ =0 (a), 1 (b), 2 (c), and 4 (d). The notations are the same as those of Fig. 3.

Finally, we examine the role of the interband excitation. Figure 4 shows R which is calculated only in the presence of the interband excitation, i.e., Eq. (8). Compared with Fig. 3, R in the absence of intraband excitation is reduced. The most region for $R(\pi/2,0.8) > R(0,0.8)$ in Fig. 3(a) is replaced by $R(0,0.8) > R(\pi/2,0.8)$ in Fig. 4(a) suggesting that the former comes from the intraband excitation. The behavior of R in Figs. 4(b), (c) and (d) clearly shows the property of $\overline{\sigma}_1$ in Eq. (12), which gives a crossover from $R(\pi/2,0.8) > R(0,0.8)$ to $R(0,0.8) > R(\pi/2,0.8)$ with increasing $\overline{\omega}$. In the case of $\overline{\mu} = 2$ the effect of the intraband excitation is still large owing to the presence of Γ. With increasing $\overline{\mu}$ (=4 and 8), the region of $\overline{\omega}$ for the contribution of the interband excitation emerges gradually where the effect of intraband excitation is still large.

We note the crossover from $R(\pi/2, 0.8) > R(0, 0.8)$ to $R(0, 0.8) > R(\pi/2, 0.8)$ appears for large $\overline{\omega}$ being much larger than $2\overline{\mu}$. The sign of $\overline{\sigma}_2$ of the intraband excitation is opposite to that of Fig. 4 suggesting that these two excitations are incompatible for the small $\overline{\mu}$. However, in the presence of large $\overline{\mu}$, both contributions to R become compatible owing to the increase of $\overline{\sigma}_1$ where the intraband contribution is large for small $\overline{\omega}$ and the interband contribution becomes large for large $\overline{\omega}$.

4. Summary

We examined the reflectance of Dirac electrons with a tilted cone using a parameter of organic conductor α-(BEDT-TTF)$_2$I$_3$, $\eta = 0.8$ and $\alpha = \pi/2$ and 0.

The reflectance on the plane is calculated from the dielectric constant which is estimated from the in-plane dynamical conductivity of the previous calculation. In addition to the chemical potential, the characteristic energy of the reflectance is the plasma frequency which is estimated with the interlayer distance for organic conductor α-(BEDT-TTF)$_2$I$_3$. We compared a difference of the reflectance and the dielectric constant between $\alpha = 0$ and $\pi/2$, where the former (the latter) corresponds to the tilting being parallel and perpendicular to the electric field. The $\overline{\omega}$ dependence of the reflectance is examined by choosing several magnitudes of the chemical potential $\overline{\mu}$ to find a crossover from the zero doping to a finite doping. With increasing $\overline{\omega}$, the reflectance decreases monotonously from unity. For the small doping with $\mu < \Gamma$, the difference between $\alpha = 0$ and $\pi/2$ is small owing to the intraband excitation. For the large doping with $\Gamma < \mu$, the reflectance shows a visible difference in a finite region of $\omega(> \mu)$. In this case, the reflectance of $\alpha = \pi/2$ is larger than that of $\alpha = 0$ for a small frequency, while the relation is reversed for a large frequency. Such a crossover originates from the interplay the intraband excitation and the interband excitation.

We comment on the doping in organic conductor. Our result shows that such a crossover occurs for $\omega/\Gamma = 10 \sim 30$ for $\mu/\Gamma \sim 5$. These are reasonable parameters since $\Gamma/k_{\rm B} \simeq 3$ K, and $|\mu| <\simeq 40$ K is obtained for the hole doping by the PEN substrate[3].

Acknowledgments

The authors are thankful to N. Tajima for useful discussions. This work was supported by Grants-in-Aid for Scientific Research (A) (No. 24244053) and (C) (No. 26400355) from the Ministry of Education, Culture, Sports, Science and Technology of Japan. One of the authors (I. P.) acknowledges the Russian Foundation for Basic Research, Grant No. 14-02-92104.

Appendix A. Dynamical conductivity in the clean limit

In the clean limit, the dynamical conductivity $\sigma(\omega)$ is obtained as $(\delta \to +0)$ [9, 8]

$$\overline{\sigma}_{\rm intra} = \frac{e^2}{\pi h}\frac{{\rm i}\mu\pi}{\hbar\omega + {\rm i}\delta}\left\langle \frac{(\eta\cos\alpha + \theta)^2}{(1+g)^2} \right\rangle_\theta , \quad (A.1)$$

$$\overline{\sigma}_{\rm inter} = \frac{e^2}{\pi h}\frac{\pi^2}{4}\left\langle (\sin\theta)^2 \left(1 + \frac{\rm i}{\pi}\ln\frac{\hbar\omega - 2\mu/(1+g) + {\rm i}\delta}{\hbar\omega + 2\mu/(1+g) + {\rm i}\delta}\right)\right\rangle_\theta , \quad (A.2)$$

where $g = \eta\cos(\theta - \alpha)$. The intra part is calculated as

$$\overline{\sigma}_{\rm intra} = \frac{e^2}{\pi h}\frac{{\rm i}\mu\pi}{\hbar\omega + {\rm i}\delta}\left\{\frac{1-\sqrt{1-\eta^2}}{\eta^2}\cos^2\alpha + \frac{1-\sqrt{1-\eta^2}}{\eta^2\sqrt{1-\eta^2}}\sin^2\alpha\right\} . \quad (A.3)$$

Noting that the real part of the bracket in Eq. (A.2) vanishes for $\hbar\omega < 2\mu/(1+g)$, the real part of the interband contribution is calculated as [10]

$$\mathrm{Re}\{\overline{\sigma}_{\mathrm{inter}}\} = \frac{e^2}{\pi\hbar}\frac{\pi^2}{8}F_c(\alpha) \,, \tag{A.4}$$

where $F_c(\alpha) = 0$ for $\omega < 2\mu/(1+\eta)$ and $F_c(\alpha) = 1$ for $2\mu/(1-\eta) < \omega$. For $2\mu/(1+\eta) < \omega < 2\mu/(1-\eta)$,

$$F_c(\alpha) = \frac{1}{\pi}\left(\varphi_0 - \frac{1}{2}\sin(2\varphi_0)\cos(2\alpha)\right) \,, \tag{A.5}$$

with $\cos\varphi_0 = (2\mu/\hbar\omega - 1)/\eta$.

References

[1] Katayama S, Kobayashi A and Suzumura Y 2006 *J. Phys. Soc. Jpn.* **75** 054705
[2] Kondo R, Kagoshima S and Maesato M 2003 *Phys. Rev.* B **67** 134519.
[3] Kajita K, Nishio Y, Tajima N, Suzumura Y, and Kobayashi A 2009 *J. Phys. Soc. Jpn.* **83** 072002.
[4] Ando T 2005 *J. Phys. Soc. Jpn.* **74** 777
[5] Kobayashi A, Katayama S, Suzumura Y and Fukuyama H 2007 *J. Phys. Soc. Jpn.* **76**, 034711
[6] Proskurin I and Ogata M 2013 *J. Phys. Soc. Jpn.* **82** 063712
[7] Suzumura Y, Proskurin I and Ogata M, 2014 *J. Phys. Soc. Jpn.* **83** 023701
[8] Suzumura Y, Proskurin I and Ogata M 2014 *J. Phys. Soc. Jpn.* **83** 094705
[9] Ando T, Zheng Y, and Suzuura H 2002 *J. Phys. Soc. Jpn.* **71** 1318
[10] Nishine T, Kobayashi A and Suzumura Y 2010 *J. Phys. Soc. Jpn.* **79** 114715
[11] Li Z Q, Henriksen E A, Jiang Z, Hao Z, Martin M C, Kim P, Stormer H L 2008 *Nature Phys.* **4** 532
[12] Falkovsky L A 2008 *J. Phys: Conf. Ser.* **129** 012004
[13] Bastin A, Lewiner C, Betbeder-Matibet O and Nozieres P, 1971 *J. Phys. Chem. Solids* **32** 1811
[14] Streda P and Smrcka L 1975 *Phys. Status Solidi* B **70** 537
[15] Shon N H and Ando T 1998 *J. Phys. Soc. Jpn.* **67** 2421
[16] Kondo R, Kagoshima S,Tajima N and Kato R 2009 *J. Phys. Soc. Jpn.* **78** 114714

Anisotropic spin motive force in multi-layered Dirac fermion system, α-(BEDT-TTF)$_2$I$_3$

K Kubo and T Morinari

Graduate School of Human and Environmental Studies, Kyoto University, Kyoto 606-8501, Japan

E-mail: kubo.kenji.22x@st.kyoto-u.ac.jp

Abstract. We investigate the anisotropic spin motive force in α-(BEDT-TTF)$_2$I$_3$, which is a multi-layered massless Dirac fermion system under pressure. Assuming the interlayer antiferromagnetic interaction and the interlayer anisotropic ferromagnetic interaction, we numerically examine the spin ordered state of the ground state using the steepest descent method. The anisotropic interaction leads to the anisotropic spin ordered state. We calculate the spin motive force produced by the anisotropic spin texture. The result quantitatively agrees with the experiment.

1. Introduction

An organic conductor α-(BEDT-TTF)$_2$I$_3$ is a multi-layered massless Dirac fermion system, in which conduction layers of BEDT-TTF molecules and I$_3$ anions stack alternatively. In each conduction layers, the massless Dirac fermion system is realized. The valence band and the conduction band contact at two inequivalent points in the Brillouin zone. The energy dispersion is linear in the vicinity of the contact points. The Fermi energy coincides with the energy of the contact point. Therefore, the system is called Dirac fermion system.

Under the magnetic field B, the energy of the n-th Landau level of the Dirac fermion is described by $E = \mathrm{sgn}(n)C\sqrt{|n|B}$ with a constant C. The energy difference between the first and the zeroth Landau levels is $C\sqrt{B}$. Therefore, in relatively high magnetic field, we may consider only the Landau level with $n = 0$.

α-(BEDT-TTF)$_2$I$_3$ has unique spin ordered states [1]. In each layer, the quantum Hall ferromagnetic state is realized by the intralayer ferromagnetic interaction due to the exchange interaction. On the other hand, the interlayer ferrimagnetic state is possible by the interlayer antiferromagnetic interaction.

α-(BEDT-TTF)$_2$I$_3$ has anisotropic crystal structure and transport property. In figure 1, the direction of the interlayer tunneling is shown. The tunneling is tilted in the direction of b-axis. For the experimental data, the angle between the vertical direction and the interlayer tunneling direction is about 28°.

At low temperature and under magnetic field, anisotropic voltage is observed in α-(BEDT-TTF)$_2$I$_3$ [2]. At high temperature the voltage disappears. The voltage increases with increasing the magnetic field. Therefore, one possible scenario is that the voltage is caused by spin motive force. The anisotropic crystal structure can lead to in the anisotropic voltage. The spin motive force produces the voltage. However, this is not a equilibrium situation. The spin motive force

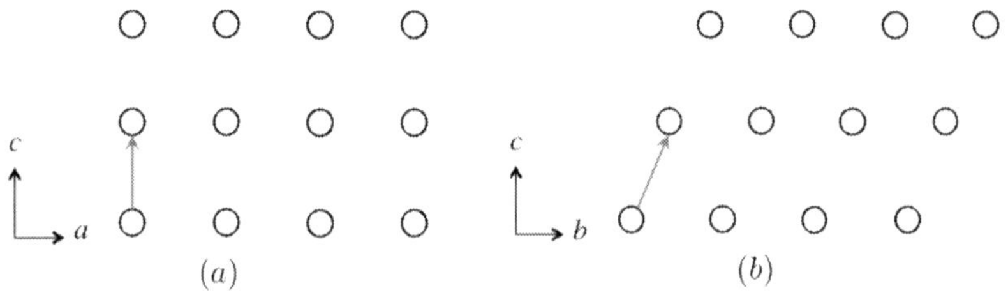

Figure 1. The red arrow shows the interlayer tunneling direction. The BEDT-TTF molecules are described by the circles. a- and b-axis is in the conducting plane. c-axis is perpendicular to it. (a) The interlayer tunneling direction in a-c plane. (b) The interlayer tunneling in b-c plane is tilted in the direction to the b-axis.

is transient due to the Gilbert damping. Charge accumulates at the edge of the system by the spin motive force and produces a electric field opposite to that created by spin motive force. The voltage will vanish due to the electric field created by the charge accumulation. Hereafter, for simplicity, we assume that the necessary time of the disappearance of the voltage is long and the voltage is constant in this time scale.

In this paper, we show that anisotropic spin ordered states are realized in α-(BEDT-TTF)$_2$I$_3$ because of the interlayer anisotropic interaction and anisotropic spin motive forces are produced under magnetic field.

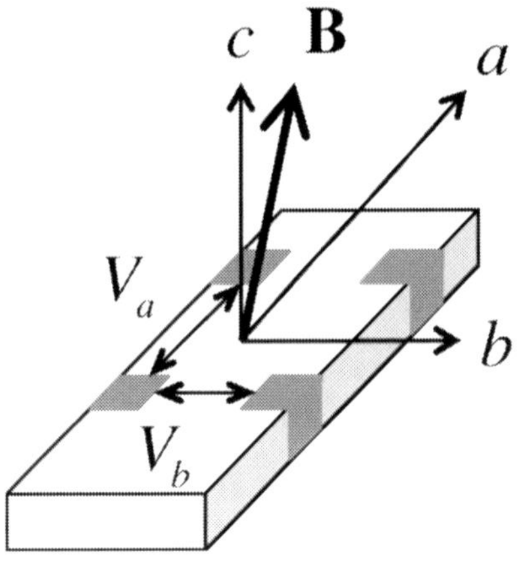

Figure 2. The anisotropic voltage in α-(BEDT-TTF)$_2$I$_3$. The voltage in the direction of a, b are defined by V_a, V_b.

2. Model

We assume the square lattice for the conducting plane and the ferromagnetic interaction $J_{\rm intra}$ is assumed between nearest neighbor molecules. We define b- and a-axis as x- and y-axis, respectively. We take the direction of interlayer tunneling as z-axis. The interlayer interaction in the direction of tunneling is denoted by $J_{\rm inter} < 0$. In this system, the lattice is distorted in x-z plane since the molecular arrangement is tilted in the direction of x-axis and there are two different next nearest neighbor sites in x-z plane. When the distance is different, the interaction should be different. For this reason, we assume a finite interaction $J'_{\rm inter}$ in one of the next nearest neighbor sites in x-z plane. On the other hand, for simplicity, we take a square lattice in x-z plane in this calculation and assume the magnetic field is parallel to the z-axis.

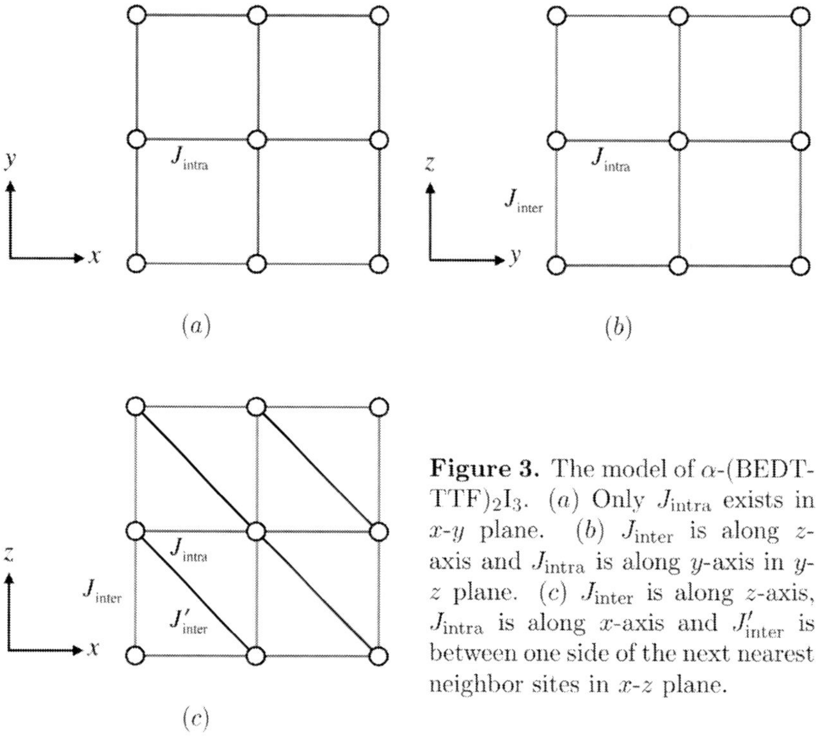

Figure 3. The model of α-(BEDT-TTF)$_2$I$_3$. (a) Only J_{intra} exists in x-y plane. (b) J_{inter} is along z-axis and J_{intra} is along y-axis in y-z plane. (c) J_{inter} is along z-axis, J_{intra} is along x-axis and J'_{inter} is between one side of the next nearest neighbor sites in x-z plane.

The Hamiltonian under the magnetic field is written as

$$\begin{aligned} H &= J_{\text{intra}} \sum_{i,j,k} (\mathbf{S}_{i+1,j,k} + \mathbf{S}_{i,j+1,k} + \mathbf{S}_{i-1,j,k} + \mathbf{S}_{i,j-1,k}) \cdot \mathbf{S}_{i,j,k} \\ &+ J_{\text{inter}} \sum_{i,j,k} (\mathbf{S}_{i,j,k+1} + \mathbf{S}_{i,j,k-1}) \cdot \mathbf{S}_{i,j,k} + J'_{\text{inter}} \sum_{i,j,k} (\mathbf{S}_{i,j+1,k+1} + \mathbf{S}_{i,j-1,k-1}) \cdot \mathbf{S}_{i,j,k} \\ &- \frac{1}{2} g \mu_B \sum_{i,j,k} \mathbf{B} \cdot \mathbf{S}_{i,j,k}, \end{aligned} \tag{1}$$

where $\mathbf{S}_{i,j,k}$ is a spin at the lattice point (i, j, k). g is g factor and μ_B is the Bohr magneton.

In this model, we investigate the spin ordered state of the ground state using the steepest descent method. At first, we put a spin with random direction on each lattice point. Next, we update spins with

$$\begin{aligned} \mathbf{n}_{i,j,k} &= -\frac{J_{\text{intra}}}{4} (\mathbf{n}_{i+1,j,k} + \mathbf{n}_{i,j+1,k} + \mathbf{n}_{i-1,j,k} + \mathbf{n}_{i,j-1,k}) \\ &- \frac{J_{\text{inter}}}{2} (\mathbf{n}_{i,j,k+1} + \mathbf{n}_{i,j,k-1}) - \frac{J'_{\text{inter}}}{2} (\mathbf{n}_{i,j+1,k+1} + \mathbf{n}_{i,j-1,k-1}) + \frac{1}{2} g \mu_B \mathbf{B}. \end{aligned} \tag{2}$$

Here, $\mathbf{n}_{i,j,k} = 2\mathbf{S}_{i,j,k}$ is the normalized localize moment. The resulting converged state is an approximate for the ground state. J_{intra} is proportional to the $\sqrt{B}$ [1], but, for simplicity, we assume $J_{\rm intra}$ is constant. We numerically calculate with the number of sites, $12 \times 12 \times 12$ and under the periodic boundary condition.

The result with the parameters $J_{\rm intra} = -5$ K, $J_{\rm inter} = 4$ K, $J'_{\rm inter} = -2$ K and $B = 5$ T is shown in figure 4. Here, we assume relatively large value for $J_{\rm inter}$, which is the same order of magnitude as the interlayer hopping estimated in a related organic compound [3]. A ferromagnetic state is realized along y-axis. On the other hand, a periodic spin texture is created along x-axis due to the interlayer next nearest neighbor interaction.

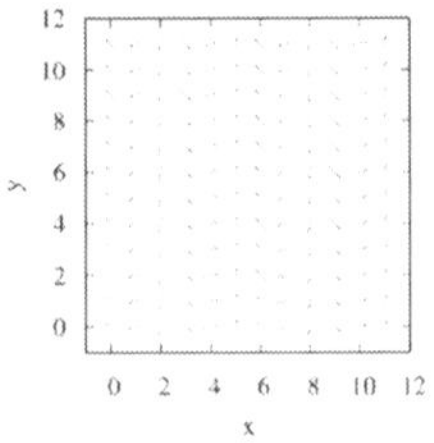

(a)

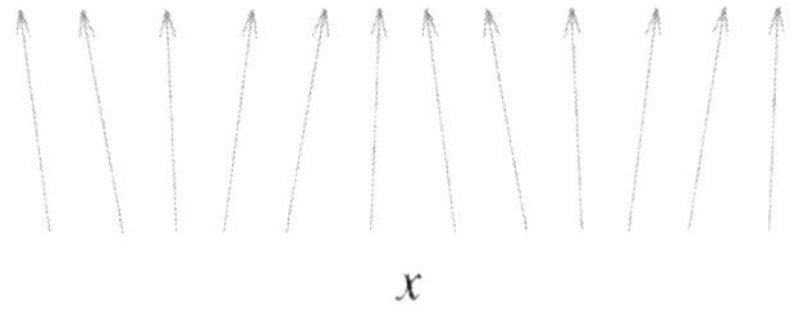

(b)

Figure 4. The resulting spin ordered state of α-(BEDT-TTF)$_2$I$_3$ with $J_{\rm intra} = -5$ K, $J_{\rm inter} = 4$ K, $J'_{\rm inter} = -2$ K and $B = 5$ T. (a) The projection of spins in a layer to x-y plane. A periodic spin ordered state is realized along x-axis, although a ferromagnetic structure is along y-axis. (b) The projection of spins in a layer to x-z plane.

Assuming this spin ordered state, we calculate the spin motive force [4] in the strong coupling limit. In this limit, the spins are polarized and there are no conduction electron with minority spin. Therefore, the spin motive force is identical with the voltage. We make indices j, k implicit below since we consider spin motive force in uniaxial direction. The localized moment is represented as

$$\mathbf{n}_i = (\cos\theta_i \cos\phi_i, \cos\theta_i \sin\phi_i, \sin\theta_i), \tag{3}$$

where θ_i is the angle between the direction of the spin and the z-axis. ϕ_i is the angle between the direction of the spin projected in the x-y plane and the positive direction of the x-axis. The electric field due to the localized moment is written as

$$\mathbf{E}_i = \frac{\hbar}{2e} \sin\theta_i \left[(\partial_t \theta_i)(\nabla\phi_i) - (\partial_t \phi_i)(\nabla\theta_i) \right], \tag{4}$$

where e is the charge of the electron [5]. Here, the condition $\partial_t \theta_i = 0$ and $\partial_t \phi_i = g\mu_B B/\hbar$ are always satisfied since the magnetic field is along z-axis. The electric field is rewritten as

$$\mathbf{E}_i = -\frac{g\mu_B}{2e} B \sin\theta_i \left(\nabla\theta_i\right). \tag{5}$$

In the numerical calculation with the periodic boundary condition, the electric field is canceled if we take the average over the spatial period and so the spin motive force does not appear. However, in the real system, there is the edge of the sample and a finite spin motive force is possible. On the other hand, spin motive force along y-axis does not appear at all. Therefore, the spin motive force in this model has strong anisotropy. In order to estimate the spin motive force in the real system, we calculate the spin motive force with taking a part of the periodic spin structure.

The numerical result is shown in figure 5. The spin motive force along x-axis is on the order of 0.01 mV to 0.1 mV which is about the same order compared with the experimental result. The most of electric field is canceled and only the spin motive force in the vicinity of the edge remains. The experimental result shows that V_b increases with increasing the magnetic field below 7 T. In order to investigate the magnetic field dependence of V_b, we define the maximum value of the V_b as $V_{\rm max}$. Although $V_{\rm max}$ does not always correspond to the experimental observed voltage, it is able to qualitatively evaluate the magnetic field dependence of V_b. The nurerical result of the magnetic field dependence of $V_{\rm max}$ is shown in figure 6. $V_{\rm max}$ increases with increasing magnetic field below 4 T. However, above 5 T, $V_{\rm max}$ decreases. Under high magnetic field, the Zeeman energy is larger than the intralayer and the interlayer interactions, and so $\{\theta_i\}, \{\nabla\theta_i\}$ are small since we assume the interactions are constant. Therefore, the electric fields due to the localized moments vanish under high magnetic field. If we consider the magnetic field dependence of the interactions, we would evaluate the anisotropic spin motive force more precisely.

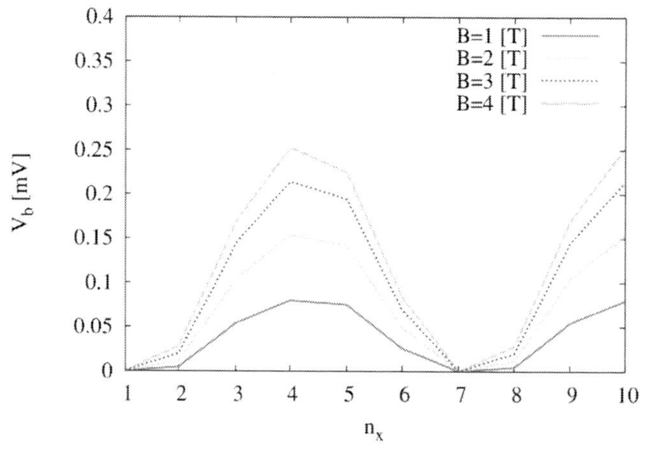

Figure 5. Spin motive forces along x-axis in α-(BEDT-TTF)$_2$I$_3$. n_x is the number of spins along x-axis from the edge of the system.

In a real system, the edge shape is different for a different layer. So, the naive expectation is that if we take the average over those edges, the spin motive force is canceled out. However, since the anisotropic spin motive force is generated by the three-dimensional structure, spins in regions without interlayer interaction should not contribute to the spin motive force. Therefore, it is justified to consider spin motive force for layers with the same edge shape.

3. Conclusion

We have calculated the spin motive force in α-(BEDT-TTF)$_2$I$_3$. The anisotropic interlayer interaction exists reflecting the anisotropic crystal structure. The anisotropic interaction leads to the unidirectional periodic spin structure. The anisotropic spin motive force is created from the anisotropic spin structure under magnetic field.

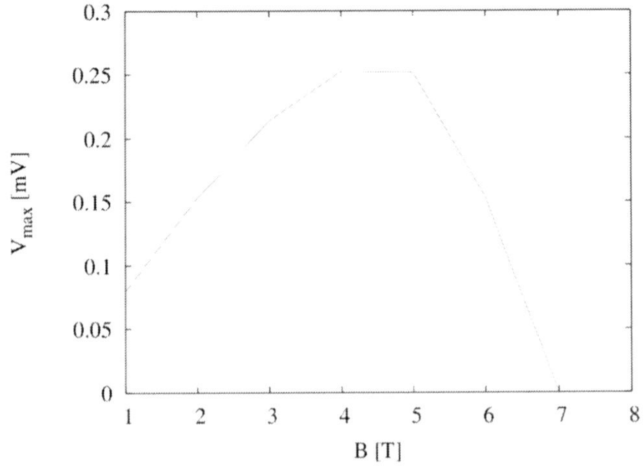

Figure 6. The magnetic field dependence of the maximum values of V_b. $V_{\max}$ increases with increasing magnetic field below 4 T and decreases above 5 T.

Acknowledgments

We would like to thank N. Tajima for discussions and sending us the experimental result. This work was financially supported in part by a Grant-in-Aid for Scientific Research (A) on " Dirac Electrons in Solids" (No. 24244053) and a Grant-in-Aid for Scientific Research (B) (No. 25287089) and (C) (No. 24540370) from the Ministry of Education, Culture, Sports, Science and Technology, Japan.

References

[1] Kubo K and Morinari T 2014 *J. Phys. Soc. Jpn.* **83** 033702
[2] Tajima N Private communication
[3] Jindo R, Sugawara S, Tajima N, Yamamoto M, Kato R, Nisino Y and Kajita K 2013 *Phys. Rev.* B **88** 075315
[4] Tatara G, Kohno H and Shibata J 2008 *Physics Reports* **468** 213
[5] Tserkovnyak Y and Mecklenburg M 2008 *Phys. Rev.* B **77** 134407

Defect-induced Kondo effect in graphene: role of localized state of π electrons

Taro Kanao, Hiroyasu Matsuura, and Masao Ogata
Department of Physics, University of Tokyo, Bunkyo, Tokyo 113-0033, Japan
E-mail: kanao@hosi.phys.s.u-tokyo.ac.jp

Abstract. We discuss a role of the localized π orbital, which exists around the defect, on the defect-induced Kondo effect in graphene by a numerical renormalization group study. We find that the localized π orbital assists this Kondo effect, and the Kondo temperature is sensitive to the broadening of the localized π orbital. Secondly, we focus on the negative magnetoresistance of this Kondo effect. In the experimental result, it has been shown that the negative magnetoresistance is ten times larger than the usual Kondo effect. In order to clarify the mechanism of the "magnetic sensitive" Kondo effect, as a first step, we study an orbital magnetic field dependence of the localized π orbital by a tight-binding model with a Peierls phase. We find that as the magnetic field increases, the spectral width of the localized π orbital increases and the local DOS at the Fermi level decreases. Since the Kondo temperature is strongly dependent of the broadening of the localized π orbital, it is expected that this Kondo effect is sensitive to the orbital magnetic field as observed in the experiment.

1. Introduction

After the fabrication of graphene, the Kondo effect in graphene has attracted much theoretical attention, because exotic Kondo effects and associated phenomena due to its Dirac conduction electrons are expected, such as multichannel Kondo effects, gate-tunable Kondo effects, and an impurity quantum phase transition between an unscreened and a screened localized moment state [1-6].As candidates of the realization of the Kondo effect in graphene, the graphene with magnetic impurities such as Fe or Co has been investigated by DFT calculations [2]. The possibility of the Kondo effect in these systems has been argued by examining the lattice positions, the spin state of the impurity atom, and the strength and symmetry of the hybridization between the impurity and the conduction electrons. Experiments on graphene with such magnetic impurities have been attempted, and spectral signatures of the Kondo screening have been observed in scanning tunneling microscope (STM) experiments [7]. In such systems, however, Kondo effect had not been observed in resistivity measurements.

In 2011, the Kondo effect was reported in the resistivity of graphene with defects [8]. The defects were introduced by the irradiation of He^+ ions with the energy of 500 eV. The temperature dependence of resistivity showed logarithmic increase with decreasing temperature and saturating behavior at low temperatures. The temperature dependence can be fitted well by the universal temperature dependence of the Kondo effect. In the ion-irradiated graphene, the negative magnetoresistance was also observed [8]. Magnetic field is applied perpendicular to the graphene plane. The contribution of the weak localization is suppressed by the magnetic field $B \sim 1$ T, and a magnetoresistance due to another mechanism occurs at higher magnetic fields. In the case of the Kondo effect, the Zeeman effect of magnetic field can cause such a negative magnetoresistance. From these observations, the Kondo effect due to magnetic origin was concluded in this system. To clarify the origin of the Kondo effect, many theoretical studies have been performed [9-15].On the other hand, it also has been proposed that the electron-electron interaction in the presence of the disorder is responsible for the logarithmic increase of the resistivity at low temperatures [16, 17]. Although there is the other possible theory, in this paper, we focus on the defect induced Kondo effect.

Until the discovery of the defect induced Kondo effect, the defects in graphene or graphite have been studied as a source of magnetism [18]. In graphite, it has been observed that magnetic moments are induced by irradiation of high-energy (2.25 MeV) protons [19]. In the graphene irradiated with ions, the magnetic moments have also been observed [20]. It has been indicated that the defect individually behaves as a localized magnetic moment. Such defects have been studied in STM experiments [21, 22]. These experiments show that the introduced defects are point defects (or point vacancies). The defects in graphene have also been investigated with transmission electron microscopes (TEMs) [23], and reconstruction and deformation of the defect are observed. Such a point defect in graphene has been studied with the DFT calculations. First, the lattice structure was investigated [24]. Two possible structures were considered: one is a threefold symmetric structure that preserves the original symmetry, and the other is a twofold symmetric structure in which one of the C-C distances is shortened. It was clarified that the latter is more stable owning to the Jahn-Teller effect, and it has an out-of-plane displacement, while the former has a planar structure. Second, the spin state was investigated by using spin-polarized DFT calculations [25, 26]. It was shown that the defect is magnetic and the dangling sp^2 bond which appears at the defect is responsible for the magnetic moment.

In graphene, effects of defect scattering on the π electrons have been studied theoretically, and it has been suggested that the point defect induces a localized state of π electrons around the defect whose energy eigenvalue coincides with the energy of the Dirac points [26-29].Such a localized π orbital has been observed experimentally with the STM. In scanning tunneling spectroscopy (STS) experiments, the localized π orbital exhibits a sharp peak in local density of states (LDOS) near the defect.

Although the origin of the Kondo effect in graphene with defects seems to be due to the localized magnetic moments on the sp^2 orbital at the defects, the role of localized π orbital has not been understood in detail. Recently, although a model where the localized state of π electrons is rather artificially added at the defect [10, 15] has been proposed, more consideration on the derivation of the model and the analysis are necessary. In addition, in this defect induced Kondo effect, the anomalous negative magnetoresistance has been observed: In the usual Kondo effect, the energy scale of a characteristic magnetic field of the negative magnetoresistance is the same order as that of the Kondo temperature. On the other hand, experimentally, the characteristic magnetic field seems to be one tenth smaller than the value expected from the Kondo temperature, which means that this Kondo effect is sensitive to the magnetic field. However, this mechanism has not been understood yet.

Thus, in this paper, we discuss the role of localized π orbital on the defect-induced Kondo effect in graphene on the basis of a numerical renormalization group (NRG) study [31]. As a result, we find that the localized π orbital assists the Kondo effect, and the Kondo temperature is sensitive to the broadening of the localized π orbital. Next, in order to clarify the mechanism of the magnetic sensitive Kondo effect, as a first step, we study an orbital magnetic field dependence of the localized π orbital by a tight-binding model with a Peierls phase. We show that the LDOS on the localized π orbital decreases as the magnetic field increases.

In §.2, we develop an effective model with the localized π orbital and transforms the effective model into a one-dimensional representation of the conduction electron states to apply NRG method. In §.3, we discuss the effect of the localized π orbital by NRG study. Finally, in §.4, we discuss an orbital magnetic field dependence of the localized state of π electrons by the tight binding model with the Peierls phase.

2. Localized moment and its hybridization with conduction electrons

2.1. Cluster model for the defect

Here, considering a cluster for the defect, we introduce a model for the system. Around the point defect, there are three sp^2 orbitals φ_i which do not form covalent bond, where i indicates one of the three sites (i = 1, 2, or 3) around the defect. There are also π orbitals on carbon atoms.

First, we consider an isolated cluster that consists of three sp^2 orbitals, as a model for the localized state of sp^2 orbitals around the defect. A tight-binding approximation is applied to this cluster. Two of the three transfer integrals between the sp^2 orbitals are $-h$, and the other is $-h'$ $(h, h' > 0)$. The threefold symmetric structure corresponds to the case with $h = h'$, and the twofold symmetric one to $h' > h$, where it is assumed that the distance between sites 2 and 3 is shorter than the others.

The energy eigenvalues and the corresponding wave functions of this cluster model are

$$\begin{aligned} E_3 &= h', & \psi_3 &= \frac{1}{\sqrt{2}}(\varphi_2 - \varphi_3), \\ E_2 &= \frac{-h' + \sqrt{8h^2 + h'^2}}{2}, & \psi_2 &= C_+\varphi_1 - \frac{1}{\sqrt{2}}C_-(\varphi_2 + \varphi_3), \\ E_1 &= \frac{-h' - \sqrt{8h^2 + h'^2}}{2}, & \psi_1 &= C_-\varphi_1 + \frac{1}{\sqrt{2}}C_+(\varphi_2 + \varphi_3), \end{aligned} \tag{1}$$

with

$$C_\pm = \left[\frac{1}{2}\left(1 \pm \frac{h'}{\sqrt{8h^2 + h'^2}}\right)\right]^{1/2}, \tag{2}$$

which are depicted in Fig. 1. There are three electrons on these sp^2 orbitals. Since the number of electrons is odd, there is an unfilled level and nonzero spin appears. Interpreting the Kohn-Sham eigenvalues of the DFT calculations [24] in terms of these tight-binding energy levels, we semiquantitatively estimate h'/h to be $h'/h \simeq 5$. When $h'/h \simeq 5$, $C_\pm$ becomes $C_+ \simeq 1$, $C_- \ll 1$, and the wave function of the lowest level ψ_1 becomes a bonding state $\psi_1 = (\varphi_2 + \varphi_3)/\sqrt{2}$, while the second state becomes $\psi_2 = \varphi_1$. Thus, φ_1 becomes the highest occupied molecular orbital of this cluster, and causes a localized magnetic moment. It should be noted that in a divacancy case, four electrons are present in the sp^2 orbitals and the localized spin does not appear.

Second, we consider the hybridization between this localized magnetic moment on the sp^2 orbital and the conduction electrons which consist of π orbitals. It depends on the lattice structure of the defect. When the defect has the planar structure, there is no hybridization between the sp^2 and π orbitals, which is schematically shown in Fig. 2(a). However, with an out-of-plane displacement, finite hybridization between the sp^2 orbital φ_i and the two π orbitals at the other sites $j \neq i$ appears as in Fig. 2(b). As mentioned above, from the DFT calculations, the twofold symmetric structure with an out-of-plane displacement is suggested to be more stable owing to the Jahn-Teller effect [24].

In the following we consider the structure shown in Fig. 3, where only one sp^2 orbital ($i = 1$) is active and it hybridizes with the π orbitals at sites 2 and 3. (The bonding state of sp^2 orbitals at sites 2 and 3 is neglected.) We assume that the hybridization term is given by

$$H_{\rm hyb} = \sum_{\sigma=\uparrow\downarrow} \left[V \left(a^\dagger_{2\sigma} + a^\dagger_{3\sigma} \right) d_\sigma + {\rm h.c.} \right], \tag{3}$$

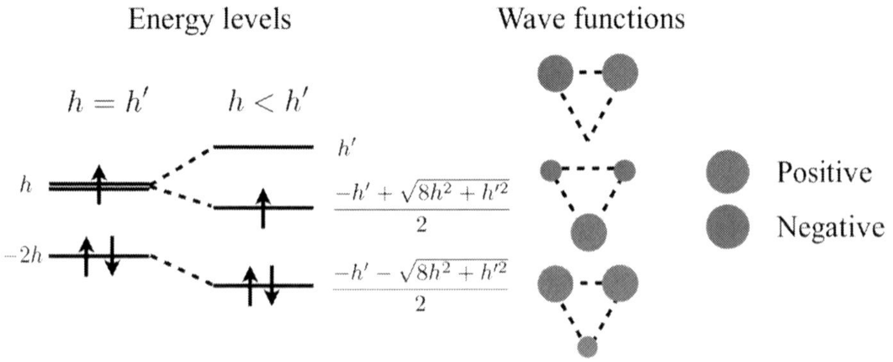

Figure 1. Energy levels and the corresponding wave functions of a cluster model for localized states of sp^2 orbitals around a point defect in graphene.

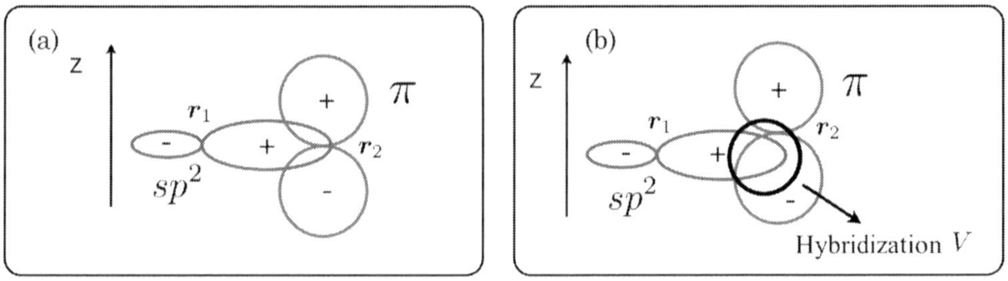

Figure 2. Schematic picture of sp^2 (blue) and π (red) orbitals on the different carbon atoms at $\boldsymbol{r}_1$ and $\boldsymbol{r}_2$. (a) Planar structure. (b) In the presence of out-of-plane displacement of the carbon atom.

where $a_{i\sigma}$ is an annihilation operator of a π electron at site i with spin $\sigma = \uparrow, \downarrow$ and d_σ is an annihilation operator of an electron on the sp^2 orbital at site $i = 1$. V is the amplitude of hybridization. Taking account of the localized nature of the active sp^2 orbital in the defect, we assume that the effective Hamiltonian for this orbital becomes

$$H_{\rm def} = \sum_\sigma (\epsilon_{sp^2} - \mu) d_\sigma^\dagger d_\sigma + U d_\uparrow^\dagger d_\uparrow d_\downarrow^\dagger d_\downarrow, \tag{4}$$

where ϵ_{sp^2}, μ, and U are the energy level, chemical potential, and Coulomb interaction on the sp^2 orbital, respectively. The Coulomb interaction term is essential for describing the Kondo effect.

The conduction electron states are formed by π orbitals in graphene, which are described by the following tight-binding model with a nearest-neighbor transfer integral $t (> 0)$,

$$H_{\rm gra} = -t \sum_{\langle ij \rangle \sigma}{}' \left(a_{i\sigma}^\dagger b_{j\sigma} + {\rm h.c.} \right) - \mu N, \tag{5}$$

where $b_{j\sigma}$ is an annihilation operator of a π electron at site j of the sublattice B, and N is the total number of π electrons. In eq. (3), only the π orbitals of the sublattice A, $a_{i\sigma}$, appear since we have assumed the missing carbon atom to be on the sublattice B. The prime in eq. (5)

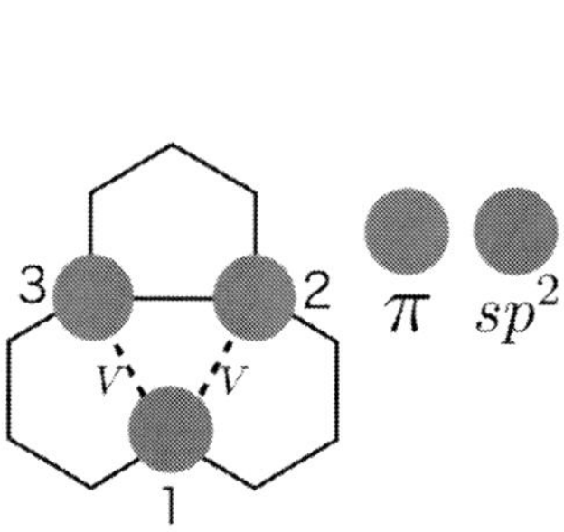

Figure 3. Schematic picture of the electronic state around the defect. Two of the three sp^2 orbitals ($i = 2, 3$) form a covalent bond. The circle at the site 1 shows the active sp^2 orbital and those at the sites 2 and 3 the π orbitals. V is the amplitude of the hybridization between the active sp^2 orbital and the two π orbitals.

denotes the presence of the point defect. The point defect in π electrons is represented by the zero transfer integrals between the missing site and the neighboring site. Here, we ignore the Coulomb interaction in the π orbitals. Although band structure may be modified by the Coulomb interaction in the bulk, the Dirac dispersion is preserved and we assume that eq. (5) represents such renormalized Dirac fermions. Finally, the effective Hamiltonian of graphene with the point defect becomes $H = H_{\rm gra} + H_{\rm def} + H_{\rm hyb}$.

2.2. One-dimensional representation of the conduction electron states

To clarify the conduction electron states which are coupled with the sp^2 orbital at the defect, and to apply the NRG method, we transform this effective Hamiltonian (3), (4), and (5) into a one-dimensional representation. In the usual Kondo problems, a magnetic impurity interacts with conduction electrons at the same site. In this case, the conduction electron states are expanded by the spherical waves, and states with particular symmetries such as s, p, or d wave symmetries are coupled with the impurity. For example, when the interaction between the impurity and the conduction electrons is spherical, only the states with s wave symmetry is coupled with the impurity. Such states are labeled by the wave vectors in the radial direction. As a result, a one-dimensional representation is obtained. However, in the present model, there are two centers: the sp^2 orbital at the defect interacts with the π orbitals of sites 2 and 3. Such a geometry has been treated in the two-impurity Kondo problem [32]. We follow the method developed for it [32, 33, 34], generalizing the method to include the effects of defect scattering on the π conduction electrons.

To express the hybridization term (3) in a one-dimensional representation, we introduce the following operator for π electrons [33, 34]:

$$a^\dagger_{\varepsilon i\sigma} = \sum_k c^\dagger_{k\sigma} \langle k|Ai\rangle \delta\left(\varepsilon - \varepsilon_k\right), \tag{6}$$

where ε_k and $|k\rangle$ are the energy eigenvalues and the wave functions of the Hamiltonian (5), and $c^\dagger_{k\sigma}$ is the corresponding creation operator with spin σ. $|Ai\rangle$ represents the π electron state localized at the site $i(= 2$ or $3)$ of sublattice A. $\delta(\varepsilon)$ is the Dirac delta function. These operators are related to the original one $a_{i\sigma}$ by

$$\int {\rm d}\varepsilon a_{\varepsilon i\sigma} = a_{i\sigma}. \tag{7}$$

The anti-commutation relations of these operators are calculated as $\left\{a_{\varepsilon i\sigma}, a^\dagger_{\varepsilon' j\sigma}\right\} =$

$\delta\left(\varepsilon-\varepsilon'\right)g_{ij}(\varepsilon)$, $\left\{a_{\varepsilon i\sigma}, a_{\varepsilon' j\sigma}\right\}=0$, and $\left\{a^{\dagger}_{\varepsilon i\sigma}, a^{\dagger}_{\varepsilon' j\sigma}\right\}=0$, where $g_{ij}(\varepsilon)$ is introduced as

$$g_{ij}(\varepsilon) = \sum_k \langle k|Ai\rangle\langle Aj|k\rangle\delta\left(\varepsilon-\varepsilon_k\right). \tag{8}$$

The diagonal element $g_{ii}(\varepsilon)$ corresponds to the LDOS at site i of sublattice A. Owning to the twofold symmetry in the present system, the LDOS at the site $i = 2$ and 3 are identical: $g_{22}(\varepsilon) = g_{33}(\varepsilon)$. The off-diagonal elements satisfy $g_{23}(\varepsilon) = g_{32}(\varepsilon)$, since the wave functions can be chosen to be real. Then, from $a_{\varepsilon 2\sigma}$ and $a_{\varepsilon 3\sigma}$, the following operators which satisfy the usual anti-commutation relations are obtained;

$$c_{\varepsilon\sigma e} = \frac{1}{N_e^{1/2}(\varepsilon)}\left(a_{\varepsilon 2\sigma}+a_{\varepsilon 3\sigma}\right), \tag{9}$$

$$c_{\varepsilon\sigma o} = \frac{1}{N_o^{1/2}(\varepsilon)}\left(a_{\varepsilon 2\sigma}-a_{\varepsilon 3\sigma}\right), \tag{10}$$

where e and o denote even and odd, respectively, and

$$\begin{aligned} N_e(\varepsilon) &= g_{22}(\varepsilon)+g_{23}(\varepsilon)+g_{32}(\varepsilon)+g_{33}(\varepsilon), \\ N_o(\varepsilon) &= g_{22}(\varepsilon)-g_{23}(\varepsilon)-g_{32}(\varepsilon)+g_{33}(\varepsilon). \end{aligned} \tag{11}$$

With these operators, the hybridization term (3) is expressed as

$$H_{\mathrm{hyb}} = \sum_\sigma \int \mathrm{d}\varepsilon N_e^{1/2}(\varepsilon)\left(V c^{\dagger}_{\varepsilon\sigma e} d_\sigma + \mathrm{h.c.}\right). \tag{12}$$

Since $N_e(\varepsilon)$ is essentially the LDOS of π electrons at the defect, this expression shows the importance of the LDOS at the defect, not the total DOS. Note that only one channel of the conduction electron $c_{\varepsilon\sigma e}$ hybridizes with the sp^2 orbital. This is because the sp^2 orbital hybridizes equally with the π orbitals at sites 2 and 3 as in Fig. 3. When the hybridization is not equal, both even and odd states are coupled to the sp^2 orbital.

Using eqs. (5), (6), and (9), it can be shown that $c_{\varepsilon\sigma e}$ satisfies the commutation relation $\left[H_{\mathrm{gra}}, c^{\dagger}_{\varepsilon\sigma e}\right] = (\varepsilon-\mu)c^{\dagger}_{\varepsilon\sigma e}$. Thus, neglecting the other channels of the conduction electrons, we assume that the conduction electron term (5) can be written as [31-33]

$$H_{\mathrm{gra}} = \sum_\sigma \int \mathrm{d}\varepsilon\left(\varepsilon-\mu\right) c^{\dagger}_{\varepsilon\sigma e} c_{\varepsilon\sigma e}. \tag{13}$$

Since decoupled channels of conduction electrons are neglected, this model can be used only when the defect contribution is concerned. The transformation above has also been justified in ref. [35] by integrating out the conduction electron degrees of freedom in the functional integral representation.

3. Role of the localized state of π electrons at the defect: numerical renormalization group study

In this section, we discuss the effect of the localized state of π electrons on the Kondo effect. The localized state of π electrons causes a peak in the LDOS on the defect as a function of energy at the charge neutrality point. To take into account this peak of the LDOS, we assume a model shown in Fig. 4,

$$N_e(\varepsilon) = |\varepsilon| + \frac{\lambda}{2w}\mathrm{e}^{-|\varepsilon|/w}, \tag{14}$$

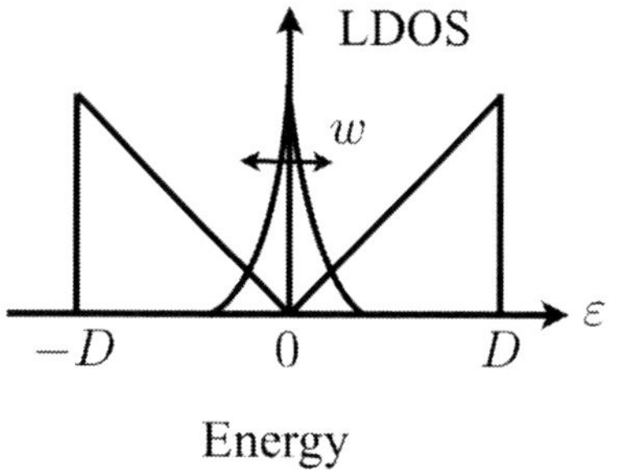

Figure 4. Local density of states of conduction electrons in the model in §3 as a function of energy ε, where D is a cut-off energy. A peak due to a localized state of π electrons is included, where w is a width of the peak.

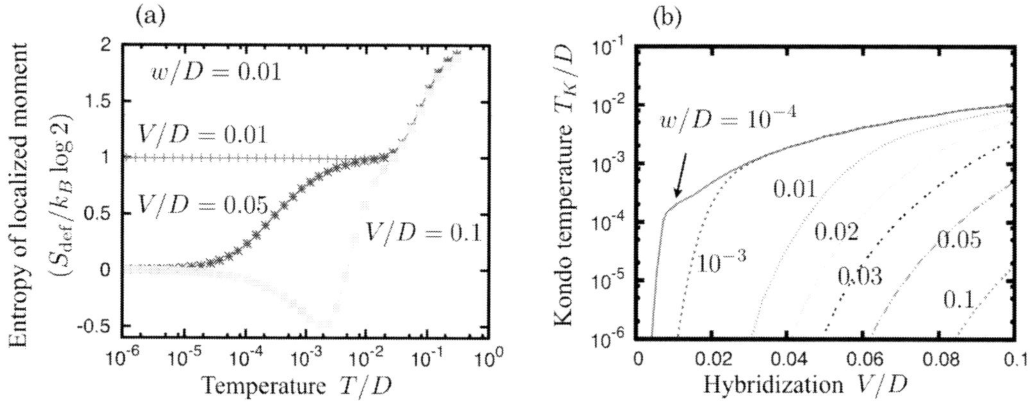

Figure 5. (a) Temperature dependence of entropy of the electron on the defect sp^2 orbital, $S_{\rm def}/\log 2$, in the presence of a peak in the local density of states of π electrons with the width $w/D = 0.01$ at the chemical potential $\mu = 0$, for the hybridization $V/D = 0.01$, 0.05, and 0.1. (b) Kondo temperature T_K/D as a function of hybridization V/D for several widths w/D. [Reprinted from Taro Kanao, Doctor Thesis, Department of physics, Graduate school of science, The university of Tokyo, Tokyo (2014).]

where λ and w are the intensity and the width of the peak, respectively. The second term becomes a Dirac delta function in the limit $w \to 0$. In the case of low density of defects, the LDOS at the defect is a Dirac delta function with zero width. On the other hand, in the case of finite concentration of defects, the LDOS can have a finite width owning to overlaps of the wave functions of the localized states on different defects [38, 39]. Assuming finite widths of the LDOS, we analyze this model by the NRG method. The $\mu = 0$ case is studied, where the effect of the peak of the LDOS is most significant.

Figure 5(a) shows a temperature dependence of entropy of the electron on the defect sp^2 orbital, $S_{\rm def}/k_B \log 2$, for $V/D = 0.01$, 0.05, and 0.1 in the presence of finite peak width $w/D = 0.01$. In this NRG calculation, the parameter of the logarithmic intervals, Λ, is set to be $\Lambda = 2$, and 300 states are retained in every step of diagonalization.

It is found that the low-energy state ($\sim 10^{-6}D$) changes from the unscreened localized

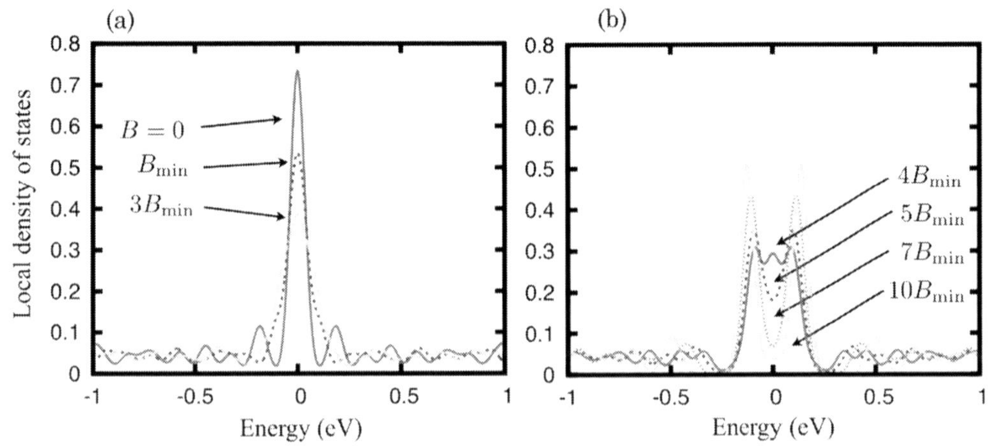

Figure 6. Local density of states of the π electrons at a point defect as a function of energy in the magnetic filed of (a) $B/B_{\rm min} = 0$, 1, and 3 and (b) $B/B_{\rm min} = 4$, 5, 7, and 10. $B_{\rm min}$ is determined by the system size. In the present calculation, $B_{\rm min} \simeq 21.2$ T. [Reprinted from Taro Kanao, Doctor Thesis, Department of physics, Graduate school of science, The university of Tokyo, Tokyo (2014).]

moment state ($S_{\rm def}/k_B \log 2 = 1$) to the Kondo-Yosida singlet state ($S_{\rm def}/k_B \log 2 = 0$) with increasing hybridization from $V/D = 0.01$ to 0.05. Since D is about 8 eV, V is of the order of 0.1 eV. Namely, this Kondo effect can occur with a realistic smaller value of hybridization between the sp^2 orbital at the defect and the π electrons.

In the case of $V/D = 0.1$, $S_{\rm def}$ is negative in a certain range of temperature. Note that $S_{\rm def}$ is defined as the deference between the entropy in the presence of defect and that without defect. This is due to the large hybridization comparing with the band width. The similar behavior has been known in the narrow-band Anderson model [40].

In order to understand the importance of localized π orbital in detail, we also discuss the width of localized π orbital dependence of the Kondo temperature. In Fig. 5(b), the Kondo temperatures T_K/D as a function of V/D for several w/D are shown. It is found that the Kondo temperature is sensitive to the width: With smaller w/D, T_K becomes larger owning to the larger LDOS at the Fermi energy.

4. Effects of magnetic field

As mentioned in the introduction, the negative magnetoresistance has been observed experimentally in this Kondo effect [8]. In the usual Kondo effect, the Zeeman effect on the localized magnetic moment causes a negative magnetoresistance, because it suppresses the formation of the Kondo-Yosida singlet. The Kondo-Yosida singlet is suppressed when the energy scales of the magnetic field and the Kondo temperature become the same order, which leads to a characteristic magnetic field in the negative magnetoresistance. In the Kondo effect in graphene with defects, however, the experimentally observed characteristic magnetic field is smaller than the energy scale of its Kondo temperature by one order of magnitude. This means that this Kondo effect (Kondo temperature) is more sensitive to the magnetic field than the usual one. However, the origin of this behavior has not been understood.

In the previous section, we have shown that the Kondo screening is assisted by the peak of the LDOS of π electrons which is caused by the defect scattering. Since the Kondo temperature sensitively depends on the height of the peak, small modulation of the localized π orbital by the magnetic field will cause appreciable change in the Kondo screening. Thus, to clarify the modulation due to the orbital motion of the π electrons in the magnetic field, as a first step, we solve the nearest-neighbor tight-binding model for the π states of graphene in the magnetic field with a single point defect. In this case, the Hamiltonian is given as

$$H'_{\rm gra} = -t \sum_{\langle ij \rangle}{}' \left(e^{i\theta_{ij}} a_i^\dagger b_j + \text{h.c.} \right), \tag{15}$$

where $'$ denotes the presence of the defect, and a_i and b_j are annihilation operators of a π electron at site i and site j on graphene, and θ_{ij} is a Peierls phase with the "string gauge" [41]. Under the periodic boundary condition, the possible magnetic fields are restricted to $B = nB_{\rm min}$ with $n = 0, 1, 2, \cdots$, where $B_{\rm min}$ is determined by the system size. In the present calculation, 61^2 unit cells are used, which corresponds to $B_{\rm min} \simeq 21.1$ T.

Figures 6 show the LDOS at the defect as a function of energy in several magnetic fields. We find that the height of the peak decreases and the width increases for relatively small B as shown in Fig. 6(a). At higher magnetic fields, the peak splits into two peaks as shown in Fig. 6(b). These results show that the localized π orbital becomes broad and the LDOS of the Dirac point (the Fermi level) decreases by the orbital magnetic field. Since the Kondo temperature is strongly sensitive to the width of the localized π orbital, it is expected that the Kondo temperature decreases drastically by the orbital magnetic field.

5. Summary

In conclusion, we discussed the role of localized π orbital on the defect-induced Kondo effect in graphene on the basis of a numerical renormalization group study. We found that the localized π orbital assists the Kondo effect due to the sp^2 orbital, and the Kondo temperature is sensitive to the broadening of the localized π orbital. Secondly, in order to clarify the mechanism of the magnetic sensitive Kondo effect, as a first step, we study the effect of the orbital magnetic field on the localized π orbital by the tight-binding model with the Peierls phase. We also found that the LDOS of the localized π orbital at the fermi level decreases as the magnetic field increases. It is suggested the the defect induced Kondo effect is strongly sensitive to the orbital magnetic field.

Acknowledgments
This work has been supported by a Grant-in-Aid for Scientific Research A on "Dirac Electrons in Solids" (No.24244053), and for Scientific Research on Innovative Areas "Ultra Slow Muon Microscope" (No.23108004) from the Ministry of Education, Culture, Sports, Science and Technology, Japan. One of authors (H.M.) is supported by a Grant-in-Aid for Scientific Research from the Japan Society for the Promotion of Science (No.25220803).

References

[1] Kotov V N, Uchoa B, Pereira V M, Guinea F, and Castro Neto A H 2012 *Rev. Mod. Phys.* **84** 1067
[2] Fritz L and Vojta M 2013 *Rep. Prog. Phys.* **76** 032501
[3] Sengupta K and Baskaran G 2008 *Phys. Rev.* B **77** 045417
[4] Vojta M, Fritz L, and Bulla R 2010 *Europhys. Lett.* **90** 27006
[5] Zhu Z G, Ding K H, and Berakdar J 2010 *Europhys. Lett.* **90** 67001
[6] Haase P, Fuchs S, Pruschke T, Ochoa H, and Guinea F 2011 *Phys. Rev.* B **83** 241408
[7] Brar V W, Decker R, Solowan H M, Wang Y, Maserati L, Chan K T, Lee H, Girit C O, Zettl A, Louie S G, Cohen M L, and Crommie M F 2011 *Nat. Phys.* **7** 43
[8] Chen J H, Li L, Cullen W G, Williams E D, and Fuhrer M S 2011 *Nat. Phys.* **7** 53

[9] Kanao T, Matsuura H, and Ogata M 2012 *J. Phys. Soc. Jpn.* **81** 063709
[10] Cazalilla M A, Iucci A, Guinea F, and Castro Neto A H 2012 *Preprint* cond-mat/1207.3135v1
[11] Mitchell A K and Fritz L 2013 *Phys. Rev.* B **88** 075104
[12] Jafari S A and Tohyama T 2014 *J. Phys.:Cond. Matt.* **26** 415601
[13] Kharitonov M and Kotliar G 2013 *Phys. Rev.* B **88** 201103
[14] Kanao T, Matsuura H, and Ogata M 2013 *J. Phys.: Conf. Ser.* **456** 012018
[15] Shirakawa T and Yunoki S 2014 *Phys. Rev.* B **90** 195109
[16] Jobst J and Weber H B 2012 *Nat. Phys.* **8** 352
[17] Chen J H, Li L, Cullen W G, Williams E D, and Fuhrer M S 2012 *Nat. Phys.* **8** 353
[18] Yazyev O V 2010 *Rep. Prog. Phys.* **73** 056501
[19] Esquinazi P, Spemann D, Höhne R, Setzer A, Han K H, and Butz T 2003 *Phys. Rev. Lett.* **91** 227201
[20] Nair R R, Sepioni M, Tsai I L, Lehtinen O, Keinonen J, Krasheninnikov A V, Thomson T, Geim A K, and Grigorieva I V 2012 *Nat. Phys.* **8** 199
[21] Ugeda M M, Brihuega I, Guinea F, and Gómez-Rodríguez J M 2010 *Phys. Rev. Lett.* **104** 096804
[22] Kondo T, Honma Y, Oh J, Machida T, and Nakamura J 2010 *Phys. Rev.* B **82** 153414
[23] Meyer J C, Kisielowski C, Erni R, Rossell M D, Crommie M F, and Zettl A 2008 *Nano Lett.* **8** 3582
[24] El-Barbary A A, Telling R H, Ewels C P, Heggie M I, and Briddon P R 2003 *Phys. Rev.* B **68** 144107
[25] Lehtinen P O, Foster A S, Ma Y, Krasheninnikov A V, and Nieminen R M 2004 *Phys. Rev. Lett.* **93** 187202
[26] Yazyev O V and Helm L 2007 *Phys. Rev.* B **75** 125408
[27] Wakabayashi K 2002 *J. Phys. Soc. Jpn.* **71** 2500
[28] Pereira V M, Guinea F, Lopes dos Santos J M B, Peres N M R, and Castro Neto A H 2006 *Phys. Rev. Lett.* **96** 036801
[29] Kumazaki H and Hirashima D S 2007 *J. Phys. Soc. Jpn.* **76** 034707
[30] Toyoda A and Ando T 2010 *J. Phys. Soc. Jpn.* **79** 094708
[31] Wilson K G 1975 *Rev. Mod. Phys.* **47** 773
[32] Jones B A and Varma C M 1987 *Phys. Rev. Lett.* **58** 843
[33] Sakai O and Shimizu Y 1992 *J. Phys. Soc. Jpn.* **61** 2333
[34] Affleck I, Ludwig A W W, and Jones B A 1995 *Phys. Rev.* B **52** 9528
[35] Bulla R, Pruschke T, and Hewson A C 1997 *J. Phys.: Condens. Matter* **9** 10463
[36] Bulla R, Costi T A, and Pruschke T 2008 *Rev. Mod. Phys.* **80** 395
[37] Gonzalez-Buxton C and Ingersent K 1998 *Phys. Rev.* B **57** 14254
[38] Peres N M R, Guinea F, and Castro Neto A H 2006 *Phys. Rev.* B **73** 125411
[39] Pereira V M, Lopes dos Santos J M B, and Castro Neto A H 2008 *Phys. Rev.* B **77** 115109
[40] Hofstetter W and Kehrein S 1999 *Phys. Rev.* B **59** R12732
[41] Hatsugai Y, Ishibashi K, and Morita Y 1999 *Phys. Rev. Lett.* **83** 2246
[42] Fritz L and Vojta M 2004 *Phys. Rev.* B **70** 214427

Dirac fermions and Kondo effect

Takashi Yanagisawa

Electronics and Photonics Research Institute, National Institute of Advanced Industrial Science and Technology (AIST), Tsukuba Central 2, 1-1-1 Umezono, Tsukuba 305-8568, Japan

E-mail: t-yanagisawa@aist.go.jp

Abstract. In this study, we investigate the Kondo effect induced by the s-d interaction where the conduction bands are occupied by Dirac fermions. The Dirac fermion has the linear dispersion and is described typically by the Hamiltonian such as $H_k = v\mathbf{k}\cdot\sigma + m\sigma_0$ for the wave number $\mathbf{k}$ where σ_j are Pauli matrices and σ_0 is the unit matrix. We derived the formula of the Kondo temperature T_K by means of the Green's function theory for Green's functions including Dirac fermions and the localized spin. The T_K was determined from a singularity of Green's functions in the form $T_K \propto \exp(-\text{const}/\rho|J|)$. The Kondo effect will disappear when the Fermi surface is point like.

1. Introduction

Recently, the Dirac electron in solid state has been investigated intensively [1-6]. It is interesting to examine how the Kondo effect occurs in a system of Dirac electrons with magnetic impurities. It is not so trivial whether the Kondo effect indeed appears there. We expect significant and interesting behaviors when the localized spin interacts with the Dirac electron through the s-d interaction. The Dirac Hamiltonian resembles the s-d model with the spin-orbit coupling of Rashba type [7]. In this paper we investigate the s-d Hamiltonian with Dirac electrons by means of the Green's function theory and evaluate the Kondo temperature T_K.

The Hamiltonian is given by $H = H_0 + H_{sd} \equiv H_m + H_K + H_{sd}$ where

$$H_m = \sum_{\mathbf{k}}(m \quad)(c^\dagger_{\mathbf{k}\uparrow}c_{\mathbf{k}\uparrow} + c^\dagger_{\mathbf{k}\downarrow}c_{\mathbf{k}\downarrow}), \tag{1}$$

$$H_K = \sum_{\mathbf{k}}[v(k_x \quad ik_y)c^\dagger_{\mathbf{k}\uparrow}c_{\mathbf{k}\downarrow} + v(k_x + ik_y)c^\dagger_{\mathbf{k}\downarrow}c_{\mathbf{k}\uparrow} + v_z k_z(c^\dagger_{\mathbf{k}\uparrow}c_{\mathbf{k}\uparrow} \quad c^\dagger_{\mathbf{k}\downarrow}c_{\mathbf{k}\downarrow})] \tag{2}$$

$$H_{sd} = \quad \frac{J}{2}\frac{1}{N}\sum_{\mathbf{k}\mathbf{k}'}[S_z(c^\dagger_{\mathbf{k}\uparrow}c_{\mathbf{k}'\uparrow} \quad c^\dagger_{\mathbf{k}\downarrow}c_{\mathbf{k}'\downarrow}) + S_+c^\dagger_{\mathbf{k}\downarrow}c_{\mathbf{k}'\uparrow} + S \;\, c^\dagger_{\mathbf{k}\uparrow}c_{\mathbf{k}'\downarrow}]. \tag{3}$$

v and v_z are velocities of conduction electrons, is the chemical potential and m is the mass of the Dirac fermion. N indicates the number of sites. We set $m = 0$ so that the chemical potential includes m. $c_{\mathbf{k}\sigma}$ and $c^\dagger_{\mathbf{k}\sigma}$ are annihilation and creation operators, respectively. S_+, S and S_z denote the operators of the localized spin. The term H_{sd} indicates the s-d interaction between the conduction electrons and the localized spin, with the coupling constant J [8, 9]. J is negative, as adopted in this paper, for the antiferromagnetic interaction.

2. Green's Functions

We define thermal Green's functions of the conduction electrons

$$G_{\mathbf{kk}'\sigma}(\ ) = \langle T_\tau c_{\mathbf{k}\sigma}(\ )c^\dagger_{\mathbf{k}'\sigma}(0)\rangle, \tag{4}$$

$$F_{\mathbf{kk}'}(\ ) = \langle T_\tau c_{\mathbf{k}\downarrow}(\ )c^\dagger_{\mathbf{k}'\uparrow}(0)\rangle, \tag{5}$$

where T_τ is the time ordering operator. We note that the spin operators satisfy the following relations:

$$S_\pm S_z = \mp\frac{1}{2}S_z, \quad S_zS_\pm = \ \frac{1}{2}S_\pm, \tag{6}$$

$$S_+S\ = \frac{3}{4} + S_z \quad S_z^2, \tag{7}$$

$$S\ S_+ = \frac{3}{4} \quad S_z \quad S_z^2. \tag{8}$$

We also define Green's functions which include the localized spins as well as the conduction electron operators. They are for example, following the notation of Zubarev [10],

$$\langle\langle S_zc_{\mathbf{k}\uparrow};c^\dagger_{\mathbf{k}'\uparrow}\rangle\rangle_\tau = \langle T_\tau S_zc_{\mathbf{k}\uparrow}(\ )c^\dagger_{\mathbf{k}'\uparrow}(0)\rangle, \tag{9}$$

$$\langle\langle S\ c_{\mathbf{k}\downarrow};c^\dagger_{\mathbf{k}'\uparrow}\rangle\rangle_\tau = \langle T_\tau S\ c_{\mathbf{k}\downarrow}(\ )c^\dagger_{\mathbf{k}'\uparrow}(0)\rangle, \tag{10}$$

$$\langle\langle S_zc_{\mathbf{k}\downarrow};c^\dagger_{\mathbf{k}'\uparrow}\rangle\rangle_\tau = \langle T_\tau S_zc_{\mathbf{k}\downarrow}(\ )c^\dagger_{\mathbf{k}'\uparrow}(0)\rangle, \tag{11}$$

$$\langle\langle S\ c_{\mathbf{k}\uparrow};c^\dagger_{\mathbf{k}'\uparrow}\rangle\rangle_\tau = \langle T_\tau S\ c_{\mathbf{k}\uparrow}(\ )c^\dagger_{\mathbf{k}'\uparrow}(0)\rangle. \tag{12}$$

The Fourier transforms are defined as usual:

$$G_{\mathbf{kk}'\sigma}(\ ) = \frac{1}{\beta}\sum_n e^{\ i\omega_n\tau}G_{\mathbf{kk}'\sigma}(i\omega_n), \tag{13}$$

$$F_{\mathbf{kk}'}(\ ) = \frac{1}{\beta}\sum_n e^{\ i\omega_n\tau}F_{\mathbf{kk}'}(i\omega_n), \tag{14}$$

$$\langle\langle S_zc_{\mathbf{k}\uparrow};c^\dagger_{\mathbf{k}'\uparrow}\rangle\rangle_\tau = \frac{1}{\beta}\sum_n e^{\ i\omega_n\tau}\langle\langle S_zc_{\mathbf{k}\uparrow};c^\dagger_{\mathbf{k}'\uparrow}\rangle\rangle_{i\omega_n}, \tag{15}$$

$$\cdots\cdots.$$

Using the commutation relations the equation of motion for $G_{\mathbf{kk}'\uparrow}(i\omega_n)$ reads

$$\begin{aligned}(i\omega_n \quad v_zk_z \quad m + \ )G_{\mathbf{kk}'\uparrow}(\omega_n) &= \ _{\mathbf{kk}'} \quad \frac{J}{2N}\sum_q \ _{\mathbf{qk}'}(i\omega_n) \\ &+ \ v(k_x \quad ik_y)F_{\mathbf{kk}'}(i\omega_n).\end{aligned} \tag{16}$$

Here we have defined

$$_{\mathbf{kk}'}(\ ) = \frac{1}{\beta}\sum_n e^{\ i\omega_n}\ _{\mathbf{kk}'}(i\omega_n) = \langle\langle S_zc_{\mathbf{k}\uparrow};c^\dagger_{\mathbf{k}'\uparrow}\rangle\rangle_\tau + \langle\langle S\ c_{\mathbf{k}\downarrow};c^\dagger_{\mathbf{k}'\uparrow}\rangle\rangle_\tau. \tag{17}$$

The equation of motion for $F_{kk'} = \langle\langle c_{\mathbf{k}\downarrow};c^\dagger_{\mathbf{k}'\uparrow}\rangle\rangle_{i\omega_n}$ is also obtained in a similar way. By using the decoupling procedure for Green's functions [11-13], we can obtain a closed solution for a set of Green's functions. The Green's function $G_{\mathbf{kk}'\uparrow}(i\omega_n)$ is obtained as

$$G_{\mathbf{kk}'\uparrow}(\omega) = \ _{kk'}G_k^0(\omega) \quad \frac{J}{2N}G_k^0(\omega)\frac{J}{2}\ (\omega)G_k^0(\omega)\frac{1}{1 + JG(\omega) + \ \frac{J}{2}\ ^2\ (\omega)F(\omega)}, \tag{18}$$

where the analytic continuation $i\omega_n \to \omega$ is carried out. We defined the following functions:

$$G^0_k(\omega) = \frac{\omega + }{(\omega + \ )^2 \quad v^2(k_x^2 + k_y^2) \quad v_z^2 k_z^2}, \tag{19}$$

$$F(\omega) = \frac{1}{N}\sum_k G^0_k(\omega), \tag{20}$$

$$(\omega) = \frac{1}{N}\sum_k \quad m_k \quad \frac{3}{4} \quad G^0_k(\omega), \tag{21}$$

$$G(\omega) = \frac{1}{N}\sum_k \quad n_k \quad \frac{1}{2} \quad G^0_k(\omega), \tag{22}$$

with $m_k = 3\sum_q \langle c^\dagger_{q\uparrow} c_{k\downarrow} S \quad \rangle$ and $n_k = \sum_q \langle c^\dagger_{q\uparrow} c_{k\uparrow}\rangle$.

3. Kondo Temperature

From the Green's function $G_{\mathbf{kk}'\uparrow}(i\omega_n)$, the Kondo temperature T_K is determined from a zero of the denominator in this formula. We consider

$$1 + JG(\omega) = 0 \tag{23}$$

in the limit $\omega \to 0$ by neglecting higher-order term being proportional to $(J/2)^2$. Let us adopt for simplicity that $v_z = v$ and then the dispersion is $\xi_k = \quad v\sqrt{k_x^2 + k_y^2 + k_z^2} \qquad$ in three dimensions. We neglect the term of the order of J in $n_k = \langle c^\dagger_{k\uparrow} c_{k\uparrow}\rangle$. The equation $1 + JG(0) = 0$ results in the Kondo temperature T_K given as

$$k_B T_K = \frac{2e^\gamma D}{\quad} \exp\left(\quad 16 \quad {}^2 \frac{v^3}{\quad {}^2} \frac{1}{|J|}\right), \tag{24}$$

where D is a cutoff and γ is Euler's constant. We assumed that $| \quad | \gg k_B T_K$ to derive the above formula. The result shows that the Kondo effect indeed occurs in a Dirac system. We can consider the Kondo effect in two dimensions by setting $v_z = 0$, and also the d-dimensional case in general. In d dimensions T_K reads

$$k_B T_K = \frac{2e^\gamma D}{\quad} \exp\left(\quad \frac{8(2 \quad)^d}{\quad _d} \quad \frac{v}{| \quad |} \quad {}^{d \quad 1} \quad \frac{v}{|J|}\right), \tag{25}$$

where $\quad _d$ is the solid angle in d-dimensional space, namely, the area of the $(d \quad 1)$-sphere $S^{d \quad 1}$.

In the limit $| \quad | \to 0$, the equation $1 + JG(0) = 0$ has no solution. Hence, when $| \quad |$ is small, T_K is reduced and vanishes. This measn that, when the Fermi surface is point like, the Kondo effect never appears because of the weak scattering by the localized spin. T_K shows an algebraic behavior $T_K \propto | \quad | \quad$ with a constant α. In three dimensions, T_K is proportional to $\quad ^2$ for small $| \quad |$:

$$k_B T_K \simeq \frac{1}{\quad {}^2}\sqrt{\frac{14\zeta(3)}{32}} \quad {}^2 \frac{1}{v}\sqrt{\frac{|J|}{v}}, \tag{26}$$

where $\zeta(3)$ is the Riemann zeta function at argument 3. In d-dimensional space, this is generalized as

$$k_B T_K \simeq \frac{1}{\quad}\sqrt{\frac{7\zeta(3) \quad _d}{8(2 \quad)^d}} | \quad |^{(d+1)/2}\sqrt{\frac{|J|}{v^d}}. \tag{27}$$

4. Summary

We investigated the s-d Hamiltonian with the localized spin interacting with the Dirac fermions. The Green's function method is applied to examine a singularity in Green's functions to determine the Kondo temperature T_K. The singularity leads to the formula of T_K as usual being proportional to exp(const/ $|J|$) with the density of states . The Kondo effect indeed occurs in the band of Dirac fermions and will produce singular properties in physical quantities. T_K vanishes when the Fermi surface is point like and shows the algebraic behavior $T_K \propto |\;|$ for small $|\;|$. The term being proportional to $_z$ like the magnetic field, which we did not have considered in this paper, is also important in the Kondo effect.

References

[1] Novoselov K S, Geim A K, Morozov S V, Jiang D, Katsnelson M I, Grigorieva S V, Dubonos S V and Firsov A A 2005 *Nature* **438** 197
[2] Zhang Y, Tan Y-W, Stormer H L and Kim P 2005 *Nature* **438** 201
[3] Hsieh D, Qian D, Wray K, Xia Y, Hor Y S, Cava R J and Hasan M Z 2008 *Nature* **452** 970
[4] Mitchell A K, Schuricht D, Vojta M and Fritz L 2013 *Phys. Rev.* B**87** 075430
[5] Beri B and Cooper N R 2012 *Phys. Rev. Lett.* **109** 156803
[6] Kanao T, Matsuura H and Ogata M 2012 *J. Phys. Soc. Jpn.* **81** 063709
[7] Yanagisawa T 2012 *J. Phys. Soc. Jpn.* **81** 094713
[8] Kondo J 1964 *Prog. Theor. Phys.* **32** 37
[9] Kondo J 2012 *The Physics of Dilute Mgnetic Alloys* (Cambridge University Press, Cambridge)
[10] Zubarev D 1960 *Sov. Phys. Uspekhi* **3** 320
[11] Nagaoka Y 1965 *Phys. Rev.* **138** A1112
[12] Hamann D R 1967 *Phys. Rev.* D**158** 570
[13] Zittartz J and Müller-Hartmann E 1968 *Z. Phys.* **212** 380

Local moment formation in Dirac electrons

M Mashkoori[1], **I Mahyaeh**[1], **S A Jafari**[1,2]

[1] Department of Physics, Sharif University of Technology, Tehran 11155-9161, Iran

[2] Center of excellence for Complex Systems and Condensed Matter (CSCM), Sharif University of Technology, Tehran 1458889694, Iran

E-mail: jafari@sharif.edu

Abstract. Elemental bismuth and its compounds host strong spin-orbit interaction which is at the heart of topologically non-trivial alloys based on bismuth. These class of materials are described in terms of 4×4 matrices at each $\vec{k}$ point where spin and orbital labels of the underlying electrons are mixed. In this work we investigate the single impurity Anderson model (SIAM) within a mean field approximation to address the nature of local magnetic moment formation in a generic Dirac Hamiltonian. Despite the spin-mixing in the Hamiltonian, within the Hartree approximation it turns out that the impuritys Green function is diagonal in spin label. In the three dimensional Dirac materials defined over a bandwidth D and spin-orbit parameter γ, that hybridizes with impurity through V, a natural dimensionless parameter $\frac{V^2 D}{2\pi\gamma^3}$ emerges. So neither the hybridization strength, V, nor the spin-orbit coupling γ, but a combination thereof governs the phase diagram. By tuning chemical potential and the impurity level, we present phase diagram for various values of Hubbard U. Numerical results suggest that strong spin-orbit coupling enhances the local moment formation both in terms of its strength and the area of the local moment region. In the case that we tune the chemical potential in a similar way as normal metal we find that magnetic region is confined to $\mu \geq \varepsilon_0$, in sharp contrast to 2D Dirac fermions. If one fixes the chemical potential and tunes the impurity level, phase diagram has two magnetic regions which corresponds to hybridization of impurity level with lower and upper bands.

1. Introduction

The low-energy features of the band structure in some materials are hyperbolic rather than parabolic. Graphene being an extreme example in two dimensions where the effective single-particle energy levels are described by a cone. Other examples in three dimensions have emerged a large class of which are based on elemental bismuth. In these class of materials, the low-energy effective theory around the Fermi level is described in terms of Dirac theory, namely the Hamiltonian at each point is of the form

$$H = v_F\vec{\alpha}.\vec{p} + \beta m \tag{1}$$

where matrices α_i, β square to unit matrix and anti-commute among themselves. The above properties constrain the matrices to be even-dimensional. In three space dimension where we need three matrices α_i, $i = 1, 2, 3$, one needs at least 4×4 matrices to satisfy the above algebra.

It was shown by Wolff [1] that the effective theory of the electronic states of bismuth near the L point of the Brillouin zone can be described by a Dirac theory. Within the isotropic approximation this equatin can be written as [2]

$$H_B = \sum_{\vec{k}} \Psi^{\dagger}_{\vec{k}} \begin{pmatrix} \Delta & i\gamma\vec{k}.\sigma \\ -i\gamma\vec{k}.\sigma & -\Delta \end{pmatrix} \Psi_{\vec{k}}, \tag{2}$$

where the Hamiltonian matrix corresponds to the Dirac representation of the gamma matrices and the four component spinor is defined as $\Psi^{\dagger}_{\vec{k}} = \left(c^{\dagger}_{\vec{k}\uparrow}, c^{\dagger}_{\vec{k}\downarrow}, d^{\dagger}_{\vec{k}\uparrow}, d^{\dagger}_{\vec{k}\downarrow}\right)$. The above Hamiltonian describes two bands of positive and negative energy states corresponding to $c^{\dagger}_{k\sigma}$ and $d^{\dagger}_{k\sigma}$ operators. Obviously $\vec{k}$ is the Bloch momentum and σ is the spin component. 2Δ is the gap which in the case of bismuth is approximately15.4 meV [3, 4] and $\gamma = \hbar v$ comes from spin-orbit interaction and sets the velocity scale [4], with $v/c \approx 10^{-3}$. The spectrum of this model is:

$$\varepsilon_k = \pm\sqrt{\gamma^2 k^2 + \Delta^2}. \tag{3}$$

In this paper we take the Dirac Hamiltonian (2) as the continuum of host material states, and couple it to an isolated impurity level within the single impurity Anderson model (SIAM) to investigate the nature of local moment formation in Dirac materials.

2. Anderson impurity in a Dirac host

Single impurity Anderson model for a host of Dirac fermions is given by,

$$H = H_{\rm B} + H_{\rm imp} + H_{\rm hyb}, \tag{4}$$

where $H_{\rm imp}$ is given by:

$$H_{\rm imp} = \sum_{\sigma} \varepsilon_0 f^{\dagger}_{\sigma} f_{\sigma} + U n_{\uparrow} n_{\downarrow}, \tag{5}$$

where $f^{\dagger}$ represents creation operator in impurity orbital, $n_{\sigma} = f^{\dagger}_{\sigma} f_{\sigma}$, the energy ε_0 is the impurity energy level and U is the on-site Coulomb repulsion. The hybridization between continuum states of the host and the impurity level is constructed as follows: The two bands of positive and negative energy states in the three dimensional Dirac Hamiltonian essentially comes from the combination of orbitals of two neighboring atoms in a unit cell. Therefore the even (or equivalently odd) combination of positive and negative energy states forms a state localized on lattice site. Hence we assume a local hybridization of the following form:

$$H_{hyb} = \frac{1}{\sqrt{\mathcal{V}}} \sum_{\vec{k}} [V(c^{\dagger}_{\vec{k}\sigma} + d^{\dagger}_{\vec{k}\sigma}) f_{\sigma} + V^* f^{\dagger}_{\sigma}(c_{\vec{k}\sigma} + d_{\vec{k}\sigma})]. \tag{6}$$

where V is hybridization strength between impurity level and Bloch state and $\mathcal{V}$ is the volume of the host material.

The local component of the retarded Green function is defined using the standard notation as follows:

$$\langle\langle f_{\sigma}(t)|f^{\dagger}_{\sigma'}(t')\rangle\rangle = -i\theta(t-t')\langle\{f_{\sigma}(t), f^{\dagger}_{\sigma'}(t')\}\rangle. \tag{7}$$

Writing the equation of motion for the above Green function in frequency domain gives,

$$(\omega - \varepsilon_0)\langle\langle f_{\sigma}|f^{\dagger}_{\sigma'}\rangle\rangle = \delta_{\sigma\sigma'} + U\langle\langle f_{\sigma} n_{\sigma}|f^{\dagger}_{\sigma'}\rangle\rangle + \sum_{\vec{k}} \frac{V^*}{\sqrt{\mathcal{V}}}\langle\langle b_{\vec{k}\sigma}|f^{\dagger}_{\sigma'}\rangle\rangle, \tag{8}$$

in which we define $b_{\vec{k}\sigma} \equiv c_{\vec{k}\sigma} + d_{\vec{k}\sigma}$. Applying the Hartree approximation to Coulomb term amounts to the replacement $\varepsilon_0 \to \varepsilon_\sigma = \varepsilon_0 + U\langle n_{\bar{\sigma}}\rangle$. To close the equation of motion we write the equation of motion for $\langle\langle b_{\vec{k}\sigma}|f^\dagger_{\sigma'}\rangle\rangle$ that gives,

$$\langle\langle b_{\vec{k}\sigma}|f^\dagger_{\sigma'}\rangle\rangle = \frac{V}{\sqrt{\mathcal{V}}}\frac{2\omega}{\omega^2 - \varepsilon_k^2}\langle\langle f_\sigma|f^\dagger_{\sigma'}\rangle\rangle. \tag{9}$$

This result shows that within Hartree mean-field approximation, the Green function reads:

$$(\omega - \varepsilon_\sigma - \Sigma_f(\omega))\langle\langle f_\sigma|f^\dagger_{\sigma'}\rangle\rangle = \delta_{\sigma\sigma'}, \tag{10}$$

where the self-energy at the Hartree level is:

$$\Sigma_f(\omega) = \frac{|V|^2}{\mathcal{V}}\sum_{\vec{k}}\frac{2\omega}{\omega^2 - \varepsilon_k^2}. \tag{11}$$

Equation (10) indicates that within the present Hartree decoupling, in spite of spin-mixing in the Hamiltonian, the impurity's Green function remains diagonal in spin label; or equivalently the spin-flip correlation function is zero, i.e. $\langle\langle f_\sigma|f^\dagger_{\bar{\sigma}}\rangle\rangle = 0$. Doing the summation over $\vec{k}$ in Eq. (10), the self-energy as a function ω becomes:

$$\Sigma(\omega) = \frac{\tilde{v}}{\pi D}\sqrt{\omega^2 - \Delta^2}\Big[2\omega\ln\Big(\frac{\sqrt{D^2 - \Delta^2} + \sqrt{\omega^2 - \Delta^2}}{\sqrt{D^2 - \Delta^2} - \sqrt{\omega^2 - \Delta^2}}\Big) - \frac{4\omega\sqrt{D^2 - \Delta^2}}{\sqrt{\omega^2 - \Delta^2}} - i\pi|\omega|\Big]. \tag{12}$$

In above equation D is the bandwidth. In contrast to normal metal where the dimensionless parameter $\frac{\pi V^2}{DU}$ emerges in Hartree approximation, here in Eq. (12) $\tilde{v} \equiv \frac{|V|^2 D}{2\pi\gamma^3}$ emerges naturally. We have to emphasize that in previous study of SIAM in two dimensional Dirac material (graphene), the conventional dimensionless parameters $Y = \frac{\mu - \varepsilon_0}{U}$ and $X = \frac{\pi V^2}{DU}$ did not appears naturally, and they were employed based on analogy with normal metallic hosts. There the phase diagram was plotted in terms of these conventional parameters Y and X to contrast the effect of the Dirac nature of charge carriers on local moment formation and to contrast it to normal metals. As we will see in this work, the emergence of the above natural parameter leads to qualitative differences in the nature of local moment formation both compared to two dimensional ones, as well as the normal metals. Given this natural parameter, neither the hybridization strength, V, nor the spin-orbit coupling, γ is the major player, but this peculiar combination of the parameters determines the phase diagram. This helps us to study the effect of spin-orbit interaction on local moment formation. In what follows, we will calculate the occupation of spin $\uparrow$ and $\downarrow$ at zero temperature. The occupation of impurity site for spin $\uparrow$ is given as an integral over spectral function, which it depends on spin $\downarrow$ occupation. As a result the occupation and magnitude of local moment is calculated self-consistently;

$$\langle n_\uparrow\rangle = -\frac{1}{\pi}\int_{-\infty}^{\mu} d\omega\frac{Im\Sigma_f}{(Z^{-1}(\omega)\omega - \varepsilon_\downarrow)^2 + Im(\Sigma_f)^2}, \tag{13}$$

where the renormalization factor is:

$$Z^{-1}(\omega) = 1 - \frac{2\tilde{v}}{\pi D}\sqrt{\omega^2 - \Delta^2}\left[\ln\Big(\frac{\sqrt{D^2 - \Delta^2} + \sqrt{\omega^2 - \Delta^2}}{\sqrt{D^2 - \Delta^2} - \sqrt{\omega^2 - \Delta^2}}\Big) - \frac{2\sqrt{D^2 - \Delta^2}}{\sqrt{\omega^2 - \Delta^2}}\right]. \tag{14}$$

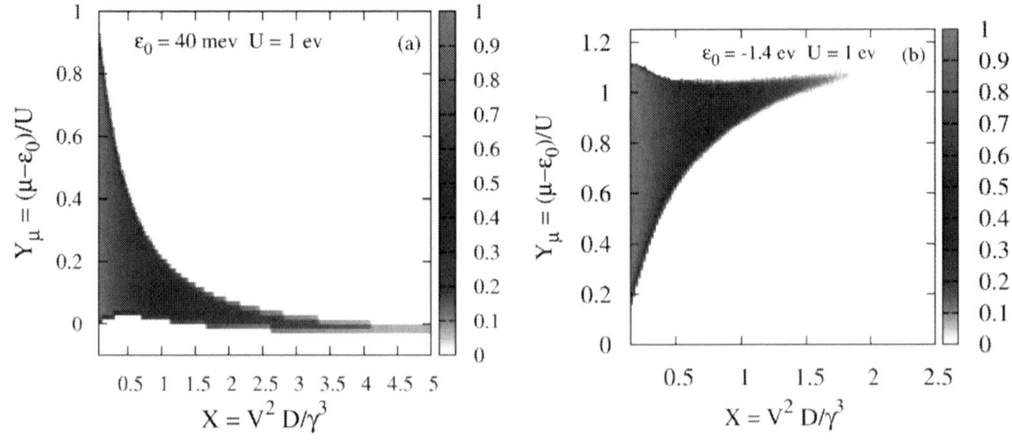

Figure 1. Phase diagram of SIAM in three dimensional Dirac matter. The intensity plot shows the magnitude of local moments. The horizontal axis represents dimensionless parameter X and the vertical axis shows effectively chemical potential energy that has been expressed as a dimensionless parameter $Y = \frac{\mu-\varepsilon_0}{U}$. The subscript μ in Y_μ indicates that we have varied μ to achieve variations of Y. In this plot Coulomb repulsion equals 1 eV, and the bandwidth D is assumed to be 2 eV. The impurity energy level is set (a) $\frac{\varepsilon_0}{D} = 0.02$ and (b) $\frac{\varepsilon_0}{D} = -0.7$. As indicated in the legend, the red color indicates fully spin polarized impurity level and white color represents the non-magnetic state. As can be seen, larger γ corresponds to larger local moments.

3. Numerical results

We solve Eq. (13) along with a similar equation obtained by replacing $\uparrow \to \downarrow$ self-consistently. Local magnetic moment is characterized by $n_\uparrow - n_\downarrow \neq 0$ the intensity plots of which are presented. We construct the phase diagram of the magnetic moments in three dimensional Dirac systems in the plane of variables $X = \tilde{v} = |V|^2 D/(2\pi\gamma^3)$ and $Y = \frac{\mu-\varepsilon_0}{U}$. The first dimensionless parameter is naturally driven from the dynamics of the SIAM, while the second dimensionless parameter is usually used in the literature following Anderson's work on local magnetic states in metals. Since in this case the variable Y does not naturally arise from the dynamics of SIAM, μ and ε_0 will be two independent variables, tuning of which gives rise to qualitatively different regions for the local magnetic states. Let us independently tune them in the sequel.

3.1. Tuning μ

Fig. 1-a shows the phase diagram for an impurity with impurity level at energy $\frac{\varepsilon_0}{D} = 0.02$ and Hubbard repulsion $U = 0.5D = 1$ eV. Note that the phase diagram which is obtained by tuning parameters μ and X does not have the particle-hole symmetry around $Y = 0.5$, as in the two-band case, particle-hole symmetry holds when in addition to $Y = 0.5$, the chemical potential μ is also at the mid-gap in which case there would be no conduction in the system. In addition, the magnetic region is limited to $\mu \geq \varepsilon_0$. In this respect the local magnetic states in 3D massive Dirac materials are akin to those in normal metals. However, for 2D massless Dirac fermions in graphene the magnetic region extends to $\mu < \varepsilon_0$. This can be understood by the broadening of impurity level which asymptotically behaves like $\frac{1}{\omega^2}$. This means that the spectral function of local impurities embedded in the Wolff Hamiltonian decreases fast enough to avoid magnetic moments for $Y < 0$ region as happens for the local moments in normal metals.

Fig. 1-b is the same as Fig. 1-a, but with a different value of $\varepsilon_0 = -1.4$ eV. In both figures it can be seen that smaller values of X that correspond to larger values of γ give rise to larger local magnetic moments. The parameter γ is controlled by the spin-orbit coupling, and also sets the velocity scale of the Dirac fermions. It is therefore curious to note that faster 3D Dirac fermions when hybridized with a local impurity level give rise to larger magnetic moments. Also it is noteworthy to mention that the sign of ε_0 determines whether the magnetic region bends upward ($\varepsilon_0 < 0$) or downward ($\varepsilon_0 > 0$) as one moves towards larger X values. In this respect the region of magnetic states in graphene also displays similar behavior, but in graphene the boundaries are round, and also extend to negative Y axis [5].

3.2. Tuning ε_0

Due to the experimental results on dielectric constant of bismuth and its compounds, which indicate a relative permeability about 10-40, the Coulomb repulsion could be substantially decreased [6]. In addition regarding the Haldane-Anderson mechanism for impurity in semiconductors, reduction of the charge accumulation on impurity site leads to a fall in on-site Coulomb repulsion up to 2 orders of magnitude [7, 8]. Therefore, we investigate the problem for smaller values of Coulomb interaction U. Since μ and ε_0 in the SIAM for Dirac materials are

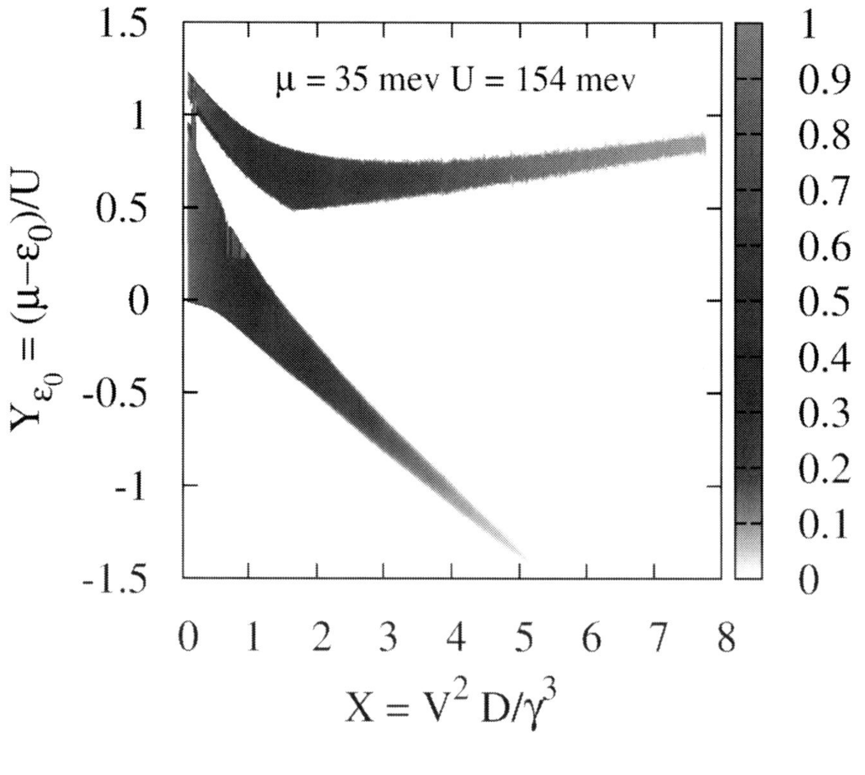

Figure 2. Phase diagram of SIAM in three dimensional massive Dirac matter. The area enclosed by the curves and the Y-axis is magnetic region. The variables are defined by $X = \frac{|V|^2 D}{2\pi\gamma^3}$ and $Y = \frac{(\mu-\varepsilon_0)}{U}$. The Coulomb repulsion and chemical potential are $U = 1$ eV and $\mu = 35$ meV.

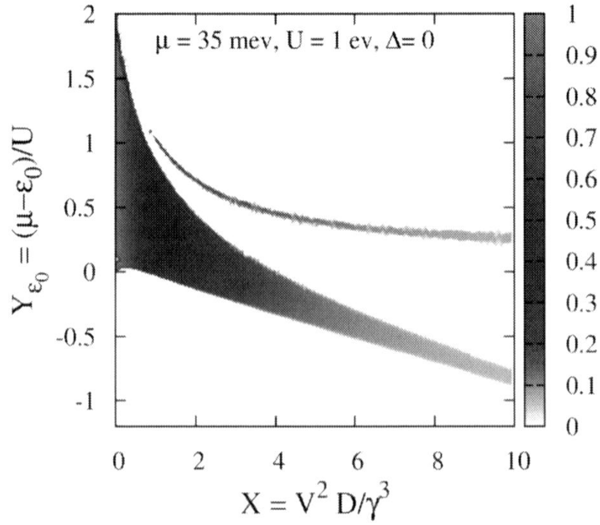

Figure 3. Phase diagram of SIAM in massless Dirac matter. The area enclosed by the curves and the Y-axis is magnetic region. The variables are defined by $X = \frac{|V|^2 D}{2\pi\gamma^3}$ and $Y = \frac{(\mu-\varepsilon_0)}{U}$. The Coulomb repulsion and chemical potential are $U = 1$ eV and $\mu = 35$ meV.

independent parameters, we choose to fix $\mu = 35$ meV which is relevant to bismuth and tune ε_0 in order to construct a phase diagram in the plane of X and Y. In general, for smaller values of Hubbard U one expects narrower magnetic region as double occupancy becomes less costly. When we tune ε_0 for a fixed μ, curiously the local moment region of the phase diagram splits into two pieces shown in Fig. 2. The two pieces correspond to the hybridization of impurity level with negative and positive energy states of the Dirac continuum. Saying it in another way, the upper magnetic region comes from the situation where the impurity level lies within the valence band ($\varepsilon_0 < -\Delta$) and the lower magnetic region arises from the case where the impurity level is above the gap ($\varepsilon_0 > \Delta$). As indicated in the legend, the red color stands for fully spin polarized impurity level and the white color represents the non-magnetic state.

3.3. Massless($\Delta = 0$)

The 3D Dirac fermions in materials such as Na_3Bi or $Bi_{1-x}Sn_x$ for $x \approx 0.04$ [9] become massless. Therefore it is interesting to check what happens to the local magnetic moments when the gap parameter of the host Dirac material vanishes. The phase diagram of local magnetic states in a host of massless Dirac fermions is depicted in Fig 3. The elbow-like upper branch of the magnetic region in the massless case is narrowed and smoothened.

Acknowledgments

SAJ thanks Professor T. Tohyama for hospitality at the Tokyo university of science (TUS).

References

[1] Wolff P A 1964 *J. Phys. Chem. Sol.***25(10)** 1057-68
[2] Fuseya Y, Ogata M, Fukuyama H 2015 J. Phys. Soc. Jpn. **84** 012001
[3] Smith G E, Baraff G A and Rowell J M 1964 *Phys. Rev.* **135(4A)**A1118
[4] Zengwei Z, Fauqué B, Fuseya Y and Behnia K 2011 *Phys. Rev. B* **84** 115137
[5] Uchoa B, Kotov V, Peres N and Castro Neto A 2008 *Phy. Rev. Lett.* **101** 026805
[6] Young K F and Frederikse H P R 1973 *J. Phys. Chem. Ref. Dat.* **2** 313-410
[7] Haldane F D M and Anderson P W 1976 *Phys. Rev B* **13**
[8] Sato K, Bergqvist L, Kudrnovsky J, Dederichs P H, Eriksson O, Turek I, Sanyal B, Bouzerar G, Katayama-Yoshida H, Dinh V A, Fukushima T, Kizaki H and Zeller R 2010 *Rev. Mod. Phys* **82**1633-90
[9] Liu Y and Allen R 1995 *Phys. Rev. B* **52** 1566-77

Intrinsic broadening of the energy levels of Dirac electrons in quantum dots

E Ahmadi[1,2], **S A Jafari**[1,3,4]

[1] Department of Physics, Sharif University of Technology, Tehran 11155-9161, Iran
[2] Novin Teyf Co., Tehran 14458-33834, Iran
[3] Center of excellence for Complex Systems and Condensed Matter (CSCM), Sharif University of Technology, Tehran 1458889694, Iran
[4] School of Physics, Institute for Research in Fundamental Sciences (IPM), Tehran 19395-5531, Iran

E-mail: jafari@physics.sharif.edu

Abstract. We theoretically study the quasi bound state of Dirac electrons in cylindrically symmetric quantum dots with sharp boundary. According to the existing picture, due to Klein tunneling "relativistic electrons" can not be localized by any confinement potential. We show however that despite of Klein tunneling, interference effects can cause the trapping of electron in quantum dots. Considering the quasi bound state as the state with complex energy, to find the energy of this state we solve the wave-equation with outgoing boundary condition at infinity. The imaginary part of complex energy determines the trapping time of electron within the quantum dot. We show that for any finite confining potential corresponding to any set of quantum numbers (n, m) where n is the principal quantum number and m the magnetic quantum number, there exists a *continuous band of states* with finite life time. Upper and lower edges of each band corresponds to *infinitely long lived states trapped inside and outside the wall of the same radius*. We term this phenomenon the *intrinsic broadening* as it is not caused by scattering from any external potential, nor by many-body effects. This broadening appears to arise from a combination of relativistic and interference effects. The imaginary part of energy which is different for energies along the energy band is controlled by the orbital angular momentum of electron and the depth of the confining potential.

1. Introduction

There is a widespread picture according to which confinement of Dirac electrons in electrostatic potential, e. g. quantum dots, is impossible due to the Klein tunneling [1-2]. Therefore instead, focus has been placed on formation of confined states within quantum dots by application of perpendicular magnetic fields [3-7] or on engineering specialized devices which introduce mass terms [8-12]. This picture limits the application of quantum dots in 2D structures with Dirac energy dispersion to trapping (not localizing) Dirac electrons for long time. If the trapping time is longer than the time required for corresponding application, then one considers the electron in such confinement potential as "practically" localized electron [13-15].

In this work we find that despite the Klein tunneling, interference effects in a cylindrically symmetric quantum dot can cause infinitely long trapping time for electrons. For such long lived states, the trapping of electrons does not depend on the angular momentum and height of electrostatic potential. The quantum dot considered in this paper has cylindrically symmetric confinement potential for 2D Dirac electrons with sharp boundaries. We study the long lived states as quasi localized states by solving the time independent wave-equation with outgoing wave boundary condition at infinity and corresponding to complex energies. The imaginary part of the energy determines the width of broadening of energy state and trapping time of the electron within the quantum dot. If E is the complex energy of localized state, then the trapping time is $\tau = \hbar/Im(E)$ where $Im(E)$ is imaginary part of the electron energy. For non-relativistic quantum dots the confining potential leads to quantization of energy, that in our case is characterized by a radial principal quantum number n. The existing picture of the relativistic quantum dots based on graphene is that, still a discrete energy levels can be labeled by some index n, but it acquires an imaginary part. To find a solution whose energy has an imaginary part is to match the wave-function at the boundary and search for solutions in the complex E plane. But the root-fining problem in complex plane is not as easy as the corresponding problem on the real axis. We revisit the problem of a relativistic quantum dot and search for complex energies by scanning a fine mesh of energies in the complex plane. We find that in addition to discrete energies with non-zero imaginary part, there are continuum of states centered around this solution. By moving away from this solution in both positive and negative real energy directions, we find that there are other solutions continuously connected to it whose imaginary part decreases in both directions and approach the real energy axis at two particular values that correspond to roots of Bessel functions that are regular at origin and infinity, respectively. This means that instead of a single discrete level we are dealing with a continuum of solutions that can be long lived at these two particular energies corresponding to relativistic particles confined inside and outside an infinitely deep well of the same radius.

Therefore the picture emerging from the present work is that, in the relativistic quantum dots based on graphene both the quasi bound states due to Klein tunneling and bound states in their usual quantum mechanical sense can co-exist.

2. Theoretical calculation

Let us first of all define what we mean by a quantum dot (QD) of Dirac electrons. The Hamiltonian of this system has a kinetic energy that is given by the Dirac Hamiltonian everywhere in the space, but the potential term is different for the interior ($\rho < R$) and exterior ($\rho > R$) regions of the space in a cylindrical geometry:

$$V(\mathbf{r}) = \begin{cases} 0 & \rho < R \\ V_0 & \rho > R \end{cases}, \qquad (1)$$

where potential height $V_0 > 0$, and R is the radius of the quantum dot. This shape of confinement potential allows us to obtain analytical expressions for the wave-functions and facilitates the numerical calculation of the trapping time of the states of the quantum dot. The confinement potential in Eq. (1) has sharp boundaries and behavior of electron should be related to the wave function interference effects within the whole region of quantum dot. The full Hamiltonian is given by,

$$H = v_F(\tau\sigma_x p_x + \sigma_y p_y) + V(\mathbf{r}), \qquad (2)$$

where $\tau = \pm 1$ stands for valley index, σ_j denotes Pauli matrices, p_j denotes the momentum operator that in the coordinate representation will be $-i\hbar\partial_j$. $v_F = 3ta/2$ is the Fermi velocity that is given in terms of the hopping parameter t and carbon-carbon bond length a.

The wave function Ψ corresponding to Hamiltonian (2) is a two component function. Here the two components correspond to two different sub-lattices of 2D honeycomb lattice. For cylindrically symmetric confinement potential the two component wave function has the following form:

$$\Psi = \begin{pmatrix} \Phi_A(\rho)e^{im\varphi} \\ \Phi_B(\rho)e^{i(m+\tau)\varphi} \end{pmatrix} \qquad (3)$$

where (ρ, φ) are cylindrical coordinates and m is an integer denoting the orbital angular momentum. With this form of wave function the wave-equation corresponding to the Hamiltonian in Eq. (2) is:

$$-i\tau\frac{d\Phi_B}{d\rho} - \frac{i(m+\tau)}{\rho}\Phi_B = (\varepsilon - \nu_0)\Phi_A, \tag{4}$$

$$-i\tau\frac{d\Phi_A}{d\rho} - \frac{im}{\rho}\Phi_A = (\varepsilon - \nu_0)\Phi_B. \tag{5}$$

We need to solve these equations to find energy E. Therefore we consider two boundary conditions: (1) Both ϕ_A and ϕ_B should be finite at $\rho = 0$, and (2) we have outgoing wave form $\propto e^{ik\rho}$ at infinity. The momentum k in general can be complex, and depends on the principal quantum number n and angular quantum number m. The angular parts of the two components are given in Eq. (3). The regular solution of equations (4) and (5) inside the quantum dot ($V_0 = 0$) is Bessel function of order $|\tau m|$ for ϕ_A and order $|\tau(m+1)|$ for ϕ_B. The general solution can be written as:

$$\begin{pmatrix} \phi_A(\rho) \\ \phi_B(\rho) \end{pmatrix} = A \begin{pmatrix} J_{|-\tau m|}(\varepsilon\rho/R) \\ iJ_{|-\tau(m+1)|}(\varepsilon\rho/R) \end{pmatrix}, \tag{6}$$

where J_m is m'th order of Bessel function and $\varepsilon = RE/at$ is dimensionless energy. Outside the quantum dot, the solutions are Hankel function of the first kind which satisfy the outgoing boundary condition at infinity. For outside of quantum dot with $\rho > R$ and $V_0 > 0$ one can write:

$$\begin{pmatrix} \phi_A(\rho) \\ \phi_B(\rho) \end{pmatrix} = C \begin{pmatrix} H^{(1)}_{|-\tau m|}((\varepsilon - \nu_0)\rho/R) \\ iH^{(1)}_{|-\tau(m+1)|}((\varepsilon - \nu_0)\rho/R) \end{pmatrix}, \tag{7}$$

where $H^{(1)}_n$ is n'th order Hankel function of first kind and $\nu_0 = RV_0/at$ is dimensionless confinement potential.

At the boundary of quantum dot the two component wave function must be continuous:

$$\frac{H^{(1)}_{|-\tau m|}(\varepsilon - \nu_0)}{H^{(1)}_{|-\tau(m+1)|}(\varepsilon - \nu_0)} = \frac{J_{|-\tau m|}(\varepsilon)}{J_{|-\tau(m+1)|}(\varepsilon)}. \tag{8}$$

From this seemingly simple equation, we can find the complex energy spectrum of graphene dot. The imaginary part of energy determines the broadening of energy state and hence the trapping of electron within quantum dot. Eq. (8) is invariant under $m \to -m$ and $\tau \to -\tau$ which means that energy spectrum is identical for an electron with a given angular momentum in a given valley and an electron in the opposite valley and opposite angular momentum. This is a different way of saying that the system is time-reversal invariant. Let us therefore focus on the positive values of m and τ.

3. Numerical Results and Discussion

To obtain energy spectrum of a Dirac electron in the quantum dot we solve the eigenvalue equation (8) for a given angular momentum and valley index. The roots of this equation belong to the complex energy plane. *The key observation is that the root finding problem in complex plane is not as straightforward as the root finding on the real axis.* Therefore it is not straightforward to apply standard algorithms that are developed to search for roots on the real axis, and expect them to work in the complex plane too. Therefore here we adopt a simple and conservative, but slightly expensive method of scanning the whole complex plane of energy in order to make sure

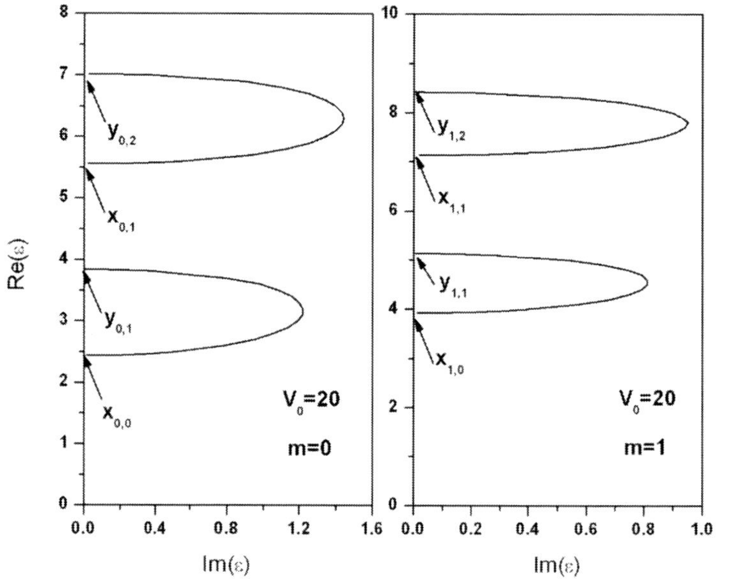

Figure 1. Two lowest energy bands of Dirac electrons in quantum dot for two angular momenta in the complex energy plane corresponding to $\nu_0 = 20$ and $\tau = 1$.

no solutions are missed. Smaller imaginary times correspond to longer trapping time and the zero imaginary time correspond to infinitely long trapping time, i.e. non-decaying state.

In Fig. 1 we show results of our calculations far from the energies near the confinement potential edge and for $\tau = 1$ and lowest values of m. As we can see from these data, instead of a single discrete point there is a continuum of solutions for the energy eigenvalue with different imaginary values. This is almost a "band" of energies which is quite counterintuitive for a quantum dot of finite size. The various "bands" corresponding to different choices of (n, m) are however separated from one another band by energy gap. Within each band every energy state has different value of imaginary part which determines the broadening of state and trapping time of electron in the quantum dot. Even more interesting are the upper and lower edged of the bands that are denoted by $x_{m,n}$ and $y_{m,n}$ in this figure: These two values are n'th roots of the Bessel functions of order m that are localized respectively inside and outside the QD. For electrons precisely tuned to these energies, one can realize the exact localization, as the imaginary part of energy vanishes at these energies. As can be seen in Fig. 1, by moving away from these "band edges" deep inside the band, the energy states at the middle of the energy bands have the typical imaginary part of the order 1 (in dimensionless units). Therefore, the trapping time for a quantum dot with R=50nm, is $\tau = \hbar R/at \sim 10^{-13}s$. This is a relatively short time and we should consider the electrons in such energy states as weakly localized in quantum dot. Using the ordinary root finding methods one can only capture the state at the middle of the present band with maximal imaginary time, while a brute force scanning captures a band of energies.

To get a feeling for the way the roots $x_{m,n}$ of the mth order Bessel function can possibly be a solution, let us examine if there is any solution around $\varepsilon = \nu_0$ where ν_0 is the dimensionless

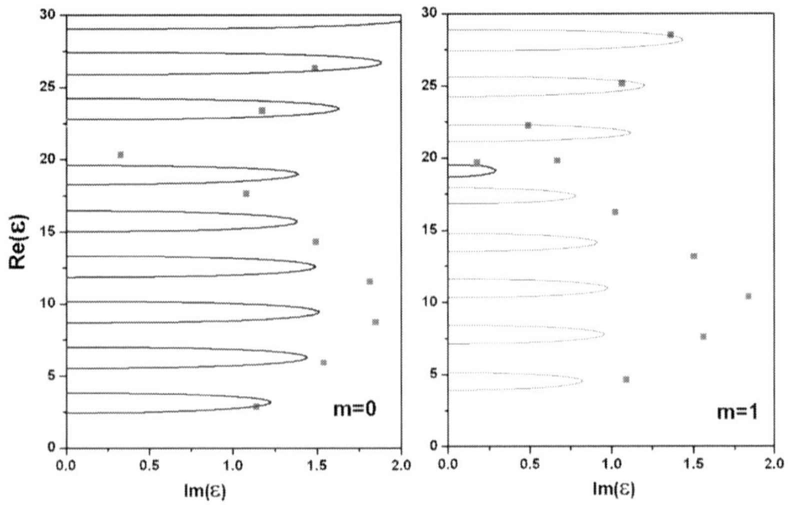

Figure 2. The energy spectrum of Dirac electrons in quantum dot for two angular momenta in complex energy plane for $\nu_0 = 20$ and $\tau = 1$.

confining potential? In this case the Eq. (8) becomes,

$$\frac{H^{(1)}_{|-\tau m|}(0)}{H^{(1)}_{|-\tau(m+1)|}(0)} = \frac{J_{|-\tau m|}(\nu_0)}{J_{|-\tau(m+1)|}(\nu_0)} \tag{9}$$

The left hand side of this equation is zero, so that $J_{|-\tau m|}(\nu_0) = 0 \quad \Rightarrow \quad \nu_0 = x_{m,n}$. As pointed out the $x_{m,n}$ is n'th root of Bessel function J_m. However, the imaginary part of the energy at this particular state is zero, which implies that the electron at this energy is exactly localized in quantum dot. This result is consistent with results of Ref. [13] where it was shown that the electron can be strongly trapped in quantum dot when the energy of electron and confinement potential satisfy special conditions.

In Fig. 2 we give a comparison of our numerical results for the energy spectrum with that of Ref. [13]. The results of this reference are denoted with a square dot, while our results are denoted with continuous solid line. As can be seen the brute force method of scanning the complex plane for finding the solutions captures at some energy eigenvalues agrees with those obtained in Ref. [13], while it also captures a continuum of solutions. This figure shows that Dirac electrons energy spectrum consist of energy bands due to the relativistic properties and interference effect in the quantum dots. In addition to quasi-bound states with finite imaginary time there are also long lived bound states possible at the edges of these bands.

It is also possible to slightly move away from this (real) root and see the qualitative behavior of the complex eigenvalues of energy as the real part deviates from the particular solution at

$\varepsilon = \nu_0$. For deviation $\delta\varepsilon$ from $\varepsilon = \nu_0$ solution the eigenvalue condition becomes,

$$\frac{H_m^{(1)}(\delta\varepsilon)}{H_{m+1}^{(1)}(\delta\varepsilon)} = \frac{J_m(x_{m,n} + \delta\varepsilon + \delta\nu)}{J_{m+1}(x_{m,n} + \delta\varepsilon + \delta\nu)} \tag{10}$$

where $\delta\varepsilon = \varepsilon - \nu_0$, $\delta\nu = \nu_0 - x_{m,n}$ (for $m > 0$ and $\tau = 1$). Taylor expansion gives,

$$\begin{cases} -\delta\varepsilon \ln \delta\varepsilon + i\frac{\pi}{2}\delta\varepsilon + (\delta\varepsilon + \delta\nu) = 0 & m = 0, \tau = 1, \\ \frac{\delta\varepsilon}{2m} + i\frac{\pi\delta\varepsilon^{2m+1}}{2^{2m+1}(m!)^2} + (\delta\varepsilon + \delta\nu) = 0 & m > 0, \tau = 1, \end{cases} \tag{11}$$

where we used the relation $J_m'(x_{m,n}) = -J_{m+1}(x_{m,n})$. By solving this equation we obtain real and imaginary parts of energy as:

$$\begin{cases} \delta\varepsilon = \frac{\delta\nu}{\ln \delta\nu} + i\frac{\pi}{2}\frac{\delta\nu}{\ln \delta\nu} & m = 0, \tau = 1, \\ \delta\varepsilon = -(\frac{2m}{2m+1})\delta\nu + i\frac{\pi}{2^{2m+1}(m!)^2}(\frac{2m}{2m+1})^{2(m+1)}\delta\nu & m > 0, \tau = 1. \end{cases} \tag{12}$$

Now we can find the intrinsic broadening of energy bans for Dirac electrons in quantum dot as:

$$\begin{cases} 1/\tau \propto \frac{\pi}{2}\frac{\delta\nu}{\ln \delta\nu} & m = 0, \tau = 1, \\ 1/\tau \propto \frac{\pi}{2^{2m+1}(m!)^2}(\frac{2m}{2m+1})^{2(m+1)}\delta\nu & m > 0, \tau = 1 \end{cases} \tag{13}$$

This equation shows that broadening of energy levels depends on the orbital angular momentum of electron, index of quasi bound state and the depth of confinement potential at least for a one particular energy band developed from the $\varepsilon = \nu_0$ infinitely long lived state.

4. Conclusion

Spectrum of relativistic Dirac electrons in a quantum well instead of single isolated complex energy points, consists of bands of complex energies. Fine tunneling the energy one could realize a discrete set of non-leaking states that are not able to make use of Klein-tunneling to escape the QD. This opens of the possibility to reconsider Dirac quantum dots for applications in various electronic devices and nano-structures.

5. References

[1] Abergel D S L, Apalkov V, Berashevich J, Ziegler K and Chakraborty T 2010 *Advances in Physics* **59** 261-482
[2] Downing C A, Stone D A and Portnoi M E 2011 *Phys. Rev. B* **84** 155437
[3] Masir M R, Matulis A and Peeters F M 2009 *Phys. Rev. B* **79** 155451
[4] Roy P, Ghosh T and Bhattacharya K 2012 *J. of Physics: Condens. Matter* **24** 055301
[5] Matulis A 2012 *Lithuanian Journal of Physics* **52** 126141
[6] M Masir P Vasilopoulos F M Peeters 2011 *J. of Physics: Condens. Matter* **23** 315301
[7] Park P S, Kim S and Yang S R E 2010 *Journal of Physics: Condens. Matter* **22** 375302
[8] Masir M R, Matulis A and Peeters F M 2011 *Phys. Rev. B* **84** 245413
[9] Giavaras G and Nori F 2011 *Phys. Rev. B* **83** 165427
[10] Giavaras G and Nori F 2012 *Phys. Rev. B* **85** 165446
[11] Pal G, Apel W and Schweitzer L 2011 *Phys. Rev. B* **84** 075446
[12] Pieper A, Heinisch R L, Wellein G and Fehske H 2014 *Phys. Rev. B* **89** 165121
[13] Hewageegana P and Apalkov V 2008 *Phys. Rev. B* **77** 245426
[14] Hewageegana P and Apalkov V 2009 *Phys. Rev. B* **79** 115418
[15] Roslyak O, Gumbs G and Mukamel S 2012 *J. Chem. Phys.* **136** 194106

International Workshop on Dirac Electrons in Solids 2015 IOP Publishing
Journal of Physics: Conference Series **603** (2015) 012017 doi:10.1088/1742-6596/603/1/012017

Weyl semimetal with strong long-range Coulomb interactions

Akihiko Sekine and Kentaro Nomura

Institute for Materials Research, Tohoku University, Sendai 980-8577, Japan

E-mail: sekine@imr.tohoku.ac.jp

Abstract. We theoretically study the stability of a Weyl semimetal against strong long-range Coulomb interactions. We consider a lattice model for a time-reversal symmetry broken Weyl semimetal with two nodes, and take into account the $1/r$ Coulomb interactions between the bulk electrons. Based on the U(1) lattice gauge theory, we investigate the system from the strong coupling limit. It is shown that spatial inversion (parity) symmetry of the system is spontaneously broken in the strong coupling limit, and a Weyl semimetal with broken time-reversal and inversion symmetries, which is different from the noninteracting one, survives in the strong coupling limit.

1. Introduction

Weyl semimetals have attracted much attention recently as a novel topological phase of matter. Weyl semimetals have three-dimensional (3D) gapless linear dispersions near the band touching points, the Weyl nodes. The low-energy effective Hamiltonian near the Weyl nodes is described by the Weyl Hamiltonian $\mathcal{H}(\boldsymbol{k}) = \pm v_{\mathrm{F}}\boldsymbol{k}\cdot\boldsymbol{\sigma}$ with $\pm$ denoting the chirality, v_{F} being the Fermi velocity, and σ_i $(i = 1, 2, 3)$ being the Pauli matrices [1, 2]. Breaking of at least time-reversal or spatial inversion symmetry is required to realize Weyl semimetal phases [3, 4]. Topological nature of Weyl semimetals is understood by the fact that each Weyl node can be regarded as a "monopole" with its charge given by chirality [5], and by the anomalous Hall effect originating from the theta term [6, 7, 8]. Further, single Weyl node cannot be gapped out, since all the three Pauli matrices are used. The bulk energy gap opens only if pairs of Weyl nodes with opposite chirality meet and annihilate each other. From this topological property, it is expected that Weyl semimetals are stable against perturbations.

In this paper, we focus on the stability of a time-reversal symmetry broken Weyl semimetal with two nodes against strong $1/r$ long-range Coulomb interactions. This is because screening effects are expected to be weak in Dirac fermion systems due to the vanishing density of states near the Fermi level. As a powerful non-perturbative method which enables us to treat strong $1/r$ Coulomb interactions properly, we employ the U(1) lattice gauge theory. Based on the U(1) lattice gauge theory with the mean-field approximation, we analyze the system from the strong coupling limit.

2. Theoretical Model

Let us consider a lattice model that describes a time-reversal symmetry broken Weyl semimetal with two nodes separated in momentum space. Theoretically, such a situation can be realized by

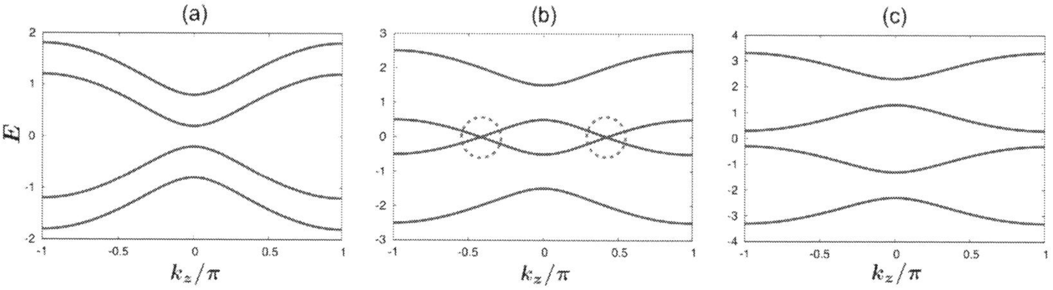

Figure 1. k_3 dependence of Eq. (3) with $k_1 = k_2 = 0$, $m_0 = -0.5$ and $r = 1$ in the case of (a) $b = 0.3$, (b) $b = 1.0$, and (c) $b = 1.8$. When the strength of b is intermediate, the two bands near the Fermi level $E_\mathrm{F} = 0$ touch at two points (Weyl nodes). Note that we have set $v_\mathrm{F} = 1$.

the case where magnetic impurities are doped to a 3D topological insulator. The noninteracting single-particle Hamiltonian is given by

$$\mathcal{H}_0(\boldsymbol{k}) = v_\mathrm{F} \sum_{j=1}^{3} \alpha_j \sin k_j + m(\boldsymbol{k})\alpha_4 + b\Sigma_3, \tag{1}$$

where v_F is the Fermi velocity, and $m(\boldsymbol{k}) = m_0 + r\sum_{j=1}^{3}(1 - \cos k_j)$. The Hamiltonian (1) with $b = 0$ and $0 > m_0/r > -2$ is known as the effective Hamiltonian for 3D topological insulators such as Bi_2Se_3 [9]. The matrices α_μ and Σ_3 are given by the Dirac representation:

$$\alpha_j = \begin{bmatrix} 0 & \sigma_j \\ \sigma_j & 0 \end{bmatrix}, \qquad \alpha_4 = \begin{bmatrix} 1 & 0 \\ 0 & -1 \end{bmatrix}, \qquad \Sigma_3 = \begin{bmatrix} \sigma_3 & 0 \\ 0 & \sigma_3 \end{bmatrix} \tag{2}$$

with σ_j being the Pauli matrices. The spinor of the Hamiltonian (1) is written in the basis of $\psi_{\boldsymbol{k}} = [c_{\boldsymbol{k}+\uparrow}, c_{\boldsymbol{k}+\downarrow}, c_{\boldsymbol{k}-\uparrow}, c_{\boldsymbol{k}-\downarrow}]^T$, where c is the annihilation operator of an electron, $+, -$ denote two orbitals, and $\uparrow$ ($\downarrow$) denotes up-(down-)spin. Then we see that the term $b\Sigma_3$ describes a ferromagnetic coupling between the bulk electrons and magnetic impurities [10]. The energy eigenvalues of the Hamiltonian (1) is easily obtained as

$$E(\boldsymbol{k}) = \pm\sqrt{v_\mathrm{F}^2(\sin k_1^2 + \sin k_2^2) + \left[\sqrt{v_\mathrm{F}^2 \sin k_3^2 + m(\boldsymbol{k})^2} \pm b\right]^2}. \tag{3}$$

The Weyl nodes appear where the wave vector $\boldsymbol{k}$ satisfies the condition $b^2 = [m_0 + r\sum_j(1 - \cos k_j)]^2 + v_\mathrm{F}\sin^2 k_3$ and $\sin k_1 = \sin k_2 = 0$. k_1 and k_2 can take the value 0 or π. In this study, we set $|m_0|$ small to restrict the existence of the Weyl nodes on the $(k_1, k_2) = (0, 0)$ line. The k_3 dependence of the energy spectrum (3) for some values of b is shown in Fig. 1.

Next we take into account the $1/r$ long-range Coulomb interactions between the bulk electrons, by introducing a scalar potential A_0. This procedure enables us to describe the system by the U(1) lattice gauge theory. The Euclidean action of the system is then given by [11]

$$S = S_F^{(0)} + (m_0 + 3r + r_\tau)\sum_n \bar{\psi}_n\psi_n + b\sum_n \bar{\psi}_n\gamma_0\Sigma_3\psi_n + S_G, \tag{4}$$

where $S_F^{(0)}$ is the fermionic part without the mass term

$$S_F^{(0)} = -\sum_n \left[\bar{\psi}_n P_0^- U_{n,0}\psi_{n+\hat{0}} + \bar{\psi}_{n+\hat{0}} P_0^+ U_{n,0}^\dagger \psi_n\right] - \sum_{n,j}\left[\bar{\psi}_n P_j^- \psi_{n+\hat{j}} + \bar{\psi}_{n+\hat{j}} P_j^+ \psi_n\right], \tag{5}$$

and S_G is the pure U(1) gauge part

$$S_G = \beta \sum_n \sum_{\mu>\nu} \left[1 - \frac{1}{2} \left(U_{n,\mu} U_{n+\hat{\mu},\nu} U^\dagger_{n+\hat{\nu},\mu} U^\dagger_{n,\nu} + \text{H.c.} \right) \right]. \tag{6}$$

Here $\bar{\psi} = \psi^\dagger \gamma_0$, $n = (n_0, n_1, n_2, n_3)$ denotes a spacetime lattice site on a 4D hypercubic lattice, $\hat{\mu}$ denotes the unit vector along the μ direction, and $P^\pm_\mu = (r_\mu \pm \gamma_\mu)/2$ with $r_0 = r_\tau$ and $r_1 = r_2 = r_3 = r$. $U_{n,\mu}$ are the U(1) link variables with $U_{n,j} = 1$ and $U_{n,0} = e^{iA_{0,n}}$ $(-\pi \le A_{0,n} \le \pi)$. Note that we have rewritten the scalar potential $A_{0,n}$ as $eA_{0,n}/v_\text{F} \to A_{0,n}$ with e the electric charge. Taking the noninteracting limit ($A_{0,n} \to 0$) in the action (4), we obtain the Hamiltonian (1). r_τ is introduced to eliminate fermion doublers resulting from the lattice regularization of the timelike action.

The parameter β which represents the strength of $1/r$ Coulomb interactions is given by

$$\beta = \frac{v_\text{F}\epsilon}{e^2} = \frac{v_\text{F}\epsilon}{4\pi c\alpha}, \tag{7}$$

where ϵ is the dielectric constant of the system, c is the speed of light in vacuum, and $\alpha(\simeq 1/137)$ is the fine-structure constant. Here note that $v_\text{F}/c \sim 10^{-3}$ in condensed matter. Namely, small Fermi velocity makes the $1/r$ Coulomb interactions effectively strong. In the following, we consider the case with $\beta \ll 1$, i.e., the case with small dielectric constant.

3. Possible Orders in the Strong Coupling Limit

In this section, we derive the effective action in the strong coupling limit $\beta = 0$, and then consider the possible orders in the strong coupling limit. The effective action up to arbitrary order in β can be obtained by integrating out the variable $U_{n,0}$ as

$$Z = \int \mathcal{D}[\psi, \bar{\psi}, U_0] e^{-S} = \int \mathcal{D}[\psi, \bar{\psi}] e^{-S_\text{eff}}. \tag{8}$$

In the strong coupling limit $\beta = 0$, we only have to evaluate the integral $\int \mathcal{D}U_0 e^{-S_F^{(0)}} = \prod_n \int_{-\pi}^{\pi} \frac{dA_{0,n}}{2\pi} e^{-S_F^{(0)}}$, since $U_{n,0}$ is contained only in $S_F^{(0)}$. After a calculation, we obtain the effective action in the strong coupling limit given by

$$\begin{aligned} S_\text{eff} = &- \sum_{n,j} \left[\bar{\psi}_n P_j^- \psi_{n+\hat{j}} + \bar{\psi}_{n+\hat{j}} P_j^+ \psi_n \right] + (m_0 + 3r + r_\tau) \sum_n \bar{\psi}_n \psi_n + b \sum_n \bar{\psi}_n \gamma_0 \Sigma_3 \psi_n \\ &+ \sum_n \text{tr} \left[N_n P_0^+ N_{n+\hat{0}} P_0^- \right], \end{aligned} \tag{9}$$

where we have defined $(N_n)_{\alpha\beta} = \bar{\psi}_{n,\alpha} \psi_{n,\beta}$. The subscripts α and β denote the component of the spinor.

Next we decouple the interaction term [the last term in Eq. (9)] to fermion bilinear form with the use of the mean-field approximation. We apply the extended Hubbard-Stratonovich transformation [12] to the trace of the product of two arbitrary matrices A and B:

$$e^{\kappa \text{tr} AB} = (\text{const.}) \times \int \mathcal{D}[Q, Q'] \exp\left\{ -\kappa \left[Q_{\alpha\beta} Q'_{\alpha\beta} - A_{\alpha\beta} Q_{\beta\alpha} - B^T_{\alpha\beta} Q'_{\beta\alpha} \right] \right\}, \tag{10}$$

where κ is a positive constant, and the superscript T denotes the transpose of a matrix. This integral is approximated by the saddle point values $Q_{\alpha\beta} = \langle B^T \rangle_{\beta\alpha}$ and $Q'_{\alpha\beta} = \langle A \rangle_{\beta\alpha}$. Then we should set $(\kappa, A, B) = (1, N_n P_0^+, -N_{n+\hat{0}} P_0^-)$.

Table 1. Transformation properties under time-reversal $\mathcal{T}$ and spatial inversion (parity) $\mathcal{P}$ of 16 independent matrices consisting of α_μ with $\{\alpha_\mu, \alpha_\nu\} = 2\delta_{\mu\nu}\mathbf{1}$ $(\mu, \nu = 1, 2, 3, 4, 5)$.

Matrices	$\mathbf{1}$	α_j	α_4	α_5	Σ_j	Σ'_j	Π_j	Π_0
Time-reversal $\mathcal{T}$	+	−	+	−	−	−	+	+
Inversion $\mathcal{P}$	+	−	+	−	+	+	−	−

Here let us consider the resulting mean-field Hamiltonian. The mean-field action of the interaction term takes the form $\bar{\psi}_n\langle N_n\rangle\psi_n$. Hence the mean-filed Hamiltonian can be written as

$$\begin{aligned}\mathcal{H}(\boldsymbol{k}) &= \mathcal{H}_0(\boldsymbol{k}) + \gamma_0\langle N_n\rangle \\ &= \alpha_j \sin k_j + m(\boldsymbol{k})\alpha_4 + b\Sigma_3 + \gamma_0\langle N_n\rangle.\end{aligned} \tag{11}$$

Note that $\gamma_0 = \alpha_4$. In Dirac fermion systems, the most general form of $\gamma_0\langle N_n\rangle$ is written in terms of 16 independent matrices which consist of α_μ. Such 16 matrices are given explicitly by $\mathbf{1}$ (the identity matrix), α_μ $(\mu = 1, 2, 3, 4, 5)$, and $\alpha_{\mu\nu} = -\frac{i}{2}[\alpha_\mu, \alpha_\nu]$ $(\mu < \nu)$. The matrices $\mathbf{1}$ and α_μ are trivial. Let us closely look at the 10 matrices $\alpha_{\mu\nu}$. In the Dirac representation, they are written explicitly as [11]

$$\begin{aligned}&\Sigma_j \equiv \alpha_{ik}\epsilon_{ikj} = \begin{bmatrix}\sigma_j & 0\\ 0 & \sigma_j\end{bmatrix}, \qquad \Sigma'_j \equiv \alpha_{j5} = \begin{bmatrix}\sigma_j & 0\\ 0 & -\sigma_j\end{bmatrix},\\ &\Pi_j \equiv \alpha_{j4} = \begin{bmatrix}0 & -i\sigma_j\\ i\sigma_j & 0\end{bmatrix}, \qquad \Pi_0 \equiv \alpha_{45} = \begin{bmatrix}0 & 1\\ 1 & 0\end{bmatrix},\end{aligned} \tag{12}$$

where $j = 1, 2, 3$. Note that these are Hermitian matrices. It is easily shown that Σ_j and Σ'_j are odd under time-reversal but even under parity:

$$\mathcal{T}\Sigma_j\mathcal{T}^{-1} = \mathcal{T}\Sigma'_j\mathcal{T}^{-1} = -1, \qquad \mathcal{P}\Sigma_j\mathcal{P}^{-1} = \mathcal{P}\Sigma'_j\mathcal{P}^{-1} = +1, \tag{13}$$

where the time-reversal and parity operators are given by $\mathcal{T} = \mathbf{1} \otimes (-i\sigma_2)\mathcal{K}$ ($\mathcal{K}$ is a complex conjugation operator) and $\mathcal{P} = \sigma_3 \otimes \mathbf{1}$, respectively. On the other hand, Π_μ are even under time-reversal but odd under parity:

$$\mathcal{T}\Pi_\mu\mathcal{T}^{-1} = +1, \qquad \mathcal{P}\Pi_\mu\mathcal{P}^{-1} = -1. \tag{14}$$

The properties of the 16 matrices are summarized in Table 1.

Let us consider the effects of the term $\gamma_0\langle N_n\rangle$. From Eq. (11), we can see the followings. First, the identity matrix shifts the energy level, and α_j $(j = 1, 2, 3)$ shifts the location of the Weyl nodes. Second, the term Σ_3 is contained in the noninteracting Hamiltonian. As mentioned before, $b\Sigma_3$ is regarded as the ferromagnetic coupling between the bulk electrons and magnetic impurities. Then the terms Σ_j $(j = 1, 2, 3)$ can be neglected. Third, the terms Σ'_j can be regarded as a kind of "antiferromagnetic" order, as is understood from its explicit form. Namely, the spins at the two orbitals point to opposite directions to each other. Here note that the spinor of the Hamiltonian is written in the basis of $\psi_{\boldsymbol{k}} = [c_{\boldsymbol{k}+\uparrow}, c_{\boldsymbol{k}+\downarrow}, c_{\boldsymbol{k}-\uparrow}, c_{\boldsymbol{k}-\downarrow}]^T$. Thus the terms Σ'_j $(j = 1, 2, 3)$ can also be neglected. To summarize, we only have to take into account the terms α_4, α_5, Π_μ $(\mu = 0, 1, 2, 3)$ as the possible orders. Here note that α_4 modifies the mass of Dirac fermions m_0. Namely, the presence of α_4 means the renormalization of the mass of Dirac fermions.

4. Free Energy in the Strong Coupling Limit

To obtain the ground state in the strong coupling limit $\beta = 0$, we need to derive the free energies for each instability at zero temperature. The free energies are derived from the usual formula $\mathcal{F} = -\frac{T}{V}\ln Z$ with T and V being the temperature and volume of the system, respectively. The partition function Z is calculated by the Grassmann integral formula $Z = \int D[\psi,\bar{\psi}]e^{-\bar{\psi}\mathcal{M}\psi} = \det\mathcal{M}$. As was shown in the previous section, we need to obtain the free energies for the α_5 and Π_μ ($\mu = 0,1,2,3$) instabilities. In what follows, we show the explicit expression of the free energy for the Π_1 (and Π_2) instability. Here note that these two instabilities are degenerate, since there is a spin degree of freedom in the xy-plane.

We assume that $\langle N_n\rangle = -\sigma\mathbf{1} + \rho_1\gamma_0\Pi_1$. With the use of the extended Hubbard-Stratonovich transformation [Eq. (10)], the mean-field decoupling of the interaction term is done to be

$$e^{-\sum_n \mathrm{tr}[N_n P_0^+ N_{n+\hat{0}} P_0^-]} \sim \exp\left\{-\sum_n \left[(1-r_\tau^2)\sigma^2 + (1+r_\tau^2)\rho_1^2 + \bar{\psi}_n\Gamma\psi_n\right]\right\}, \tag{15}$$

where $\Gamma = \frac{1}{2}\left[(1-r_\tau^2)\sigma + (\gamma_0\Pi_1)^T(1+r_\tau^2)\rho_1\right]$. Substituting Eq. (15) into Eq. (9) and doing the Fourier transform, we obtain the effective action in terms of σ and ρ_1 as

$$S_{\mathrm{eff}}(\sigma,\rho_1) = \frac{V}{T}\left[(1-r_\tau^2)\sigma^2 + (1+r_\tau^2)\rho_1^2\right] + \sum_k \bar{\psi}_k\mathcal{M}(\boldsymbol{k};\sigma,\rho_1)\psi_k. \tag{16}$$

By calculating the determinant of $\mathcal{M}$, the free energy is obtained as [11]

$$\mathcal{F}(\sigma,\rho_1) = (1-r_\tau^2)\sigma^2 + (1+r_\tau^2)\rho_1^2 - \int_{-\pi}^{\pi}\frac{d^3k}{(2\pi)^3}\ln\left\{\frac{I(\boldsymbol{k},b,\tilde{\rho}_1)I(\boldsymbol{k},-b,-\tilde{\rho}_1) - J(\boldsymbol{k},b,\tilde{\rho}_1)}{[\tilde{m}(\boldsymbol{k}) + r_\tau]^2 - b^2}\right\}, \tag{17}$$

where $\tilde{\rho}_1 = \frac{1}{2}(1+r_\tau^2)\rho_1$ and $\tilde{m}(\boldsymbol{k}) = m_0 + \frac{1}{2}(1-r_\tau^2)\sigma + r\sum_j(1-\cos k_j)$. Two functions $I(\boldsymbol{k},b,\tilde{\rho}_1)$ and $J(\boldsymbol{k},b,\tilde{\rho}_1)$ are given by $I(\boldsymbol{k},b,\tilde{\rho}_1) = [\tilde{m}(\boldsymbol{k}) + r_\tau - b]\times\{[\tilde{m}(\boldsymbol{k}) + r_\tau]^2 + \sin^2 k_3 - b^2\} + [\tilde{m}(\boldsymbol{k}) + r_\tau + b]\times[\sin^2 k_1 + (\sin k_2 + \tilde{\rho}_1)^2]$ and $J(\boldsymbol{k},b,\tilde{\rho}_1) = 4\sin^2 k_3(\{\tilde{\rho}_1[\tilde{m}(\boldsymbol{k}) + r_\tau] + b\sin k_2\}^2 + b^2\sin^2 k_1)$. The ground state is determined by the stationary condition $\partial\mathcal{F}(\sigma,\rho_1)/\partial\sigma = \partial\mathcal{F}(\sigma,\rho_1)/\partial\rho_1 = 0$.

5. Numerical Results

It is found that the α_5, Π_0, and Π_3 instabilities do not appear in the strong coupling limit $\beta = 0$, and that the Π_1 (or Π_2) order is realized as the ground state in the strong coupling limit. Namely, spatial inversion symmetry is spontaneously broken in the strong coupling limit. The b dependences of σ and ρ_1 are shown in Fig. 2(a). It can be seen that the bandgap renormalization by σ and the inversion symmetry breaking by ρ_1 occur simultaneously.

Let us consider the single-particle mean-field Hamiltonian in the phase with the Π_1 order, which is given by

$$\mathcal{H}_{\mathrm{MF}}(\boldsymbol{k}) = \alpha_j\sin k_j + \tilde{m}(\boldsymbol{k})\alpha_4 + b\Sigma_3 + \rho_1\Pi_1. \tag{18}$$

We can obtain the energy spectrum analytically as

$$E(\boldsymbol{k}) = \pm\left\{s^2(\boldsymbol{k}) + [\tilde{m}(\boldsymbol{k})]^2 + b^2 + \rho_1^2 \pm 2\sqrt{[\tilde{m}(\boldsymbol{k})b - \rho_1\sin k_2]^2 + (b^2+\rho_1^2)\sin^2 k_3}\right\}^{1/2}, \tag{19}$$

where $s^2(\boldsymbol{k}) = \sum_{i=1}^3 \sin^2 k_i$. Substituting the obtained values of σ and ρ_1 into Eq. (19), it is found that the Weyl semimetal phase with broken time-reversal and inversion symmetries exists in the strong coupling limit. The Weyl semimetal phase in the strong coupling limit has two Weyl nodes, as in the noninteracting phase. However, due to the inversion symmetry breaking, the two Weyl nodes do not appear on the k_3 axis but appear on the k_2-k_3 plane. The phase diagram in the strong coupling limit is shown in Fig. 2(b). For comparison, the phase diagram in the noninteracting limit ($\beta = \infty$), which is obtained from Eq. (3), is shown in Fig. 2(c).

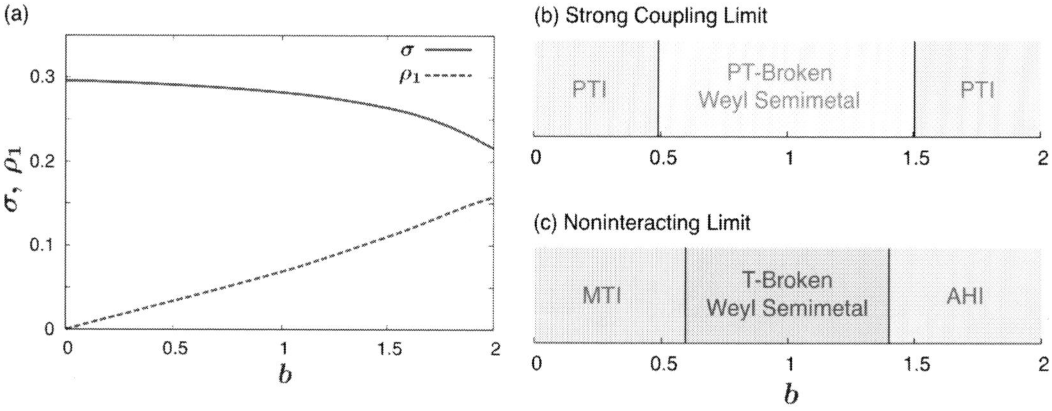

Figure 2. (a) b dependences of σ and ρ_1 in the strong coupling limit $\beta = 0$. (b) Phase diagram in the strong couplig limit. PTI and PT-broken Weyl semimetal represent the insulator with broken time-reversal and spatial inversion (parity) symmetries, and the Weyl semimetal with broken time-reversal and inversion symmetries, respectively. These phases are characterized by nonzero ρ_1. (c) Phase diagram in the noninteracting limit ($\beta = \infty$). MTI, T-broken Weyl semimetal, and AHI represent the magnetic topological insulator, the Weyl semimetal with broken time-reversal symmetry, and the anomalous Hall insulator, respectively. In the above three figures, parameters are set to be $m_0 = -0.6$, $r = 1$ and $r_\tau = 0.5$.

6. Discussions and Summary

Here it should be noted that no symmetry is spontaneously broken in the strong coupling limit when $b = 0$. As mentioned in Sec. 2, the system with $b = 0$ represents 3D topological insulators of Bi_2Se_3 family. Therefore, in this study, it was also shown that 3D topological insulators of Bi_2Se_3 family are also stable against strong $1/r$ Coulomb interactions.

In the noninteracting limit, the phase boundaries where the Weyl semimetal phase appears and disappears are determined by the analytical condition such that $b = |m_0|$ and $b = 2r + m_0$. In the strong coupling limit, we determined the phase boundaries by plotting Eq. (19) numerically, and found that they can be approximated by $b \simeq |m_{\rm eff}|$ and $b \simeq 2r + m_{\rm eff}$ with $m_{\rm eff} = m_0 + \frac{1}{2}(1 - r_\tau^2)\sigma$. Note that $\sigma > 0$ and $r_\tau \leq 1$. The condition $r_\tau \leq 1$ comes from the reflection positivity of lattice gauge theories with Wilson fermions [13]. Then it can be said that the broadening of the Weyl semimetal phase, i.e., the stabilization of the Weyl semimetal phase, in the presence of $1/r$ Coulomb interactions is due to the bandgap renormalization.

In summary, we have studied the effects of strong $1/r$ long-range Coulomb interactions in a time-reversal symmetry broken Weyl semimetal. In the mean-field decoupling process of the interaction term in the strong coupling limit, we have taken into account all the possible 16 instabilities allowed in our model. It was found that spatial inversion (parity) symmetry of the system is spontaneously broken in the strong coupling limit. Nevertheless, the Weyl semimetal phase with broken time-reversal and inversion symmetries, which is different from the noninteracting phase, survives in the strong coupling limit.

Acknowledgments

A.S. is supported by the JSPS Research Fellowship for Young Scientists. This work was supported by Grant-in-Aid for Scientific Research (No. 25103703, No. 26107505 and No. 26400308) from the Ministry of Education, Culture, Sports, Science and Technology, Japan.

Journal of Physics: Conference Series **603** (2015) 012017 doi:10.1088/1742-6596/603/1/012017

References

[1] Volovik G E 2003 *The Universe in a Helium Droplet* (Clarendon, Oxford)
[2] Volovik G E 2007 *Lect. Notes Phys.* **718** 31
[3] Murakami S 2007 *New J. Phys.* **9** 356
[4] Burkov A A, Hook M D and Balents L 2011 *Phys. Rev.* B **84** 235126
[5] Wan X, Turner A M, Vishwanath A and Savrasov S Y 2011 *Phys. Rev.* B **83** 205101
[6] Burkov A A and Balents L 2011 *Phys. Rev. Lett.* **107** 127205
[7] Grushin A 2012 *Phys. Rev.* D **86** 045001
[8] Zyuzin A A and Burkov A A 2012 *Phys. Rev.* B **86** 115133
[9] Zhang H, Liu C-X, Qi X-L, Dai X, Fang Z and Zhang S-C 2009 *Nat. Phys.* **5** 438
[10] Kurebayashi D and Nomura K 2014 *J. Phys. Soc. Jpn.* **83** 063709
[11] Sekine A and Nomura K 2014 *J. Phys. Soc. Jpn.* **83** 094710
[12] Miura K, Nakano T Z, Ohnishi A and Kawamoto N 2009 *Phys. Rev.* D **80** 074034
[13] Menotti P and and Pelissetto A 1987 *Commun. Math. Phys.* **113** 369

Study of spin transport in Dirac systems

Nobuyuki Okuma and Masao Ogata
Department of Physics, The University of Tokyo, Hongo 7-3-1, Tokyo 113-0033, Japan
E-mail: okuma@hosi.phys.s.u-tokyo.ac.jp

Abstract. We investigate spin Hall effect in a three-dimensional Dirac electron system. We derive diffusion equations in the Keldysh formalism and define a spin current from a spin diffusion equation. We solve the obtained diffusion equations in a steady state in the bulk and find that the bulk spin Hall conductivity is non-zero, though that for the conventional spin current is zero.

1. Introduction

Spintoronics is one of the most important fields in recent condensed matter physics. In particular, studies of spin currents in materials with spin-orbit coupling (SOC) led to discoveries such as intrinsic [1, 2] (and extrinsic [3]) spin Hall effect (SHE) and topological insulators [4, 5, 6].

In such materials, however, spin currents are often not conserved currents because SOC breaks spin rotational symmetry. Conventionally, a spin current is defined as a Hermitian operator

$$\hat{j}^{S_i} = \frac{\{\hat{S}_i, \hat{j}\}}{2}, \tag{1}$$

where $\hat{S}_i$ $(i = 1, 2, 3)$ is the spin operator and $\hat{j}$ is the electric current devided by $|e|$. The spin current defined in eq.(1) is a conserved Noether current in the systems with spin rotational symmetry. In the systems with SOC, however, this conventional spin current does not satisfy the continuity equation and sometimes leads crucial problems [7] . Hence it is not clear whether or not we can use the definition to each system.

Instead of using the conventional definition, we define a spin current from a spin diffusion equation [8] in this paper. We derive the spin diffusion equation from a quantum kinetic equation [9]. In this formalism, we do not need any assumption for the spin current operator. In order to investigate intrinsic SHE in a spin-non-conserved system, we choose a three-dimensional semi-metallic Dirac electron system with spin-mixing terms. We find that the spin Hall conductivity (SHC) can be non-zero even when the conventional SHC is zero.

This paper is organized as follows. In section 2 we introduce a three-dimensional Dirac Hamiltonian and derive its quantum kinetic equation. Then we outline the derivation to obtain diffusion equations. In section 3 we define 16 charges which describe the transport of the system and choose a specific form of alpha matrices. In section 4 we define a spin current and calculate the SHC. In section 5 we briefly summarize this paper.

2. Model and Formalism

In this section, we introduce a three-dimensional Dirac Hamiltonian. Then we derive its quantum kinetic equation and diffusion equations in the same formalism as previous works [10]. We consider only the effect of nonmagnetic impurities and ignore electron-electron interactions.

We investigate a 4×4 three-dimensional massless Dirac Hamiltonian with impurities

$$\begin{aligned}
\mathrm{H} &= \mathrm{H}_0 + \mathrm{H}_{imp}, \\
\mathrm{H}_0 &= \int d\boldsymbol{x}\psi^\dagger(\boldsymbol{x})v\left(-i\partial_i - eA_i\right)\hat{\alpha}_i\psi(\boldsymbol{x}), \\
\mathrm{H}_{imp} &= \int d\boldsymbol{x}\psi^\dagger(\boldsymbol{x})V(\boldsymbol{x})\psi(\boldsymbol{x}) = \int d\boldsymbol{x}\psi^\dagger(\boldsymbol{x})\left[u_0\sum_i \delta(\boldsymbol{x}-\boldsymbol{x}_i)\right]\psi(\boldsymbol{x}),
\end{aligned} \tag{2}$$

where v is the Fermi velocity, A_i is a vector potential, $V(\boldsymbol{x})$ is a nonmagnetic impurity potential, ψ is a four-component spinor filed, and alpha matrices $\hat{\alpha}_i$ obey the anticommutation relations

$$\{\hat{\alpha}_i, \hat{\alpha}_j\} = 2\delta_{ij}\hat{1}, \;(i, j = 1, 2, 3). \tag{3}$$

In order to investigate a non-equilibrium state of this system, we work on the Keldysh formalism. We start from the following equations:

$$\hat{G}^{R(A)} = \left[\left(\hat{G}^{(0)}\right)^{-1} - \hat{\Sigma}^{R(A)}\right]^{-1}, \tag{4}$$

$$\left[\hat{G}^R\right]^{-1}\hat{G}^K - \hat{G}^K\left[\hat{G}^A\right]^{-1} = \hat{\Sigma}^K\hat{G}^A - \hat{G}^R\hat{\Sigma}^K, \tag{5}$$

where $\hat{G}^{R(A)}$ is the retarded (advanced) Green's function, $\hat{G}^K$ is the Keldysh Green's function, and $\hat{\Sigma}'s$ are their self-energies.

First, we consider the system without an electromagnetic field. Ignoring the weak-localization effects, we perform the disorder average in the Born approximation and obtain the foliowing relations [10]:

$$\hat{G}^{R,A}(\boldsymbol{p}, \epsilon) = \frac{1}{(\epsilon + \mu) - vp_i\hat{\alpha}_i \pm \frac{i}{2\tau}}, \tag{6}$$

$$\hat{\Sigma}^K(\boldsymbol{x}, \boldsymbol{x}') = \frac{1}{\pi\nu\tau}\delta(\boldsymbol{x}-\boldsymbol{x}')\hat{G}^K(\boldsymbol{x}, \boldsymbol{x}), \tag{7}$$

$$\tau = \frac{1}{\pi\nu u_0^2 n_i}, \tag{8}$$

where μ is the chemical potential, ν is the density of states of one band at the Fermi energy, and n_i is the density of impurities. In this approximation, eq.(5) becomes

$$\left[\epsilon, \hat{G}^K\right] + \left[\hat{G}^K, \hat{H}\right] + \frac{i}{\tau}\hat{G}^K = \hat{\Sigma}^K\hat{G}^A - \hat{G}^R\hat{\Sigma}^K. \tag{9}$$

All we have to do is determine the Keldysh Green's function and its self-energy in eq.(9).

In order to obtain the quantum kinetic equation, we consider the Wigner transformation of eq.(9). The Wigner transformation of a function $A(x + \frac{\delta x}{2}, x - \frac{\delta x}{2})$ is defined as

$$A_{\boldsymbol{p}\epsilon}(\boldsymbol{x}, t) = \int d(\delta\boldsymbol{x})d(\delta t)e^{-i\boldsymbol{p}\cdot\delta\boldsymbol{x}+i\epsilon\delta t}A\left(x + \frac{\delta x}{2}, x - \frac{\delta x}{2}\right), \tag{10}$$

where $x = (\boldsymbol{x}, t)$. Applying the semi-classical approximation (See AppendixA), the Wigner transformation of eq.(9), or the quantum kinetic equation, can be written as

$$\frac{\partial}{\partial t}\hat{g}_{\boldsymbol{p}\epsilon}(\boldsymbol{x},t) - i\left[\hat{g}_{\boldsymbol{p}\epsilon}(\boldsymbol{x},t), v\hat{\boldsymbol{\alpha}}\cdot\boldsymbol{p}\right] + \frac{1}{2}\left\{\nabla\hat{g}_{\boldsymbol{p}\epsilon}(\boldsymbol{x},t), v\hat{\boldsymbol{\alpha}}\right\} + \frac{1}{\tau}\hat{g}_{\boldsymbol{p}\epsilon}(\boldsymbol{x},t)$$
$$= \frac{i}{\tau}\left(\hat{G}^R_{\boldsymbol{p}\epsilon}\hat{\rho}_\epsilon(\boldsymbol{x},t) - \hat{\rho}_\epsilon(\boldsymbol{x},t)\hat{G}^A_{\boldsymbol{p}\epsilon}\right). \tag{11}$$

Here we define the non-equilibrium distribution function $\hat{g}_{\boldsymbol{p}\epsilon}(\boldsymbol{x},t)$ and the density matrix $\hat{\rho}_\epsilon(\boldsymbol{x},t)$ as

$$\hat{g}_{\boldsymbol{p}\epsilon}(\boldsymbol{x},t) = \frac{1}{2i}\hat{G}^K(\boldsymbol{x}t, \boldsymbol{p}\epsilon),$$
$$\hat{\rho}_\epsilon(\boldsymbol{x},t) = \frac{1}{\pi\nu}\int\frac{d\boldsymbol{p}}{(2\pi)^3}\hat{g}_{\boldsymbol{p}\epsilon}(\boldsymbol{x},t). \tag{12}$$

The details of the derivation is included in AppendixA.

The effect of a uniform static electric field can be taken by a replacement

$$\nabla \to \tilde{\nabla} \equiv \nabla + e\boldsymbol{E}\partial_\epsilon \tag{13}$$

and a modification of the Wigner transformation as

$$A_{\boldsymbol{p}\epsilon}(\boldsymbol{x},t) = \int d(\delta\boldsymbol{x})d(\delta t)e^{-i(\boldsymbol{p}+e\boldsymbol{A}(t))\cdot\delta\boldsymbol{x}+i\epsilon\delta t}A\left(x+\frac{\delta x}{2}, x-\frac{\delta x}{2}\right) \tag{14}$$

in eq.(11).

In the following, we outline the derivation to solve eq.(11) by iterations. We consider the Fourier transformation of eq.(11) with respect to time. Using the following notations:

$$\Omega \equiv \omega + \frac{i}{\tau},\ E_p \equiv vp,\ \hat{\alpha}_{\theta\phi} \equiv \sin\theta\cos\phi\hat{\alpha}_1 + \sin\theta\sin\phi\hat{\alpha}_2 + \cos\theta\hat{\alpha}_3,$$
$$\hat{F}_0 \equiv \frac{i}{\tau}\left(\hat{G}^R_{\boldsymbol{p}\epsilon}\hat{\rho}_\epsilon - \hat{\rho}_\epsilon\hat{G}^A_{\boldsymbol{p}\epsilon}\right),\ \hat{F}_{grad} \equiv -\frac{1}{2}\left\{\nabla\hat{g}_{\boldsymbol{p}\epsilon}, v\hat{\boldsymbol{\alpha}}\right\}, \tag{15}$$

we can rewrite eq.(11) as

$$\Omega\ \hat{g}_{\boldsymbol{p}\epsilon}(\boldsymbol{x},\omega) = \hat{F}_0 + \hat{F}_{grad} - E_p\left[\hat{g}_{\boldsymbol{p}\epsilon}(\boldsymbol{x},\omega), \hat{\alpha}_{\theta\phi}\right]. \tag{16}$$

Equation (16) can be formally solved as

$$\hat{g}_{\boldsymbol{p}\epsilon} = A_{\boldsymbol{p}\omega}\hat{F}_0 + B_{\boldsymbol{p}\omega}\hat{\alpha}_{\theta\phi}\hat{F}_0\hat{\alpha}_{\theta\phi} + C_{\boldsymbol{p}\omega}\left[\hat{\alpha}_{\theta\phi}, \hat{F}_0\right] + A_{\boldsymbol{p}\omega}\hat{F}_{grad} + B_{\boldsymbol{p}\omega}\hat{\alpha}_{\theta\phi}\hat{F}_{grad}\hat{\alpha}_{\theta\phi} + C_{\boldsymbol{p}\omega}\left[\hat{\alpha}_{\theta\phi}, \hat{F}_{grad}\right]$$
$$\equiv \hat{g}^{(0)}_{\boldsymbol{p}\epsilon} + \hat{G}_{grad}\left[\hat{g}_{\boldsymbol{p}\epsilon}\right], \tag{17}$$

where

$$A_{\boldsymbol{p}\omega} = i\frac{2E_p^2 - \Omega^2}{\Omega(4E_p^2 - \Omega^2)},\ B_{\boldsymbol{p}\omega} = i\frac{2E_p^2}{\Omega(4E_p^2 - \Omega^2)},\ C_{\boldsymbol{p}\omega} = i\frac{-E_p}{4E_p^2 - \Omega^2}. \tag{18}$$

Solving eq.(17) with respect to $\hat{g}_{\boldsymbol{p}\epsilon}$ by the second order iteration and performing momentum integration, we obtain the diffusion equation of $\hat{\rho}_\epsilon$, which is the momentum-integral of $\hat{g}_{\boldsymbol{p}\epsilon}$ (eq.(12)). The inverse Fourier transformation with respect to ω can be performed when we assume $\Omega \cong i/\tau$ in eq.(18).

3. Definition of charges

The density matrix $\hat{\rho}_\epsilon$ is a 4×4 Hermitian matrix. Every 4×4 Hermitian matrix can be decomposed into the following 16 matrices:

$$\begin{aligned}
&\mathbf{1},\\
&\hat{\alpha}_0,\\
&\hat{\alpha}_i\ (i=1,\ 2,\ 3),\\
&\hat{\alpha}_5 \equiv \hat{\alpha}_0\hat{\alpha}_1\hat{\alpha}_2\hat{\alpha}_3,\\
&\hat{\alpha}_{\mu\nu} \equiv i\hat{\alpha}_\mu\hat{\alpha}_\nu\ (\mu<\nu \text{ and } \mu,\nu=0,1,2,3,5).
\end{aligned} \tag{19}$$

Therefore the density matrix $\hat{\rho}_\epsilon$ can be witten as

$$\hat{\rho}_\epsilon = \frac{1}{4}\left[n_\epsilon \mathbf{1} + \rho_\epsilon^\mu \hat{\alpha}_\mu + \rho_\epsilon^{\mu\nu}\hat{\alpha}_{\mu\nu}\right]. \tag{20}$$

Here we define 16 "charges" as coefficients corresponding to 16 matrices. Using these charges, we can derive 16 diffusion equations (to be published). For example, diffusion equations for n_ϵ, ρ_ϵ^1, and ρ_ϵ^{12} can be written as

$$\begin{aligned}
\frac{\partial n_\epsilon}{\partial t} =& \frac{v^2\tau}{3}\nabla^2 n_\epsilon - \frac{v}{3}\partial_x\rho_\epsilon^1 - \frac{v}{3}\partial_y\rho_\epsilon^2 - \frac{v}{3}\partial_z\rho_\epsilon^3,\\
\frac{\partial \rho_\epsilon^1}{\partial t} =& \frac{v^2\tau}{5}\partial_x^2\rho_\epsilon^1 + \frac{v^2\tau}{15}\partial_y^2\rho_\epsilon^1 + \frac{v^2\tau}{15}\partial_z^2\rho_\epsilon^1 + \frac{2v^2\tau}{15}\partial_x\partial_y\rho_\epsilon^2 + \frac{2v^2\tau}{15}\partial_x\partial_z\rho_\epsilon^3\\
&- \frac{v}{3}\partial_x n_\epsilon - \frac{v}{6\mu\tau}\partial_y\rho_\epsilon^{12} - \frac{v}{6\mu\tau}\partial_z\rho_\epsilon^{13} - \frac{\rho_\epsilon^1}{\left(\frac{3\tau}{2}\right)},\\
\frac{\partial \rho_\epsilon^{12}}{\partial t} =& \frac{v^2\tau}{15}\partial_x^2\rho_\epsilon^{12} + \frac{v^2\tau}{15}\partial_y^2\rho_\epsilon^{12} + \frac{v^2\tau}{5}\partial_z^2\rho_\epsilon^{12} - \frac{2v^2\tau}{15}\partial_y\partial_z\rho_\epsilon^{13} + \frac{2v^2\tau}{15}\partial_x\partial_z\rho_\epsilon^{23}\\
&- \frac{v}{6\mu\tau}\partial_x\rho_\epsilon^2 + \frac{v}{6\mu\tau}\partial_y\rho_\epsilon^1 - \frac{v}{3}\partial_z\rho_\epsilon^{05} - \frac{\rho_\epsilon^{12}}{\left(\frac{3\tau}{2}\right)}.
\end{aligned} \tag{21}$$

We use these equations in the next section, which describe z-component SHE.

In the following, we assume that alpha matrices can be witten as

$$\hat{\alpha}_i = \hat{\sigma}_i \times \hat{\tau}_1,\ (i=1,2,3), \tag{22}$$

where σ_i and τ_1 are Pauli matrices in spin and orbital space, respectively. In this system, the particle density N, the spin density S_i, and other charge density ρ^a can be calculated as

$$\begin{aligned}
N(\boldsymbol{x},t) &= \nu\int d\epsilon\ n_\epsilon(\boldsymbol{x},t),\\
S_1(\boldsymbol{x},t) &= \nu\int d\epsilon\ \frac{1}{2}\left[-\rho_\epsilon^{23}(\boldsymbol{x},t)\right],\\
S_2(\boldsymbol{x},t) &= \nu\int d\epsilon\ \frac{1}{2}\left[\rho_\epsilon^{13}(\boldsymbol{x},t)\right],\\
S_3(\boldsymbol{x},t) &= \nu\int d\epsilon\ \frac{1}{2}\left[-\rho_\epsilon^{12}(\boldsymbol{x},t)\right],\\
\rho^a(\boldsymbol{x},t) &= \nu\int d\epsilon\ \left[\rho_\epsilon^a(\boldsymbol{x},t)\right].
\end{aligned} \tag{23}$$

Journal of Physics: Conference Series **603** (2015) 012018 doi:10.1088/1742-6596/603/1/012018

4. Spin Hall effect

In this paper, we focus on z-component SHE when the electric field $\boldsymbol{E} = (E_x, 0, 0)$ is applied. We assume that the system size is infinite in x and z directions and finite in y direction. Under this condition, we can drop x and z derivatives. Rewriting diffusion equations (21) with eqs.(13), (14) and (23), we obtain the following equations:

$$\begin{aligned}\frac{\partial N}{\partial t} =& \frac{v^2\tau}{3}\partial_y^2 N - \frac{v}{3}\partial_y \rho^2_{,}\\ \frac{\partial \rho^1}{\partial t} =& \frac{v^2\tau}{15}\partial_y^2\rho^1 + \frac{ve(2\nu)}{3}E_x + \frac{v}{3\mu\tau}\partial_y S_3 - \frac{\rho^1}{\left(\frac{3\tau}{2}\right)},\\ \frac{\partial S_3}{\partial t} =& \frac{v^2\tau}{15}\partial_y^2 S_3 - \frac{v}{12\mu\tau}\partial_y\rho^1 - \frac{S_3}{\left(\frac{3\tau}{2}\right)}.\end{aligned} \tag{24}$$

Here we have kept only the first order with respect to the electric field and neglected

$$eE_x\partial_\epsilon\rho_\epsilon^\alpha, (\alpha = \mu, \mu\nu), \tag{25}$$

which are the second order contributions.

When $\boldsymbol{j}^{S_3}$ is the z-component spin current and τ_s is the spin relaxation time,

$$\frac{\partial S_3}{\partial t} = -\nabla\cdot\boldsymbol{j}^{S_3} - \frac{S_3}{\tau_s}. \tag{26}$$

Comparing eqs.(24) and (26), we obtain

$$j_y^{S_3} = -\frac{v^2\tau}{15}\partial_y S_3 + \frac{v}{12\mu\tau}\rho^1, \tag{27}$$

$$\tau_s = \frac{3}{2}\tau. \tag{28}$$

This is our definition of the spin current which is naturally defined from the obtained spin diffusion equation.

In the steady state in the bulk, we can drop time and space derivatives in eq.(24). In this case, we have

$$\rho^1 = \tau v e \nu E_x. \tag{29}$$

Combining eqs.(27) and (29), we obtain non-zero bulk spin current:

$$j_y^{S_3} = \frac{e\mu}{24\pi^2 v}E_x \equiv \sigma_{xy}^{S_3}E_x. \tag{30}$$

Here we have used $\nu = \mu^2/2\pi^2 v^3$ and defined the z-component SHC $\sigma_{xy}^{S_3}$. In the system with $\hat{\alpha}_i = \hat{\sigma}_i \times \hat{\tau}_1$, the conventional spin current operator can be defined as

$$j_y^{S_3} = \frac{\{\hat{S}_3, \hat{j}_y\}}{2} = 0. \tag{31}$$

Therefore the SHC of this system is 0 for the conventional definition. In our definition, however, we obtain non-zero SHC

It is important to note that we do not assume the form of the spin current operator, while the conventional spin current has no justification.

5. Summary

In this paper, we have derived diffusion equations of a three-dimensional massless Dirac electron system from the quantum kinetic equation. Assuming the specific form of alpha matrices $\hat{\alpha}_i = \hat{\sigma}_i \times \tau_1$, we have investigated diffusion equations of the Dirac electron system in terms of spin transport. We have defined the spin current from diffusion equations and found that the system has a finite SHC, though the conventional SHC is zero.

Acknowledgement

This work was supported by Japan Society for the Promotion of Science through Program for Leading Graduate Schools (MERIT) and Grants-in-Aid for Scientific Research (A) (No. 24244053).

Appendix A. Semi-classical approximation

In this appendix, we derive the quantum kinetic equation in the semi-classical approximation.

The Wigner transformation of the convolution $\hat{A}\hat{B}$ is defined as

$$\left[\hat{A}\hat{B}\right](\boldsymbol{x}t,\boldsymbol{p}\epsilon) \equiv A(\boldsymbol{x}t,\boldsymbol{p}\epsilon) * B(\boldsymbol{x}t,\boldsymbol{p}\epsilon), \tag{A.1}$$

where the Moyal product $*$ is

$$* = \exp\left[\frac{i\hbar}{2}\left(\overleftarrow{\partial}_{\boldsymbol{x}}\cdot\overrightarrow{\partial}_{\boldsymbol{p}} - \overleftarrow{\partial}_{\boldsymbol{p}}\cdot\overrightarrow{\partial}_{\boldsymbol{x}} - \overleftarrow{\partial}_t\overrightarrow{\partial}_\epsilon + \overleftarrow{\partial}_\epsilon\overrightarrow{\partial}_t\right)\right]. \tag{A.2}$$

In the semi-classical approximation, we have kept only zeroth and first order terms with respect to $\hbar$. Up to this order, eq.$(A.1)$ becomes

$$A(\boldsymbol{x}t,\boldsymbol{p}\epsilon) * B(\boldsymbol{x}t,\boldsymbol{p}\epsilon) \cong A(\boldsymbol{x}t,\boldsymbol{p}\epsilon)B(\boldsymbol{x}t,\boldsymbol{p}\epsilon) + \frac{i\hbar}{2}\left\{A(\boldsymbol{x}t,\boldsymbol{p}\epsilon), B(\boldsymbol{x}t,\boldsymbol{p}\epsilon)\right\}_{Poisson}, \tag{A.3}$$

where

$$\left\{A(\boldsymbol{x}t,\boldsymbol{p}\epsilon), B(\boldsymbol{x}t,\boldsymbol{p}\epsilon)\right\}_{Poisson} = \frac{\partial A}{\partial \boldsymbol{x}}\frac{\partial B}{\partial \boldsymbol{p}} - \frac{\partial A}{\partial \boldsymbol{p}}\frac{\partial B}{\partial \boldsymbol{x}} - \frac{\partial A}{\partial t}\frac{\partial B}{\partial \epsilon} + \frac{\partial A}{\partial \epsilon}\frac{\partial B}{\partial t} \tag{A.4}$$

is the Poisson bracket.

In the following, we consider the Wigner transformation of eq.(9) in the semi-classical approximation. First and second terms of LHS can be written as

$$\begin{aligned}
\left[\hat{\epsilon},\hat{G}^K\right](\boldsymbol{x}t,\boldsymbol{p}\epsilon) &\cong \left[\epsilon, G^K(\boldsymbol{x}t,\boldsymbol{p}\epsilon)\right] + \frac{i\hbar}{2}\left\{\epsilon, G^K\right\}_{Poisson} - \frac{i\hbar}{2}\left\{G^K,\epsilon\right\}_{Poisson} \\
&= i\hbar\frac{\partial G^K}{\partial t}, \\
\left[\hat{G}^K,\hat{H}\right](\boldsymbol{x}t,\boldsymbol{p}\epsilon) &\cong \left[G^K(\boldsymbol{x}t,\boldsymbol{p}\epsilon), v\hat{\boldsymbol{\alpha}}\cdot\boldsymbol{p}\right] + \frac{i\hbar}{2}\left\{G^K, v\hat{\boldsymbol{\alpha}}\cdot\boldsymbol{p}\right\}_{Poisson} - \frac{i\hbar}{2}\left\{v\hat{\boldsymbol{\alpha}}\cdot\boldsymbol{p}, G^K\right\}_{Poisson} \\
&= \left[G^K(\boldsymbol{x}t,\boldsymbol{p}\epsilon), v\hat{\boldsymbol{\alpha}}\cdot\boldsymbol{p}\right] - \frac{i\hbar}{2}\left\{\nabla G^K, v\hat{\boldsymbol{\alpha}}\right\}.
\end{aligned} \tag{A.5}$$

The first term of RHS can be written as

$$\begin{aligned}\left[\hat{\Sigma}^K \hat{G}^A\right](\boldsymbol{x}t, \boldsymbol{p}\epsilon) \cong & \Sigma^K(\boldsymbol{x}t, \boldsymbol{p}\epsilon) G^A(\boldsymbol{x}t, \boldsymbol{p}\epsilon) \\ = & \int d(\delta \boldsymbol{x}) d(\delta t) e^{-i\boldsymbol{p}\cdot\delta\boldsymbol{x} + i\epsilon\delta t} \Sigma^K\left(x + \frac{\delta x}{2}, x - \frac{\delta x}{2}\right) G^A_{\boldsymbol{p}\epsilon} \\ = & \int d(\delta \boldsymbol{x}) d(\delta t) e^{-i\boldsymbol{p}\cdot\delta\boldsymbol{x} + i\epsilon\delta t} \frac{\delta(\delta \boldsymbol{x})}{\pi\nu\tau} G^K\left(\boldsymbol{x}, t - \frac{\delta t}{2}; \boldsymbol{x}, t + \frac{\delta t}{2}\right) G^A_{\boldsymbol{p}\epsilon} \\ = & \frac{1}{\pi\nu\tau} \int d(\delta t) e^{i\epsilon\delta t} \frac{i}{\pi} \int \frac{d\epsilon' d\boldsymbol{p}'}{(2\pi)^3} g_{\boldsymbol{p}'\epsilon'}(\boldsymbol{x}, t) e^{-i\epsilon'\delta t} G^A_{\boldsymbol{p}\epsilon} \\ = & \frac{2i}{\tau} \hat{\rho}_\epsilon(\boldsymbol{x}, t) G^A_{\boldsymbol{p}\epsilon}. \end{aligned} \tag{A.6}$$

Here we have used eq.(7) in the third line. We also obtain

$$\left[\hat{G}^R \hat{\Sigma}^K\right](\boldsymbol{x}t, \boldsymbol{p}\epsilon) \cong \frac{-2i}{\tau} G^R_{\boldsymbol{p}\epsilon} \hat{\rho}_\epsilon(\boldsymbol{x}, t) \tag{A.7}$$

in the same procedure. Finally, we obtain eq.(11).

References

[1] Murakami S, Nagaosa N, and Zhang S C, 2003 *Science.* **301** 1348
[2] Sinova J, et al., 2004 *Phys. Rev. Lett.* **92** 126603
[3] Hirsch J E, 1999 *Phys. Rev. Lett.* **83** 1834
[4] Kane C L and Mele E J, 2005 *Phys. Rev. Lett.* **95** 226801
[5] Fu L, Kane C L, and Mele E J, 2007 *Phys. Rev. Lett.* **98** 106803
[6] Qi X L and Zhang S C, 2011 *Rev. Mod. Phys.* **83** 1057
[7] Rashba E, 2003 *Phys. Rev.* B **68** 241315(R)
[8] Burkov A A, Núñez Alvaro S, and MacDonald A H, 2004 *Phys. Rev.* B **70** 155308
[9] Kadanoff L P, Baym G, *Quantum Statistical Mechanics* (Benjamin, New York, 1962)
[10] Mishchenko E G, Shytov A V, and Halperin B I, 2004 *Phys. Rev. Lett.* **93** 226602

Distribution of Chern number by Landau level broadening in Hofstadter butterfly

Nobuyuki Yoshioka, Hiroyasu Matsuura and Masao Ogata
Department of Physics, University of Tokyo, Hongo, Bunkyo-ku, 113-0033 Japan
E-mail: yoshioka@cams.phys.s.u-tokyo.ac.jp

Abstract. We discuss the relationship between the quantum Hall conductance and a fractal energy band structure, Hofstadter butterfly, on a square lattice under a magnetic field. At first, we calculate the Hall conductance of Hofstadter butterfly on the basis of the linear responce theory. By classifying the bands into some groups with a help of continued fraction expansion, we find that the conductance at the band gaps between the groups accord with the denominators of fractions obtained by aborting the expansion halfway. The broadening of Landau levels is given as an account of this correspondance.

1. Introduction

Through the ages, the problem of 2D Bloch electrons under an influence of vertical uniform magnetic field has attracted many interests of physicists. Under two limiting conditions, i.e. only with either magnetic field or potential, discrete Landau levels (LLs) or continuous Bloch bands are given as solutions [1, 2], both of which are fundamental knowledge in solid state physics. A typical problem in the intermediate, or two-combined case is as follows: a single electron in a very strong potential under a weak magnetic field in an isotropic square lattice, where the Hamiltonian reduces to a proper equation named "Harper equation" [3]. While analytic solution does not exist for general B, the system shows two extremely intriguing properties: integer quantum Hall effect (IQHE) and fractal energy band structure. The latter combines the adverse nature of the LLs and the Bloch bands, which is expressed in the energy diagram known as "Hofstadter's butterfly" [4].

The IQHE is described as a quantization of Hall conductance σ, which usually occurs when the Fermi energy lies between the LLs. First discovered by von Klitzing *et al.* [5], this strikingly odd phenomenon has contributed to the development of condensed matter physics both theoretically and experimentally. For instance, since the theoretical explanation of IQHE was given first by Laughlin [6] based on the gauge invariance, physicists were further encouraged to demonstrate the validity of the method of gauge theory. Thouless *et al.* (or TKNN) suggested that a noninteracting two-dimensional electron system always shows IQHE [7], which was later understood as the consequence of topological invariance of σ. Namely, the Hall conductance can be identified as a Berry phase in the Brillouin zone (BZ) [8, 9], which is gauge invariant, so that σ is stable unless the band gaps open and close [10, 11]. Therefore, one may expect that the label of each gap is related to the Hall conductance. As for the square lattice, a very simple yet powerful diophantic equation to associate σ with the gap label was discovered by TKNN. However, the multivalency of the solution requires another rule to give the authentic σ, whose reasonable justification is still an open question. Note that, of course one may compute σ directly from TKNN formula if needed. Other methods are Streda formula [12] and the adoption of lattice gauge theory techniques [13, 14].

While the quantum Hall effect (QHE) has been a research topic, not so many works have focused on the fractality, an apparently interesting nature first confirmed by Hofstadter with a help of continued fraction expansion [4]. In this paper, we clarify the relation of the IQHE and the fractal structure. As a result, we find that the Hall conductance at gaps between "families" (the definition of a family is given in §5) can be determined without numerical calculation, whose uniqueness is supported by a physical background.

The organization of this paper is as follows. §2 gives a brief derivation of Harper equation, which is the proper equation in our system. In §3, we estimate the conductance on the basis of the linear responce theory, and we describe the correspondence between the Hall conductance and the energy spectrum. In §4, we divide the energy spectrum into "subcells", employing the notation used by Hofstadter. This enables us to state the grouping rules for families in §5. We discuss the relationship between two nature in §6, and the conclusion is given in §7.

2. Harper equation

The tight-binding Hamiltonian on an isotropic square lattice under a vertical uniform magnetic field is written as

$$\mathscr{H} = -\sum_{\langle i,j\rangle} t c_i^\dagger c_j e^{i\theta_{ij}} + (h.c.), \tag{1}$$

$$\theta_{ij} = \frac{e}{\hbar}\int_{\mathbf{r}_i}^{\mathbf{r}_j} \mathbf{A}\cdot \mathrm{d}\mathbf{l}, \tag{2}$$

where $c_i^\dagger$ (c_i) is a creation (annihilation) operator on the i site, and t_{ij} is a transfer integral between i and j sites. Peierls phase θ_{ij} satisfies $\sum_{\text{plaquette}} \theta_{ij} = 2\pi\phi$, with the right-hand side representing the magnetic flux per plaquette in units of magnetic flux quantum, i.e. $\phi = \frac{BS_{\text{plaquette}}}{h/e}$. When we choose the Landau gauge $\mathbf{A} = (0, Bx, 0)$, only the transfer integrals along the y axis acquire nonzero Peierls phase. Using the Cartesian coordinate (m, n) of site i (m, n=integer), the Peierls phase becomes $\theta_{ij} = 2\pi\phi n$ for the link between $i = (m, n)$ and $j = (m, n+1)$.

Let us assume ϕ is a rational number p/q where p and q are coprime integers. In this case [15][16], the magnetic unit cell becomes q times larger in the x-direction, and correspondingly the BZ is q times smaller than the original one. By a straightfoward calculation, the Hamiltonian is transformed to

$$\mathscr{H} = -t\sum_{k_x,k_y} \tilde{\mathbf{c}}^\dagger(\mathbf{k})\tilde{\mathscr{H}}(\mathbf{k})\tilde{\mathbf{c}}(\mathbf{k}), \tag{3}$$

where each component is defined by

$$\tilde{\mathscr{H}}(\mathbf{k}) = \begin{pmatrix} 2\cos(k_y) & 1 & 0 & \cdots & e^{-iqk_x} \\ 1 & 2\cos(k_y+2\pi\phi) & 1 & \cdots & 0 \\ 0 & 1 & 2\cos(k_y+4\pi\phi) & \cdots & 0 \\ \vdots & & & \ddots & \\ e^{iqk_x} & 0 & 0 & \cdots & 2\cos(k_y+2(q-1)\pi\phi) \end{pmatrix}, \tag{4}$$

with

$$\tilde{\mathbf{c}}(\mathbf{k}) = {}^t(\tilde{c}_0(\mathbf{k}), \tilde{c}_1(\mathbf{k}), ..., \tilde{c}_{q-1}(\mathbf{k})). \tag{5}$$

Here, $\widetilde{c_n}(\mathbf{k})$ denotes the fermion operator in the reciprocal space,

$$c_{m,n} = c_{qm'+m'',n} = \frac{1}{\sqrt{L_x/q}}\frac{1}{\sqrt{L_y}}\sum_{k_x,k_y} e^{iqk_x m' + ik_y n}\widetilde{c}_{m''}(\mathbf{k}). \tag{6}$$

Equation (3), which is called Harper equation, is known to have q bands when $\phi = p/q$.
Figure 1 shows the energy spectrum obtained by Hofstadter [4]. It was shown [18] that two

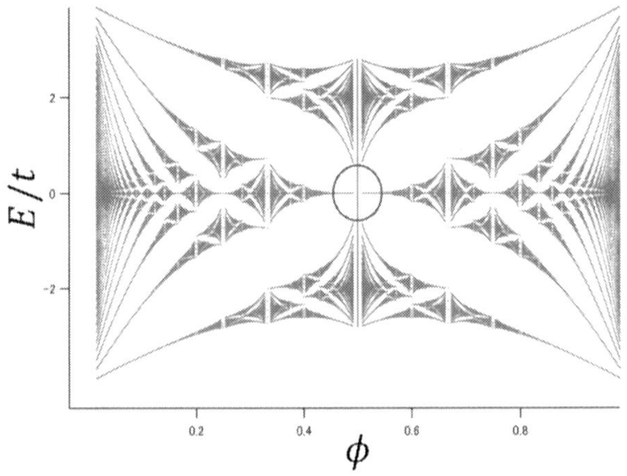

Figure 1. Energy spectrum of the tight-binding Hamiltonian of a square lattice under a magnetic field whose flux per plaquette is ϕ. The blue circle indicates a Dirac point.

bands happen to cross linearly at zero energy, i.e. form a Dirac cone, when q is an even number. The blue circle shown in Fig. 1 indicates one of the Dirac cone at $\phi = 1/2$. Further more, the magnetic translation symmetry leads to a q-fold degeneracy of the Brillouin zone in the direction of k_y, which means that each LL is q-fold degenerate [18]. In the vicinity of this Dirac point, we can see that the vertical uniform magnetic field measured from the Dirac point (i.e. $\phi - 1/2$) induces LLs with their energy in proportion to $\sqrt{|n|}$ and $\sqrt{|B|}$ with n being the LL number. We refer these LLs as "the Dirac LLs", discriminating from "the Fermi LLs" with the ordinary linear energy spectrum [14].

3. Hall Conductance

The Hall conductance of the n-th band is computed in the linear response theory as

$$\sigma_n = -\frac{1}{L_x L_y}\frac{ie^2}{\hbar}\sum_{\mathbf{k}}\sum_{m\neq n} f(E_{n,\mathbf{k}})\left[\frac{\langle u_{n,\mathbf{k}}|\frac{\partial \mathscr{H}(\mathbf{k})}{\partial k_x}|u_{m,\mathbf{k}}\rangle\langle u_{m,\mathbf{k}}|\frac{\partial \mathscr{H}(\mathbf{k})}{\partial k_y}|u_{n,\mathbf{k}}\rangle - (k_x \leftrightarrow k_y)}{(E_{n,\mathbf{k}} - E_{m,\mathbf{k}})^2}\right]. \tag{7}$$

where $f(E)$ denotes the Fermi distribution function. Assuming zero temperature and sufficiently large system size, $L_x, L_y >> 1$, (7) is reduced to

$$\sigma_n = -\frac{e^2}{h}\int_{\mathrm{MBZ}}\frac{\mathrm{d}^2\mathbf{k}}{2\pi}\sum_{m\neq n}\left[\frac{\langle u_{n,\mathbf{k}}|\frac{\partial \mathscr{H}(\mathbf{k})}{\partial k_x}|u_{m,\mathbf{k}}\rangle\langle u_{m,\mathbf{k}}|\frac{\partial \mathscr{H}(\mathbf{k})}{\partial k_y}|u_{n,\mathbf{k}}\rangle - (k_x \leftrightarrow k_y)}{(E_{n,\mathbf{k}} - E_{m,\mathbf{k}})^2}\right] \tag{8}$$

$$= \gamma_n \frac{e^2}{h}, \tag{9}$$

where MBZ means magnetic Brillouin zone and γ_n is regarded as the Berry phase in the Brillouin zone [9]. Since the MBZ is two-torus, γ_n is integer for arbitrary n.

As an example, we calculate the Hall conductance as a function of Fermi energy E_f at $\phi = 9/26$ as shown in Fig. 2. Since there is a particle-hole symmetry, we discuss only

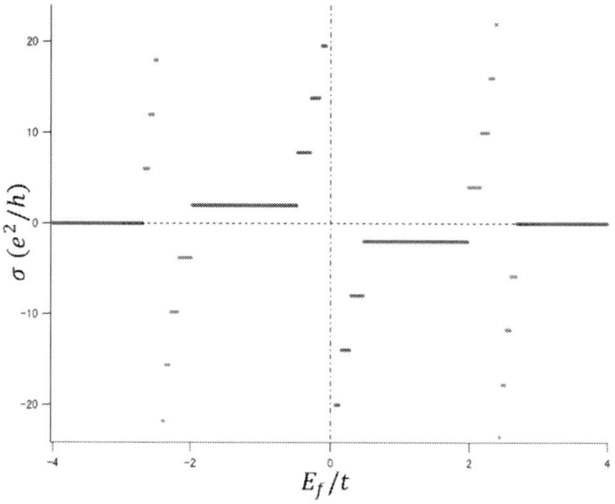

Figure 2. Hall conductance as a function of the Fermi energy for $\phi = 9/26$. Plateaux in family gaps, whose definitions are given in §5, are shown in blue lines and others in red lines.

the region of $E_f > 0$ below. As E_f decreases from $E_f/t = 4$, the conductance changes as $\sigma = 0, \cdots, 22, 16, 10, 4, -2, -8, -14, -20$ as shown in Fig. 2. We can see that the Hall conductance is constant between band gaps in the energy spectrum. In the following, we clarify the relationship between the energy spectrum and the value of the conductance.

4. Fractal structure of Hofstadter butterfly

In order to understand the behavior of Hall conductance, we find it necessary to classify the magnetic Bloch bands. First, we review minimum but sufficient information of the notion "subcell" which was first proposed by Hofstader. We restrict our argument to $0 \leq \phi < 1/2$ since the identical diagram is given for the other ϕ's.

There are three types of subcells: L, R and C (See Fig. 3). The L and R subcells are the outermost bands as shown in Fig. 3(a). L_n and R_n $(n = 0, 1, 2 \cdots)$ are in the regions of

$$\frac{1}{n+3} < \phi < \frac{1}{n+2}. \tag{10}$$

C subcells consist of the center bands as shown in Fig. 3(a) and C_n $(n = 0, 1, 2 \cdots)$ are in the regions of

$$\frac{n}{2n+1} < \phi < \frac{n+1}{2n+3}. \tag{11}$$

Note that the existence of gaps between the L, C and R subcells enable us to determine the label of each band uniquely.

From Fig. 3(a) we can see that each subcell, L_n, R_n, C_n, contains the self-similar butterfly pattern as in the original one for $0 \leq \phi \leq 1$. Mathematically, as first claimed by Hofstadter and

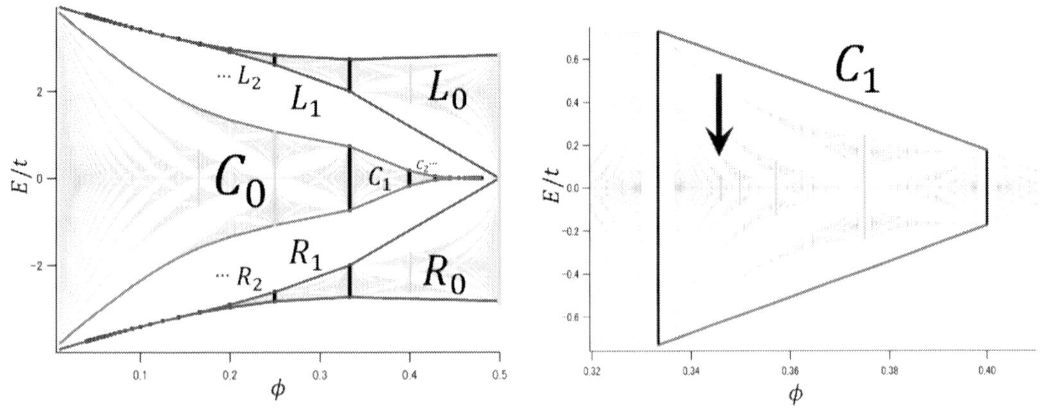

(a) (b)

Figure 3. (a) Butterfly divided into subcells. Red and blue lines stand for the boundary of C subcells and that of L, R subcells. (b) Expanded energy spectrum inside the C_1 subcell. Yellow straight lines denote the centermost bands which have self-similar structure.

proved by MacDonald [17], an appropriate linear stretching of the local variable $\beta \equiv 1/\phi - [1/\phi]$ gives nearly the same diagram as the original butterfly pattern. Here the term 'nearly' reflects the fact that the streched diagram of L_n and R_n subcells differ from the original butterfly in that the gaps do not close at $\beta = 1/2, 1/4$ etc, although there are no gap in the original butterfly at $\phi = 1/2, 1/4$. However, we do not need to discuss this difference further because in the following we restrict our argument on C subcells in which gap closes at $\beta = 1/2, 1/4$ etc. The substructure inside C_n can be classified again by $\widetilde{L}_m, \widetilde{R}_m$ and $\widetilde{C}_m$. In the following, we denote $\widetilde{C}_m$ in C_n as C_nC_m etc; which means that "C_m subcell inside C_n." Note that repetition of such operation determines the division of the butterfly spectrum uniquely.

5. Grouping rule for "families"

Next we introduce the concept of "families" in the Hofstadter butterfly. Figure 4 shows the energy spectrum obtained for $\phi = 9/26$. The energy levels in C_1 is expanded in the r.h.s. of Fig. 4 and the energy bands in C_1 are named in the method described in the previous section. We define these bands as families.

In order to understand these families, we use $\Gamma_m(\beta)$ and $\Lambda_l(\beta)$ ($m, l =$ integer) [4, 17]:

$$\Gamma_m(\beta) = \frac{1}{2 + (m+\beta)^{-1}}, \quad 0 \le \beta < 1 \tag{12}$$

$$\Lambda_l(\beta) = \frac{1}{(l+2)+\beta}, \tag{13}$$

which were introduced by Hofstadter to describe the continued fraction expansion. For example, $\phi = 9/26$ can be expressed in various ways of continued fraction expansions as follows:

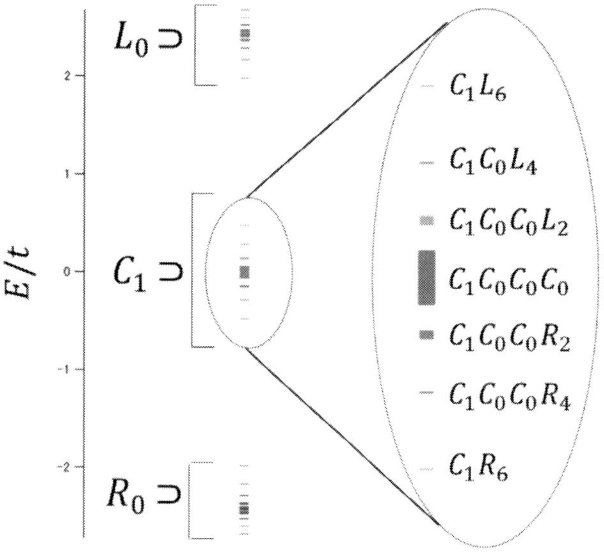

Figure 4. Energy spectrum at $\phi = \frac{9}{26}$.

$$\frac{9}{26} = \begin{cases} \frac{1}{(2+0)+\frac{8}{9}} = \Lambda_0(\frac{8}{9}) \\ \frac{1}{2+(1+\frac{1}{8})^{-1}} = \Gamma_1(\frac{1}{8}) = \begin{cases} \frac{1}{2+(1+\frac{1}{(2+6)+0})^{-1}} = \Gamma_1\Lambda_6(0) \\ \frac{1}{2+(1+\frac{1}{2+(0+\frac{1}{6})^{-1}})^{-1}} = \Gamma_{1,0}(\frac{1}{6}) = \begin{cases} \Gamma_{1,0}\Lambda_4(0) \\ \Gamma_{1,0,0}(\frac{1}{4}) = \begin{cases} \Gamma_{1,0,0}\Lambda_2(0) \\ \Gamma_{1,0,0,0}(\frac{1}{2}). \end{cases} \end{cases} \end{cases} \end{cases} \tag{14}$$

Here $\Gamma_{m_1,m_2\cdots m_N}(\beta)$ represents the continued fraction $\Gamma_{m_1}(\Gamma_{m_2}(\cdots(\Gamma_{m_N}(\beta))\cdots))$ and $\Gamma_{m_1,m_2\cdots m_N}\Lambda_l(\beta)$ is $\Gamma_{m_1}(\Gamma_{m_2}(\cdots(\Gamma_{m_N}(\Lambda_l(\beta)))\cdots))$.

In the general case, $\phi = p/q$ is expressed in various ways as

$$\begin{aligned} \phi = \Lambda_l\left(\frac{p^{(1)}}{q^{(1)}}\right) = \Gamma_{m_1}(\beta^{(2)}) &= \Gamma_{m_1}\Lambda_{l_1}\left(\frac{p^{(2)}}{q^{(2)}}\right) = \Gamma_{m_1 m_2}(\beta^{(3)}) = \\ \cdots &= \Gamma_{m_1\cdots m_{N-1}}\Lambda_{l_N}\left(\frac{p^{(N)}}{q^{(N)}}\right) = \Gamma_{m_1\cdots m_N}(\beta^{(N+1)}), \qquad (15) \\ \beta^{(N+1)} &= \begin{cases} 0 & \text{for odd } q \\ 1/2 & \text{for even } q \end{cases}. \end{aligned}$$

The relation between the denominators is [4, 17]

$$\begin{aligned} q &= 2q' + 2q'' + \cdots + 2q^{(i)} + \gamma, \qquad (16) \\ \gamma &= \begin{cases} 1 & \text{for odd } q \\ 2 & \text{for even } q \end{cases}. \end{aligned}$$

6. Relationship between conductance and energy spectrum

We discuss the relationship between the Hall conductance and the energy spectrum on the basis of the "families". As shown in §3, the Hall conductance are, -2, -8, -14, -20 when $0 < E_f \leq 2$. We find that the absolute values of σ are equivalent to the denominators of 1/2, $\Gamma_1(\frac{1}{2}) = 3/8$, $\Gamma_{1,0}(\frac{1}{2}) = 5/14$, $\Gamma_{1,0,0}(\frac{1}{2}) = 7/20$. In the general case, when the magnetic flux is $\phi = \Gamma_{m_1 \cdots m_N}(\beta^{(N+1)})$, we calculate $\Gamma_{m_1}(\frac{1}{2}) = P_1/Q_1, \Gamma_{m_1 m_2}(\frac{1}{2}) = P_2/Q_2, \cdots, \Gamma_{m_1 \cdots m_{N-1}}(\frac{1}{2}) = P_{N-1}/Q_{N-1}$. Then the Hall conductance when E_f is located in the gap between $(l+1)$th and l-th family from the top is given by

$$\sigma(l) = \frac{1}{2} sgn\left(\phi - \frac{P_l}{Q_l}\right) \cdot 2Q_l. \tag{17}$$

It is widely known that the Hall conductance is quantized in unconventional way in the gaps of the Dirac LLs (e.g. $\sigma = \frac{e^2}{h}4(N+1/2)$ in graphene) [19, 20]. Therefore, we can expect the Hall conductance to be quantized as $\sigma = 2q(N+1/2)$ when infinitesimal magnetic flux $\Delta\phi$ additionarily is imposed to $\phi = p/q$. In our case $\phi = 9/26$, the family $C_1C_0C_0C_0$ corresponds to the $n = 0$ Dirac LL originated from $\phi = 7/20$. As antisipated from Fig. 3(b), this $n = 0$ Dirac LL is broadened as ϕ decreases from 7/20, and at $\phi = 9/26$, it becomes the family of $C_1C_0C_0C_0$. Therefore the conductance when E_f is just above this family becomes -20 which is determined for $\phi = 7/20$. Similarly, the central three families correspond to the broadened $n = 0$ Dirac LL originated from $\phi = 5/14$. As a result, the conductance when E_f is just above $C_1C_0C_0L_2$ becomes -14. Details would be discussed elsewhere [21].

7. Conclusion

We have shown that the bands in 2D electron system under a vertical uniform magnetic field can be classified into families, which is a notion introduced to describe the recursive structure of the Hofstadter butterfly. The Hall conductance when E_f is located in the gaps between families is given by the denominator of the aborted continued fraction expansion. This can be physically understood from the two types of LLs.

References

[1] Landau L 1930 *Z. Phys.* **64** 629
[2] Bloch F 1929 *Z. Phys.* **52** 555
[3] Harper P G 1955 *Proc. R. Soc. London, Ser. A* **68** 874
[4] Hofstadter D R 1976 *Phys. Rev. B* **14** 2239
[5] von Klitzing K, Dorda G, and Pepper M 1980 *Phys. Rev. Lett.* **45** 494
[6] Laughlin R B 1981 *Phys. Rev. B* **23** 5632
[7] Thouless D J, Kohmoto M, Nightingale M P, and den Nijs M 1982 *Phys. Rev. Lett.* **49** 405
[8] Berry M V 1984 *Proc. R. Soc. London, Ser. A* **392** 45
[9] Xiao D, Chang M C, and Niu Q 2010 *Rev. Mod. Phys.* **82** 1959
[10] Avron J E, Seiler R, Simon B 1983 *Phys Rev. Lett.* **51** 51
[11] Oshikawa M 1994 *Phys. Rev. B* **50** 17357
[12] Streda P 1982 *J. Phys. C* **15** L718
[13] Fukui T, Hatsugai Y, and Suzuki H 2005 *J. Phys. Soc. Jpn.* **74** 1674
[14] Hatsugai Y, Fukui T, and Aoki H 2006 *Phys. Rev. B* **74** 205414
[15] Zak J 1964 *Phys. Rev.* **134** A1602
[16] Chang M C and Niu Q 1996 *Phys. Rev. B* **53** 7010
[17] MacDonald A H 1983 *Phys. Rev. B* **28** 6713
[18] Kohmoto M 1989 *Phys. Rev. B* **39** 11943
[19] Gusynin V P and Sharapov S G 1956 *Phys. Rev.* **104** 666
[20] Watanabe H, Hatsugai Y, and Aoki H 2010 *Phys. Rev. B* **82** 241403
[21] Yoshioka N, Matsuura H, and Ogata M *to be submitted*

International Workshop on Dirac Electrons in Solids 2015 IOP Publishing
Journal of Physics: Conference Series **603** (2015) 012020 doi:10.1088/1742-6596/603/1/012020

Anomalous Hall effect in the Dirac electron system with a split term

Shota Suetsugu, Hiroyasu Matsuura, and Masao Ogata

Department of Physics, University of Tokyo, Hongo, Bunkyo-ku, 113-0033 Japan

E-mail: 6363846388@mail.ecc.u-tokyo.ac.jp

Abstract. We study an anomalous Hall effect in the massive Dirac electron system with broken time reversal symmetry. Using the model Hamiltonian with the spin-orbit interaction and a split term which breaks time reversal symmetry, we calculate the energy band, the Berry curvature and the intrinsic Hall conductivity in an analytical way. We show that the non-zero Berry curvature appears and thus an intrinsic Hall conductivity occurs. This anomalous Hall effect can be observed in such systems as a ferromagnetic Dirac electron system or a ferromagnet-coated Dirac electron system.

1. Introduction

Anomalous Hall effect is a Hall effect in a ferromagnet without magnetic fields. Its mechanism has been controversial issues [1]. The main mechanisms of the anomalous Hall effect are the intrinsic contribution [2] which is related to the conception of Berry phase [3-5], skew scattering and side jump scattering, and a unified theory of the issues has been revealed [6].

It has been known that unusual quantum Hall effect occurs in graphene which has a massless Dirac Hamiltonian [7,8]. This comes from the existence of non-zero Berry phase around the Dirac point. In addition, the quantized anomalous Hall effect occurs in a N-layer ABC-stacked graphene system with a term which breaks time-reversal symmetry [9]. Moreover, a valley Hall effect occurs in graphene with broken inversion symmetry. [10]

On the other hand, the topological insulators have recently been focused because of the interesting properties of these materials [11,12]. The topological insulator has the gapless band on the surface of a bulk insulator. This band has Dirac-like dispersion and helical spin polarized. The quantized anomalous Hall effect also occurs in the magnetic topological insulator, which is topological insulator doped with transition metal elements such as Cr or Fe [13-15].

Although the anomalous Hall effects in the usual dispersion systems have been well studied, those in the linear dispersion systems like Dirac electron system have not been studied yet. Moreover, the quantized anomalous Hall effect in N-layer graphene depends on a pseudospin which has nothing to do with a real spin, and the magnetic topological insulator is massless Dirac electron system. Therefore, we study in this paper the anomalous Hall effect in the massive Dirac electron system with a real spin.

2. Method

In this paper, we discuss the anomalous Hall effect in a massive Dirac electron system. Generally, the anomalous Hall effect occurs in the systems whose Hamiltonian contains a spin-orbit interaction and a split term due to the internal magnetic field [6]. Therefore, in this paper we use the Wolff Hamiltonian [16] with an isotropic approximation. Cohen and Blount proposed such a Hamiltonian as an effective model of the electrons in Bi [17] and later Wolff found that this Hamiltonian can be rewritten as the 4×4 Dirac Hamiltonian. This model contains the spin-orbit interaction in an essential way. In order to study the anomalous Hall effect, we add to this Hamiltonian a split term, $\Delta\sigma_z$. Therefore, the Hamiltonian becomes

$$H = \begin{pmatrix} m + \Delta\sigma_z & i\gamma\mathbf{k}\cdot\boldsymbol{\sigma} \\ -i\gamma\mathbf{k}\cdot\boldsymbol{\sigma} & -m - \Delta\sigma_z \end{pmatrix}$$
$$= \begin{pmatrix} m+\Delta & 0 & i\gamma k_z & i\gamma(k_x - ik_y) \\ 0 & m-\Delta & i\gamma(k_x + ik_y) & -i\gamma k_z \\ -i\gamma k_z & -i\gamma(k_x - ik_y) & -m-\Delta & 0 \\ -i\gamma(k_x + ik_y) & i\gamma k_z & 0 & -m+\Delta \end{pmatrix}$$

where m represents the mass gap and we use typical set of parameters, $m = 10$ (meV) and $\gamma = 10^{-7}$ (meV m). Because the sign of a magnetic moment is opposite between conduction and valence bands [18] , the split terms have opposite signs. Note that this split term breaks time reversal symmetry.

The intrinsic Hall conductivity of this system is given by

$$\sigma_{xy} = \sum_n \frac{e^2}{\hbar} \int \frac{d\mathbf{k}}{(2\pi)^d} \Omega^n_{k_x k_y}(\mathbf{k}) f(E(\mathbf{k})),$$

where $f(E(\mathbf{k}))$ is the Fermi distribution function and $\Omega^n_{k_x k_y}$ given by

$$\Omega^n_{k_x k_y}(\mathbf{k}) = i\left(\left\langle \frac{\partial n(\mathbf{k})}{\partial k_x} \middle| \frac{\partial n(\mathbf{k})}{\partial k_y} \right\rangle - (x \leftrightarrow y)\right)$$

is the Berry curvature of the n-th band [3]. This formula is derived from Kubo formula. Although the bands in real systems have a cutoff, the $\mathbf{k}$ integral is taken over all $\mathbf{k}$ values. When the non-zero Berry curvature exists, the velocity becomes

$$\mathbf{v}_n = \frac{\partial E_n(\mathbf{k})}{\hbar \partial \mathbf{k}} + \frac{e}{\hbar}\mathbf{E} \times \boldsymbol{\Omega}^n(\mathbf{k})$$

where $\mathbf{E}$ is a electric field. This contribution from the Berry curvature known as an anomalous velocity is related to a Hall current. In the following, we calculate the energy band, the Berry curvature and the intrinsic Hall conductivity at $T = 0$ by

$$\sigma_{xy} = \sum_n \frac{e^2}{\hbar} \int_{E_n(\mathbf{k})<E_f} \frac{d\mathbf{k}}{(2\pi)^d} \Omega^n_{k_x k_y}(\mathbf{k})$$

in an analytical way. Here, E_f is the Fermi energy. Note that the whole Hall conductivity has other contributions such as the extrinsic contribution which comes from impurities.

In the same way as in the three-dimensional case, we also study the two-dimensional case which is obtained by ignoring k_z dependence. The integration in the Hall conductivity formula is also carried out in the two-dimensional $\mathbf{k}$-space.

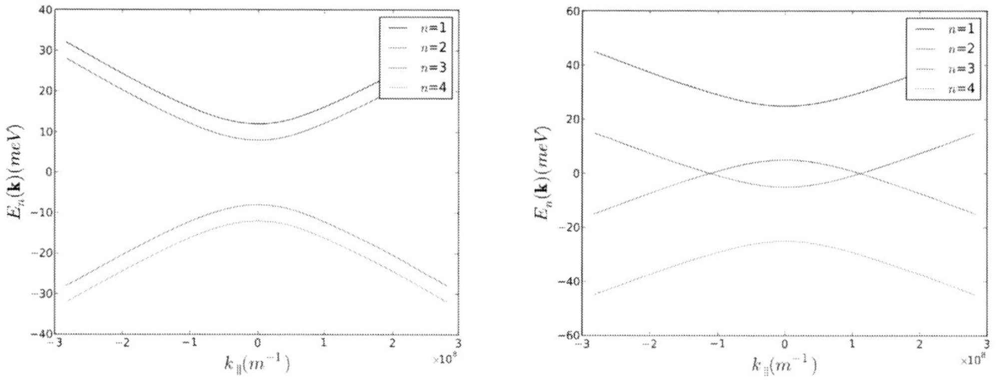

(a) (b)

Figure 1. Energy Band of this system. Here $k_{\|}^2 = k_x^2 + k_y^2$. (a) $m \geq \Delta$ case. ($\Delta = 2$(meV)) (b) $m \leq \Delta$ case. ($\Delta = 15$(meV)) In this case, the second- and the third-band avoid crossing.

3. Result

By diagonalizing the Hamiltonian, we obtain the energy band as

$$E_n = \pm\sqrt{m^2 + \Delta^2 + \gamma^2 k^2 \pm 2\Delta\sqrt{m^2 + \gamma^2(k_x^2 + k_y^2)}} \qquad (n = 1, \cdots 4),$$

where n is the band index in descending order. Figure 1 shows examples of this band dispersion for typical cases.

Using the eigenfunctions, we calculate Berry curvature of this model,

$$\Omega^n_{k_x k_y} = \pm\frac{m\gamma^2}{2(m^2 + \gamma^2(k_x^2 + k_y^2))^{3/2}} \qquad (+ : n = 2 \text{ or } 3, - : n = 1 \text{ or } 4)$$

Opposite signs of the Berry curvature mean that the bands $n = 1$ and 2 ($n = 3$ and 4) have the opposite signs anomalous velocity. Note that the Berry curvatures vanishes when the bands are not split and the bands $n = 1$ and 2 ($n = 3$ and 4) are degenerate. Since the energy bands and the Berry curvatures are symmetric with respect to energy, we consider the cases with $E_f > 0$ and calculate the first- and the second-band contributions in the following.

Owing to the population difference in the spin-up and spin-down electrons, the intrinsic Hall

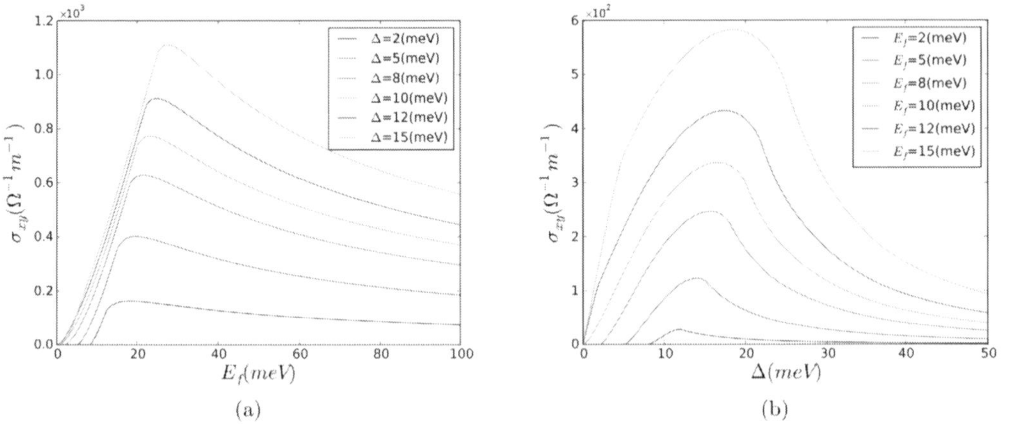

Figure 2. The intrinsic Hall conductivity calculated from the Berry curvature. These are the sum of the contribution from the first- and the second-band. (a) The E_f dependence of the intrinsic Hall conductivity. (b) The Δ dependence of intrinsic Hall conductivity.

conductivities for the first- and the second-band occur and these are obtained analytically as

$$\begin{aligned}\frac{\sigma_{xy}^{(1)}}{\frac{e^2}{h}} &= -\frac{E_f}{2\pi\gamma}\sqrt{1-\left(\frac{m+\Delta}{E_f}\right)^2}+\frac{m}{2\pi\gamma}\arccos\left(\frac{m+\Delta}{E_f}\right)\\ &+\frac{m}{2\pi\gamma}\frac{\Delta}{\sqrt{E_f^2-\Delta^2}}\ln\left(\frac{\sqrt{(E_f-\Delta)(E_f+m+\Delta)}+\sqrt{(E_f+\Delta)(E_f-m-\Delta)}}{\sqrt{(E_f-\Delta)(E_f+m+\Delta)}-\sqrt{(E_f+\Delta)(E_f-m-\Delta)}}\right)\\ &(E_f>m+\Delta)\end{aligned}$$

$$\begin{aligned}\frac{\sigma_{xy}^{(2)}}{\frac{e^2}{h}} &= \frac{E_f}{2\pi\gamma}\sqrt{1-\left(\frac{m-\Delta}{E_f}\right)^2}-\frac{m}{2\pi\gamma}\arccos\left(\frac{m-\Delta}{E_f}\right)\\ &\begin{cases}+\frac{m}{2\pi\gamma}\frac{\Delta}{\sqrt{E_f^2-\Delta^2}}\ln\left(\frac{\sqrt{(E_f+\Delta)(E_f+m-\Delta)}+\sqrt{(E_f-\Delta)(E_f-m+\Delta)}}{\sqrt{(E_f+\Delta)(E_f+m-\Delta)}-\sqrt{(E_f-\Delta)(E_f-m+\Delta)}}\right) & (E_f>|m-\Delta|, E_f>\Delta)\\ +\frac{m}{2\pi\gamma}\frac{2\Delta}{\sqrt{\Delta^2-E_f^2}}\arctan\left(\sqrt{\frac{(\Delta-E_f)(E_f-m+\Delta)}{(E_f+\Delta)(E_f+m-\Delta)}}\right) & (|m-\Delta|<E_f<\Delta)\end{cases}\end{aligned}$$

$$\frac{\sigma_{xy}^{(2)}}{\frac{e^2}{h}} = \frac{1}{2\pi}\frac{m\pi}{\gamma}\left(\frac{\Delta}{\sqrt{\Delta^2-E_f^2}}-1\right) \qquad (0\le E_f\le\Delta-m)$$

where $\sigma_{xy}^{(n)}$ is the contribution from the n-th band.

There are two reasons why the analytic form of the intrinsic Hall conductivity changes depending on the parameters, E_f, m, and Δ. Firstly, since the first band has a band minimum at $E=m+\Delta$, it starts to contribute to σ_{xy} when E_f exceeds $m+\Delta$. Secondly, anticrossing of the second band changes the range of integration when $\Delta>m$.

Figure 2 (a) shows the intrinsic Hall conductivity as a function of E_f. It begins to increase at $E_f=m-\Delta$ and starts to decrease near $E_f=m+\Delta$. This is because the second-band which has the positive Berry curvature starts to contribute to σ_{xy} at $E_f=m-\Delta$ and the first-band which has the negative Berry curvature joins from $E_f=m+\Delta$. Figure 2 (b) shows that the intrinsic Hall conductivity is 0 at $\Delta=0$ because the bands are not split on this occasion.

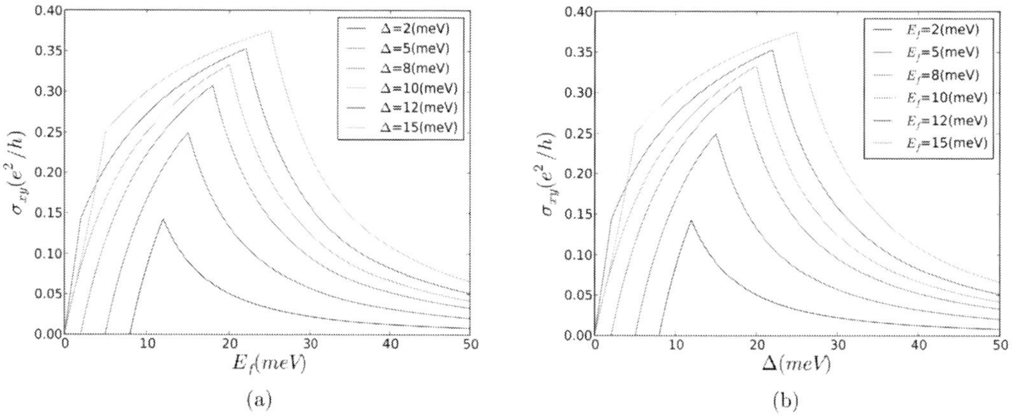

Figure 3. The intrinsic Hall conductivity in the two-dimensional case. (a) The E_f dependence of the intrinsic Hall conductivity. (b) The Δ dependence of intrinsic Hall conductivity. However, this is identical with (a).

In the two-dimensional system, the energy bands and the Berry curvatures are obtained by replacing k_z with 0. The intrinsic Hall conductivity is given by

$$\frac{\sigma_{xy}}{\frac{e^2}{h}} = \begin{cases} \frac{1}{2}\frac{E_f+\Delta-m}{E_f+\Delta} & (m-\Delta < E_f < m+\Delta, m > \Delta \quad or \quad \Delta - m < E_f < \Delta + m, \Delta > m) \\ \frac{m\Delta}{(E_f-\Delta)(E_f+\Delta)} & (m+\Delta < E_f) \\ \frac{mE_f}{(\Delta-E_f)(\Delta+E_f)} & (0 < E_f < \Delta - m, \Delta > m) \end{cases}$$

Interestingly, this formula is symmetric with respect to exchange of E_f for Δ. This is clearly shown in Figure 3.

4. Conclusion

In this paper, we showed that the anomalous Hall conductivity occurs in the massive Dirac electron system with a spin-orbit interaction and a split term. Unlike the magnetic topological insulator, the anomalous Hall conductivity is not quantized. This is because the quantized anomalous Hall effect in the magnetic topological insulator originates from the massless Dirac fermion.

Although we only studied the system with broken time reversal symmetry in this paper, it is interesting to study the systems in which other symmetries are broken. In the N-layer graphene system, it has been shown that some interesting quantum phases occur [9]. In addition, a valley Hall effect in graphene with broken inversion symmetry [10] is essentially similar to our result in the two-dimensional case except for a kind of broken symmetry. Therefore, it is interesting to study the system with broken inversion symmetry in the three-dimensional case.

Experimentally, this anomalous Hall conductivity will be observed in such systems as a ferromagnetic Dirac electron system or a ferromagnet-coated Dirac electron system. In this paper, only the intrinsic contribution was considered. Other contributions, such as the extrinsic contribution from impurities, will be observed in experiments, so further calculation is necessary for the extrinsic contribution. However, the result of this paper will help to identify the intrinsic contribution by measuring the chemical potential dependence and the split gap dependence.

References

[1] N. Nagaosa et al. 2010 *Rev. Mod. Phys.* **82**, 1539
[2] R. Karplus and J. M. Luttinger 1954 *Phys. Rev.* **95**, 1154
[3] D. Xiao et al. 2010 *Rev. Mod. Phys.* **82**, 1959
[4] M. Onoda and N. Nagaosa 2002 *J. Phys. Soc. Jpn.* **71**, 19-22
[5] T. Jungwirth et al. 2002 *Phys. Rev. Lett.* **88**, 207208
[6] S. Onoda et al. 2006 *Phys. Rev. Lett.* **97**, 126602
[7] N. M. R. Peres et al. 2006 *Phys. Rev. B* **73**, 125411
[8] Y. Zhang et al. 2005 *Nature Phys.* **438**, 201-204
[9] F.Zhang et al. 2011 *Phys. Rev. Lett.* **106**, 156801
[10] D. Xiao et al. 2007 *Phys. Rev. Lett.* **99**, 236809
[11] L. Fu et al. 2007 *Phys. Rev. Lett.* **98**, 106803
[12] D. Hsieh et al. 2008 *Nature* **452** 970-974
[13] R. Yu et al. 2010 *Science* **329**, 61
[14] C. Chang et al. 2013 *Science* **340**, 167
[15] J. G. Checkelsky et al. 2014 *Nature Phys.* **10**, 731736
[16] P. Wolff 1964 *J. Phys. Chem. Solids* **25**, 1057
[17] M. Cohen and E. Blount 1960 *Philos. Mag.* **5**, 115
[18] Y. Fuseya et al. 2011 *J. Phys. Soc. Jpn.* **81**, 013704

Negative magnetoresistance in Ti-cleaned single-layer graphene

A Fujimoto[1], C A Joiner[2], Y Jiang[3], D Terasawa[4], A Fukuda[4], Z Jiang[3] and E M Vogel[2]

[1]Applied Physics, Osaka Institute of Technology, Osaka, Japan

[2]School of Materials Science & Engineering, Georgia Tech, Atlanta, GA, USA

[3]School of Physics, Georgia Tech, Atlanta, GA, USA

[4]Department of Physics, Hyogo College of Medicine, Nishinomiya, Hyogo, Japan

E-mail: akira.fujimoto@oit.ac.jp

Abstract. Symmetric graphene tunneling field-effect transistors (SymFETs) consisting of two independently gated graphene layers separated by a potential barrier have been proposed as a room-temperature resonant tunneling device. In a SymFET, the inelastic-coherent length (L_φ) of the electrons is considered to be one of the important characteristic scattering lengths. Weak localization (WL) is a good tool to estimate L_φ. In this study, the surface of the chemical-vapor-deposited single-layer graphene was Ti cleaned after performing the conventional fabrication processes for graphene transistors. We found that the charge-neutral point (V_{CNP}) shifted to a lower back-gate voltage and the mobility increased owing to Ti cleaning. Ti-cleaned Hall bars were investigated at 0.3 K under magnetic fields of up to 14 T. Negative magnetoresistance (MR) appears because of the WL effect, and the MR increases as the back-gate voltage (V_G) approaches V_{CNP}. From a fitting analysis using the theoretical formulation of WL, we found that L_φ was greater than 100 nm and that L_φ decreased as V_G approached V_{CNP} because of electron–hole puddles and electron–electron interaction.

1. Background

Graphene is a promising candidate for future electronic devices owing to the high mobility of its carriers and its perfect two-dimensionality. However, single-layer graphene transistors have a low on–off ratio due to the zero bandgap. Therefore, new transistors using double-layer graphene have been proposed, among which is the bilayer pseudospin field-effect transistor (BiSFET) [1]. In a BiSFET, electrons and holes in respective layers form excitons, and Bose–Einstein condensation occurs. Consequently, the tunnel resistance between two layers reduces and a high on-current is obtained. However, BiSFET operation has not been experimentally demonstrated. Another new type of transistor is the symmetric graphene tunneling field-effect transistor (SymFET) [2, 3], which consists of two independently gated graphene layers separated by a tunnel barrier. The SymFET has been proposed as a room-temperature resonant tunneling device because a large resonant current is produced when the Dirac points of the graphene sheet are aligned by the applied voltage, resulting in negative differential resistance (NDR). This phenomenon requires that the tunneling electrons conserve their lateral momentum, and the inelastic-coherent length (L_φ) of the electrons is considered to be one of the important characteristic scattering lengths. In our previous studies, we fabricated double-layer graphene with h-BN tunnel dielectrics formed using CVD. The conventional tunneling characteristics for these devices were equivalent to similar characteristics for exfoliated devices [4, 5]. However, devices fabricated using CVD graphene and h-BN have not yet exhibited a large resonant peak and NDR.

In comparison with exfoliated graphene, chemical-vapor-deposited (CVD) graphene is better for fabricating devices on a large scale. However, the charge-neutral point (V_{CNP}) of CVD graphene appears at a relatively high back-gate voltage because of molecular adsorption [6] and impurities on the graphene surface. One of the challenges with CVD graphene is that the transfer process typically results in contamination, which is very difficult to remove. A good method to overcome this issue is Ti cleaning [7]. Some researchers have reported p-type and n-type graphene produced using the adsorption of Ti. Li *et al.* [8] reported a *p*-type graphene FET with titanium oxide on the graphene surface. The conductance (G) is not controlled by the back-gate voltage, and the Fermi level is pinned between an electrode and graphene. Iqbal *et al.* [9] reported an *n*-type Ti-coated graphene FET and showed that the V_{CNP} shifts toward a negative back-gate voltage and electron transport becomes dominant.

In this study, by performing Ti cleaning during the graphene Hall bar fabrication process, we improved the graphene surface to observe quantum transport phenomena. In particular, weak localization (WL) measurements were performed to estimate the inelastic and intervalley scattering lengths.

2. Experimental

CVD single-layer graphene on Cu purchased from ACS Material was used. A solution-based transfer process was used to transfer the graphene onto 90 nm or 268 nm SiO_2 on Si substrates, and then annealed in forming gas [10]. Back-gated Hall bars were fabricated using conventional photolithography and etching in O_2 plasma. Electrodes ware fabricated with e-beam evaporation for the deposition of Ni contacts with Au. After patterning CVD-graphene Hall bars, 2 nm of Ti was deposited on the graphene using e-beam evaporation and allowed to oxidize to Ti oxide upon exposure to air. The Ti oxide was then removed with HF, and the graphene surface was cleaned of resist residue.

Four kinds of graphene Hall bars were employed in our measurements as shown in Table 1. Figure 1 shows one of the measured single-layer graphene Hall bars. Two Hall bars, one with (#1) and the other without Ti cleaning (#2), were measured in this study at 77 K. At low temperatures, we examined another two Ti-cleaned Hall bars on 268-nm-thick SiO_2 (#3 and #4). #3 and #4 were fabricated on the same die. We employed a ^{3}He cryostat system with a 14 T superconducting magnet for cooling to 0.3 K. At low temperatures, a standard ac lock-in technique was used with an excitation current of 50 nA to prevent Joule heating. Our measurements in the magnetic field were performed for different carrier types and densities controlled using the back-gate voltage (V_G).

Table 1. Four kinds of graphene Hall bars measured in our study. #3 and #4 were fabricated on the same die.

Sample No.	SiO_2 thickness (nm)	Ti-cleaning
#1	90	with
#2	90	without
#3	268	with
#4	268	with

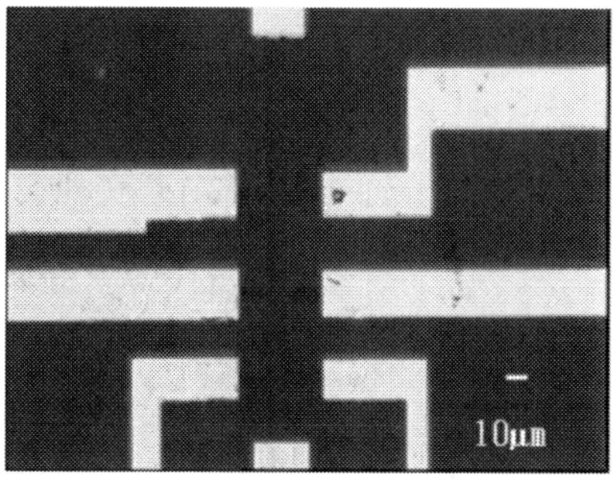

Figure 1. One of the single-layer graphene Hall bars measured in this study.

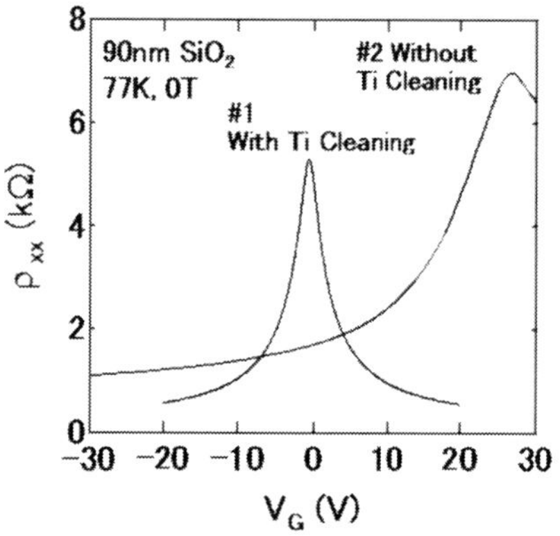

Figure 2. Back-gate voltage (V_G) dependence of longitudinal resistance (ρ_{xx}) with/without Ti cleaning at 77 K. The charge neutral point shifts to approximately $V_G = 0$ V owing to Ti cleaning.

3. Results and Discussion

Figure 2 shows the V_G dependence of longitudinal resistance (ρ_{xx}) with (#1) and without Ti cleaning (#2) at 77 K. V_{CNP} decreases to approximately $V_G = 0$ V owing to Ti cleaning, and the mobility increases with Ti cleaning. The graphene Hall bars tended to become *n*-type with Ti cleaning (#1), and this result supported the report by Iqbal *et al.* [9]. To analyze the data, we employed the constant mobility model proposed by Kim *et al.* [11]. The graphene transistor's resistance (R) consists of contact (R_{Cont}) and channel resistances ($R_{Channel}$), in which carrier concentration in the graphene channel is expressed as the square root of the sum of the squares of intrinsic carrier concentration (n_0) and carrier concentration (n_{ind}) induced by back-gate voltage (V_{bg}*), where $V_{bg}^* = V_G - V_{CNP}$. Kim *et al.* assumed that the mobility (μ) is constant and independent of V_G. The resistance is expressed as follows:

$$R = R_{Cont} + R_{Channel} \quad (1)$$

$$= R_{Cont} + \frac{L / W}{\sqrt{n_0^2 + n_{ind}^2(V_{bg}^*)}\; e\mu}. \quad (2)$$

Here, e is the electron charge and L and W are the length and width of the measured device, respectively. Table 2 lists the parameters extracted with a fitting analysis using Kim's model. By performing Ti cleaning, μ increases and n_0 decreases because of the improvement of the graphene surface. The most important advantage of Ti cleaning is that the V_{CNP} shifts to a lower V_G. On the other hand, R_{Cont} increases; as we perform the Ti-cleaning after fabricating the electrodes, Ti cleaning seems to negatively affect R_{Cont}. Furthermore, we have checked the electrical properties of various CVD

Table 2. Characteristic parameters extracted with a fitting analysis using Kim's model.

	R_{cont} (Ω)	n_0 (cm^{-2})	μ (cm^2/Vs)	V_{CNP} (V)
With Ti cleaning (#1)	7.1 x 10^2	3.0 x 10^{11}	4.6 x 10^3	-0.6
Without Ti cleaning (#2)	4.9 x 10^2	1.2 x 10^{12}	7.8 x 10^2	+26

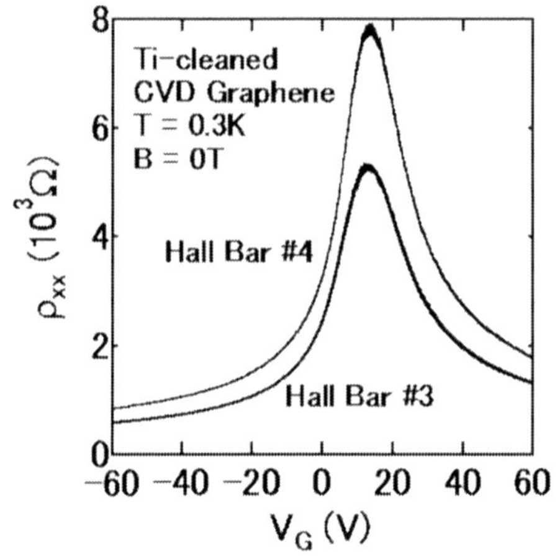

Figure 3. V_G dependence of ρ_{xx} for two Ti-cleaned single-layer graphene Hall bars with a 268-nm-thick SiO_2 film at 0.3 K and 0 T.

graphene transistors and confirmed that V_{CNP} shifts to a lower V_G and μ improves upon Ti cleaning [7]. On the other hand, Morozov *et al.* reported that resistivity of doped graphene could empirically be described by two contributions: $\rho_L \propto 1/n$ and ρ_S independent of n due to long- and short-range scatterers, respectively, where n is the carrier concentration [12]. The constant mobility model proposed by Kim *et al.* is also composed of two terms, which are dependent and independent of n. Therefore we need to measure the detailed temperature dependence of resistivity of graphene and discuss our results in the future.

Figure 3 shows the V_G dependence of ρ_{xx} for two Ti-cleaned single-layer graphene Hall bars with a 268-nm-thick SiO_2 film (#3 and #4) at 0.3 K and 0 T. V_{CNP} appears at +13 V for the two Hall bars, and its value is larger than that of Hall bar #1 because of the SiO_2 film. Figure 4 shows ρ_{xx} and Hall resistance (ρ_{xy}) of Hall bar #3 at 0.3 K and a magnetic field of 14 T applied perpendicular to the graphene surface. The value of ρ_{xx} appears to oscillate under the magnetic field (B) with changing V_G, and the sign of ρ_{xy} inverts at V_{CNP}. Furthermore, we calculated σ_{xx} (= ρ_{xx} / (ρ_{xx}^2+ρ_{xy}^2)) and σ_{xy} (= ρ_{xy} / (ρ_{xx}^2+ρ_{xy}^2)). Figure 5 shows the quantum Hall plateaus, which exhibits a half-integer step ($0.5 \times 4e^2/h$). This indicates that the samples are single-layer graphene [13]. However, σ_{xx} is not 0 at the Hall plateaus. In fact, there exists some mixing between σ_{xx} and σ_{xy}. At 0.3 K Hall mobility of Hall bar #3 is 3.6×10^3 cm^2/Vs at V_G = 0 V.

Figure 6 shows the V_G dependence of the magnetoresistance (MR) for magnetic fields less than 1 T at 0.3 K. The figure shows three MR curves of Hall bar #3 for V_G = -60 V (hole transport region), +13 V (charge neutral point), and +60 V (electron transport region). The vertical axis in figure 6 represents the ratio $\Delta\,\rho_{xx}(B)$ / $\rho_{xx}(0)$, which is the rate of change of MR to the resistance in the absence of a

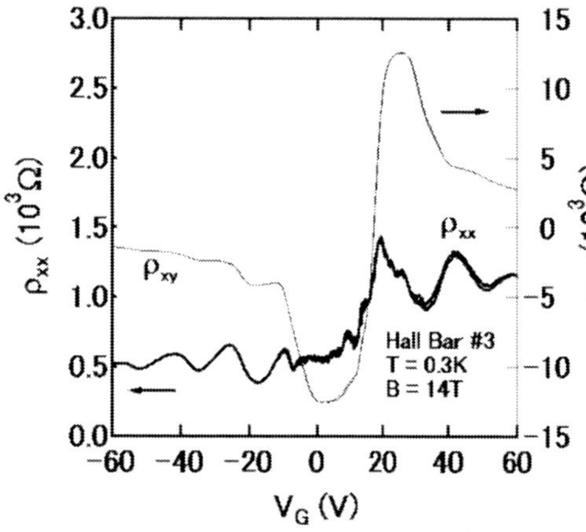

Figure 4. ρ_{xx} and ρ_{xy} as functions of V_G of Hall bar #3 at 0.3 K and 14 T.

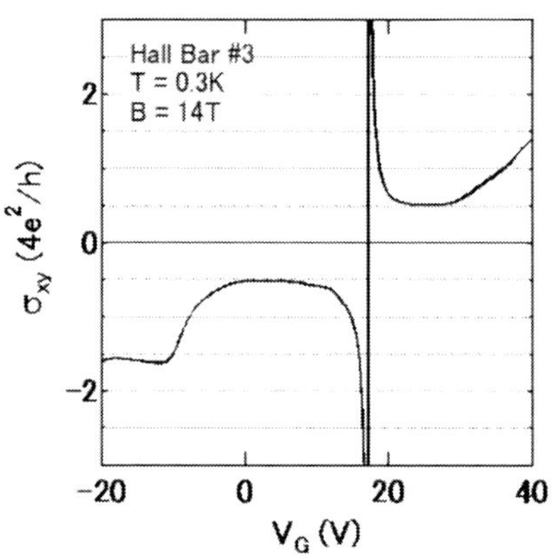

Figure 5 σ_{xy} as a function of V_G calculated with ρ_{xx} and ρ_{xy}.

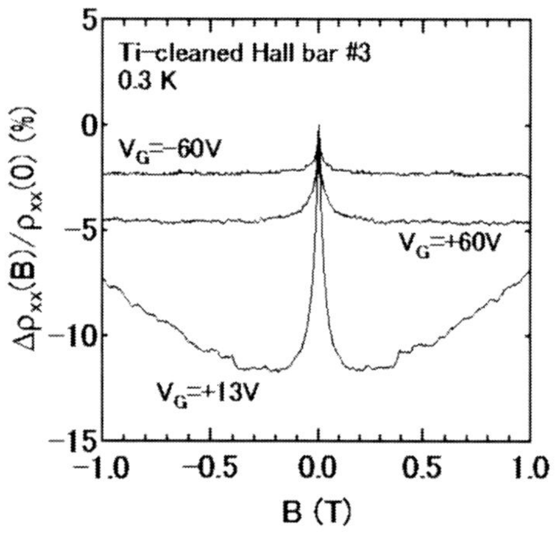

Figure 6. Back-gate voltage (V_G) dependence of MR at 0.3 K.

magnetic field. The MR shows a gradual decrease with B and increases as V_G approaches V_{CNP} (= +13 V for this Hall bar). The negative MR at V_G = +13 V is several times larger than that at V_G = -60V. The negative MR is caused by the WL effect. WL is a constructive quantum interference effect of the scattered electron waves at low temperatures. The WL states for electrons are broken by a magnetic field and inelastic scattering such as phonon scattering, which reduces resistance. The theory of MR due to the WL in single-layer graphene was proposed by McCann *et al.* [14]. The theory should hold for not only inelastic but also elastic scattering. Elastic scattering that breaks chirality will destroy the interference within each of the two graphene valleys in k space. The theory is expressed as follows:

$$\Delta\rho(B) = -\frac{e^2\rho^2}{\pi h}\left[F\left(\frac{B}{B_\phi}\right) - F\left(\frac{B}{B_\phi + 2B_i}\right) - 2F\left(\frac{B}{B_\phi + B_*}\right)\right], \qquad (3)$$

$$F(z) = \ln z + \psi\left(\frac{1}{2} + \frac{1}{z}\right), \qquad (4)$$

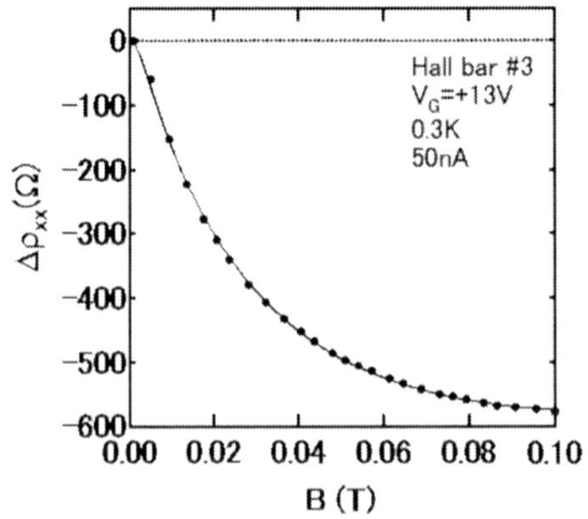

Figure 7. Fitting result for $V_G = +13$ V for Hall bar #3 below 0.1 T. Black circles represent the experimental data, and the black solid line is the fitting result.

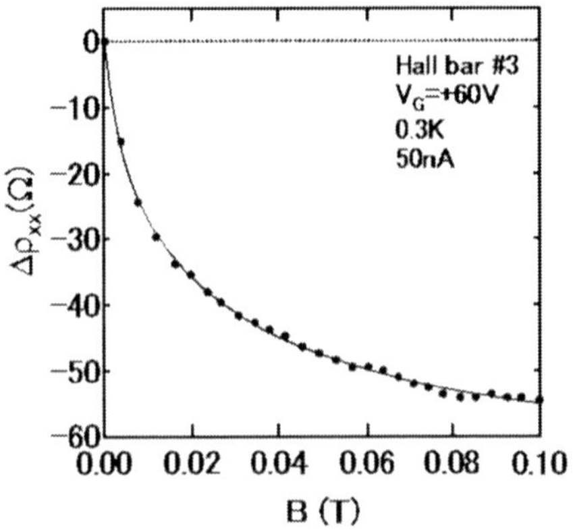

Figure 8. Fitting result for $V_G = +60$ V for Hall bar #3 below 0.1 T. Black circles represent the experimental data, and the black solid line is the fitting result.

$$B_\varphi = \frac{\hbar}{4De\tau_\varphi}, \quad B_i = \frac{\hbar}{4De\tau_i}, \quad B_* = \frac{\hbar}{4De\tau_*}. \tag{5}$$

The first term in equation (3) is responsible for WL, while the other terms express weak anti-localization (WAL). Here, ρ is the resistivity, h is the Planck's constant, $\psi(x)$ is the digamma function, D is the diffusion constant, and $\tau_*^{-1} = \tau_w^{-1} + \tau_z^{-1} + \tau_i^{-1}$. τ_ϕ, τ_i, and τ_z are the inelastic, intervalley, and intravalley scattering times, respectively. τ_w is the effect of trigonal warping. One can see that the curve for $V_G = +13$ V in figure 6 has a steeper upward turn, indicating the greater importance of the second and third terms (WAL) in equation (3) for magnetic fields above 0.3 T. On the other hand, the curves for V_G = -60 and +60V show only negative MR, indicating the importance of the first term (WL) in equation (3). In this study, we focus on the negative MR below 0.1 T. We confirmed that τ_w and τ_z did not contribute to our MR data by performing the fitting analysis. That is, the effect of trigonal warping and intravalley scattering are suppressed in our Hall bars. Therefore, we performed the fitting analysis with the assumption $B_i = B_*$. Figures 7 and 8 show the results of the fits for V_G = +13 V and +60 V for Hall bar #3, respectively. The black circles represent the experimental data, and the black solid lines represent the results of fitting using the theoretical expression for magnetic fields below 0.1 T. We could obtain the best fits and extract the inelastic-coherent and intervalley scattering lengths.

Figure 9 shows the gate-voltage dependence of the inelastic-coherent (L_φ) and intervalley scattering lengths (L_i) extracted from the MR analyses for Hall bars #3 and #4. At a temperature of 0.3 K, L_φ is greater than 100 nm. L_φ is also observed to decrease as V_G approached V_{CNP}. Although the average carrier concentration vanishes at V_{CNP}, electrons and holes may form electron–hole puddles. Electrons and holes inelastically scatter more frequently because of the electron–hole puddle and electron–electron interaction [15]. On the other hand, L_i is nearly constant with respect to V_G, but L_i is always less than L_φ. The mechanism of intervalley scattering is considered to be the significant fluctuations in potential energy in the atomically sharp defects. The atomically sharp defect is known to strengthen the intervalley scattering [16]. In our Hall bars, the atomically sharp defects might originate from dislocations and some domains in our CVD graphene.

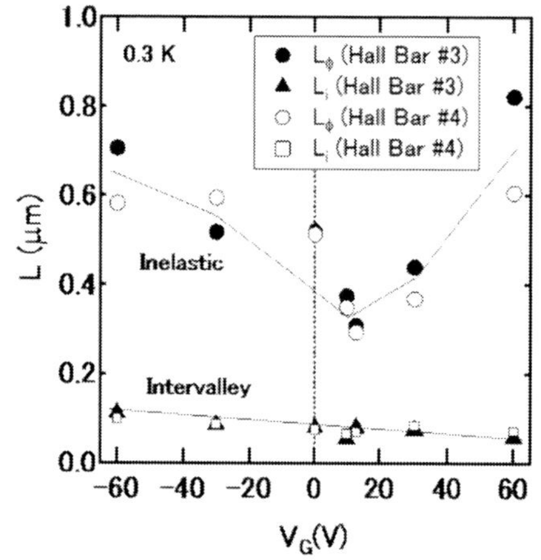

Figure 9. Gate-voltage dependence of the inelastic-coherent (L_φ) and intervalley scattering lengths (L_i) extracted from the analyses of MR for Hall bars #3 and #4 at 0.3 K. The solid lines are guides to the eyes.

4. Conclusions

Ti cleaning was performed after the conventional fabrication of CVD single-layer graphene Hall bars in order to clean the surfaces. V_{CNP} appeared at a lower back-gate voltage and the mobility increases because of the Ti cleaning. At 0.3 K, negative MR appears because of the WL effect. As V_G approaches V_{CNP}, the negative MR increases. By performing a fitting analysis using the theoretical expression, L_ϕ was observed to decrease as V_G approaches V_{CNP} because of electron–hole puddles and electron–electron interaction. On the other hand, L_i is nearly constant with respect to V_G, and L_i is always less than L_ϕ, indicating that atomically sharp defects might influence the intervalley scattering. In order to examine the origins of the inelastic scattering at V_{CNP} and the other scattering mechanisms, a detailed investigation of the V_G and the temperature dependences of the MR is required. Our results will stimulate further investigations on the physics of graphene and applications such as SymFET.

References

[1] Banerjee S K, Register L F, Tutuc E, Reddy D and MacDonald A H 2009 *IEEE Electron Dev. Lett.* **30**(2) 158-160
[2] Feenstra R M, Jena D and Gu G 2012 *J. Appl. Phys.* **111**(4) 043711
[3] Britnell L, Gorbachev R V, Geim A K, Ponomarenko L A, Mishchenko A, Greenaway M T, Fromhold T M, Novoselov K S and Eaves L 2013 *Nature Comm.* **4** 1794
[4] Roy T, Hesabi Z R, Joiner C A, Fujimoto A and Vogel E M 2013 *Microelectron. Eng.* **109** 117
[5] Roy T, Liu L, de la Barrera S, Chakrabarti B, Hesabi Z R, Joiner C A, Feenstra R M, Gu G and Vogel E M 2014 *Appl. Phys. Lett.* **104**(12) 123506
[6] Sidorov A N, Sherehiy A, Jayasinghe R, Stallard R, Benjamin D K, Yu Q, Liu Z, Wu W, Cao H, Chen Y P, Jiang Z and Sumanasekera G U, 2011 *Appl Phys Lett.* **99**(1) 013115
[7] Joiner C A, Roy T, Hesabi Z R, Chakrabarti B and Vogel E M 2014 *Appl. Phys. Lett.* **104**(22) 223109
[8] Li H, Zhang Q, Liu C, Xu S and Gao P 2011 *ACS Nano* **5**(4) 3198
[9] Iqbal M W, Singh A K, Iqbal M Z and Eom J 2012 *J. Phys. Cond. Mat.* **24**(33) 335301
[10] Chan J, Venugopal A, Pirkle A, McDonnell S, Hinojos D, Magnuson C W, Ruoff R S, Colombo L, Wallace R M and Vogel E M, 2012 *ACS Nano* **6**(4) 3224

[11] Kim S, Nah J, Jo I, Shahrjerdi D, Colombo L, Yao Z, Tutuc E and Banerjee S K 2009 *Appl. Phys. Lett.* **94**(6) 062107
[12] Morozov S V, Novoselov K S, Katsnelson M I, Schedin F, Elias D C, Jaszczak J A and Geim A K, 2008 *Phys. Rev. Lett.* **100**(1) 016602
[13] Geim A K and Novoselov K S 2007 *Nat. Mater.* **6**(3) 183
[14] McCann E, Kechedzhi K, Fal'ko V I, Suzuura H, Ando T and Altshuler B L 2006 *Phys. Rev. Lett.* **97**(14) 146805
[15] Ki D K, Jeong D, Choi J H, Lee H J and Park K S 2008 *Phys. Rev. B* **78**(12) 125409
[16] Tikhonenko F V, Horsell D W, Gorbachev R V and Savchenko A K 2008 *Phys. Rev. Lett.* **100**(5) 056802

Nontrivial contributions to the magnetoconductivity due to anomalous g-factor in the Luttinger Hamiltonian

Riki Suemasa and Yuki Fuseya

Department of Engineering Science, University of Electro-Communications, Chofu, Tokyo 182-8585, Japan

E-mail: `suemasa@kookai.pc.uec.ac.jp`

Abstract. The effective g-factor and the selection rules for the matrix elements of the velocity operator are investigated for the Luttinger Hamiltonian under a magnetic field. It is shown that the g-factor has a strong k_z-dependence, where k_z is the wave vector along the magnetic field. This anomalous k_z-dependence is due to the interband effect of the magnetic field between the light and heavy hole bands. It is found that a nontrivial contribution to the longitudinal magnetoconductivity arises also due to the interband effects by analyzing the selection rules. This nontrivial interband contribution can be a source of the longitudinal spin current.

1. Introduction

The Luttinger Hamiltonian was introduced first as an effective Hamiltonian for semiconductors such as Si or Ge [1]. In these semiconductors, three bands are degenerate at $\boldsymbol{k} = 0$ when there is no spin-orbit interaction (SOI) and effects of electron spin are neglected. If the SOI is present, these three degenerate bands (sixfold due to spins) breaks up into one fourfold and one twofold bands. The fourfold band lies higher than the twofold one and it is more important in practice. The Luttinger Hamiltonian is the effective Hamiltonian derived for this fourfold band on the basis of $\boldsymbol{k} \cdot \boldsymbol{p}$ theory. It is written in terms of 4×4 matrix. The Luttinger Hamiltonian is valid also for semiconductors such as GaAs and InSb at temperatures higher than the energy scale of the band splitting originated from the lack of the inversion symmetry [2]. Thus the Luttinger Hamiltonian is the most fundamental effective Hamiltonian in semiconductor physics. Moreover, the Luttinger Hamiltonian has attracted revived interests since the proposal of intrinsic spin Hall effect in the Luttinger Hamiltonian [3, 4].

In addition, the relationship to the Dirac Hamiltonian, which is also written in terms of 4×4 matrix, is an interesting subject. In some materials with a strong SOI, such as Bi and PbTe, the effective Hamiltonian is essentially equivalent to the Dirac Hamiltonian except for a spatial anisotropy [5, 6]. It was found that the Dirac Hamiltonian also exhibits the intrinsic spin Hall effect [7, 8]. Therefore, the Luttinger and Dirac Hamiltonian should share common features since both of them are written in terms of 4×4 matrix and the interband matrix elements due to the SOI plays crucial roles. But it has not been clearly understood yet.

The purpose of the present work is to revisit the Luttinger Hamiltonian and study the quantum transport phenomena paying special attention to the anomalous property of g-factor in the Luttinger Hamiltonian. It is well known that the effective g-factor can be greatly different from the value for bare electrons ($g = 2$) when there is a SOI [1, 9–11]. Actually, the effective g-factor of Bi can exceed 1000 due to the strong SOI [6, 12]. Furthermore, the effective g-factor can be $\boldsymbol{k}$-dependent. It was approximately shown [11] that the effective g-factor of the Luttinger Hamiltonian exhibit a linear-k_z dependence for large quantum numbers of Landau levels as $g^* \propto |k_z|$, where k_z is the wave vector along the magnetic field.

Here we propose that an interesting spin transport will emerge with such a k-dependent effective g-factor. For example, suppose the effective Hamiltonian is given as

$$\mathcal{H} = \frac{\hbar^2 k^2}{2m^*} + \frac{g^*(k_z)}{2}\mu_{\rm B}\boldsymbol{\sigma}\cdot\boldsymbol{B}, \tag{1}$$

where $\mu_{\rm B} = e\hbar/2mc$ is the Bohr magneton, $\boldsymbol{\sigma}$ is the Pauli spin matrix, and $\boldsymbol{B}$ is an external magnetic field. The second term of Eq. (1) is the effective Zeeman term originates from the orbital motion of electrons (not the bare Zeeman term). In general cases, g^* is independent from k_z, so that the current $j_i = ev_i = e\partial\mathcal{H}/\hbar\partial k_i$ is independent from g^* and the spin. (e is the elementary charge and v_i is the velocity along the i-direction.) By contrast, in the case that g^* depends on k_z, the current is different between opposite spins as

$$j_{z\uparrow,\downarrow} \;=\; ev_z(k_z) \pm \frac{e\mu_{\rm B}B_z}{2\hbar}\frac{\partial g^*(k_z)}{\partial k_z}. \tag{2}$$

This suggests that the spin current, $j_{si} = j_{i\uparrow} - j_{i\downarrow}$, can flow along the magnetic field when the g^* depends on k_z. Of course, the above argument based on the quasi-classical picture is too naive to discuss the spin transport phenomena, which is the essentially quantum phenomena. In order to examine this idea, we need to study the magnetoconductivity based on the quantum transport theory.

2. Model

The Luttinger Hamiltonian under a magnetic filed is given by

$$\mathcal{H} = \mathcal{H}_0 + \mathcal{H}_1 + \mathcal{H}_2, \tag{3}$$

$$\mathcal{H}_0 = \frac{1}{2m}\begin{pmatrix} \mathcal{H}_0' & 0 \\ 0 & \mathcal{H}_0'' \end{pmatrix}, \tag{4}$$

$$\mathcal{H}_0' = \begin{pmatrix} (\gamma_1+\bar\gamma)({\pi_x}^2+{\pi_y}^2)+3\hbar^2 s\kappa & -\sqrt{3}\bar\gamma(\pi_x-i\pi_y)^2 \\ -\sqrt{3}\bar\gamma(\pi_x+i\pi_y)^2 & (\gamma_1-\bar\gamma)({\pi_x}^2+{\pi_y}^2)-\hbar^2 s\kappa \end{pmatrix}, \tag{5}$$

$$\mathcal{H}_0'' = \begin{pmatrix} (\gamma_1-\bar\gamma)({\pi_x}^2+{\pi_y}^2)+\hbar^2 s\kappa & -\sqrt{3}\bar\gamma(\pi_x-i\pi_y)^2 \\ -\sqrt{3}\bar\gamma(\pi_x+i\pi_y)^2 & (\gamma_1+\bar\gamma)({\pi_x}^2+{\pi_y}^2)-3\hbar^2 s\kappa \end{pmatrix}. \tag{6}$$

$$\mathcal{H}_1 = \frac{1}{2m}\begin{pmatrix} (\gamma_1-2\bar\gamma){\pi_z}^2 & 0 & 0 & 0 \\ 0 & (\gamma_1+2\bar\gamma){\pi_z}^2 & 0 & 0 \\ 0 & 0 & (\gamma_1+2\bar\gamma){\pi_z}^2 & 0 \\ 0 & 0 & 0 & (\gamma_1-2\bar\gamma){\pi_z}^2 \end{pmatrix}, \tag{7}$$

$$\mathcal{H}_2 = \frac{1}{2m}\begin{pmatrix} 0 & 0 & -2\bar\gamma\sqrt{3}(\pi_x-i\pi_y)\pi_z & 0 \\ 0 & 0 & 0 & 2\bar\gamma\sqrt{3}(\pi_x-i\pi_y)\pi_z \\ -2\bar\gamma\sqrt{3}(\pi_x+i\pi_y)\pi_z & 0 & 0 & 0 \\ 0 & 2\bar\gamma\sqrt{3}(\pi_x+i\pi_y)\pi_z & 0 & 0 \end{pmatrix}, \tag{8}$$

where $\boldsymbol{\pi} = \boldsymbol{p} + (e/c)\boldsymbol{A}$ is the kinetic momentum operator, c is the light velocity, $\boldsymbol{A}$ is the vector potential of the external magnetic field, and $s = eB/\hbar c$. γ_1, $\bar{\gamma}$, and κ are the effective mass parameters [1].

Hereafter we consider the case of the magnetic field is in the z direction. When $k_z = 0$, the Hamiltonian is block diagonalized as $\mathcal{H} = \mathcal{H}_0$ and the eigenvalues can be analytically obtained just by solving the 2×2 matrix problem [10]. An interesting nontrivial contribution arises from the interband matrix elements of $\mathcal{H}_2$. The eigenvectors of $\mathcal{H}$ can be written as [1, 10]

$$\psi_{l,m} = \begin{pmatrix} c_{1,\alpha} G_{l-2} \\ c_{2,\alpha} G_l \\ c_{3,\alpha} G_{l-1} \\ c_{4,\alpha} G_{l+1} \end{pmatrix}, \tag{9}$$

in terms of the functions

$$G_l = \frac{e^{i(k_x x + k_z z)}}{\sqrt{L_x L_z}} \sqrt{\frac{\sqrt{s}}{2^l l! \sqrt{\pi}}} H_l(t) e^{-t^2/2}, \tag{10}$$

where $t = \sqrt{s}y - k_z/\sqrt{s}$ and $H_l(t)$ is the Hermite polynomials. The functions G_l satisfy following relations in terms of s and $\zeta = k_z/\sqrt{s}$:

$$(\pi_x^2 + \pi_y{}^2) G_l = \hbar^2 s(2l+1) G_l, \tag{11}$$
$$(\pi_x + i\pi_y) G_l = -\hbar\sqrt{s}\sqrt{2(l+1)} G_{l+1}, \tag{12}$$
$$(\pi_x - i\pi_y) G_l = -\hbar\sqrt{s}\sqrt{2l} G_{l-1}, \tag{13}$$
$$\pi_z G_l = \hbar\sqrt{s}\zeta G_l. \tag{14}$$

For each quantum number l, there are four different eigenvalues, whose degree of freedom is denoted by $\alpha = 1 \sim 4$. The coefficients $c_{1,\alpha} \sim c_{4,\alpha}$ are determined by solving numerically the 4×4 secular equation.

3. Anomalous g-factor

The eigenenergies of the Luttinger Hamiltonian for $\gamma_1 = 13.20$, $\bar{\gamma} = 4.92$, and $\kappa = 3.30$, which are the parameters for Ge [10], are shown in Fig. 1. Not the that the energy axis is inverted in Fig. 1, namely, the hole bands look like "electron like". The left and right panels of Fig. 1 correspond to the heavy and light holes, respectively. A non-parabolic dispersion, which was pointed out by Wallis and Bowlden [10], is obtained for the heavy holes. This non-parabolic dispersion is due to $\mathcal{H}_2$ and gives an unusual effective g-factors.

The effective g-factor is determined by [11]

$$g^* = \frac{E_{n\uparrow} - E_{n\downarrow}}{\mu_B B} = 2\frac{E_{n\uparrow} - E_{n\downarrow}}{\frac{e\hbar B}{mc}}. \tag{15}$$

According to this definition, we firstly calculated the exact effective g-factor for every energy levels of the Luttinger Hamiltonian as is shown in Fig. 2 for heavy holes. We found $g^* \propto |k_z|$ for large l, which is consistent with the approximate theory by Yafet [11]. As discussed before, this k_z-dependent g^* can generate nontrivial spin current.

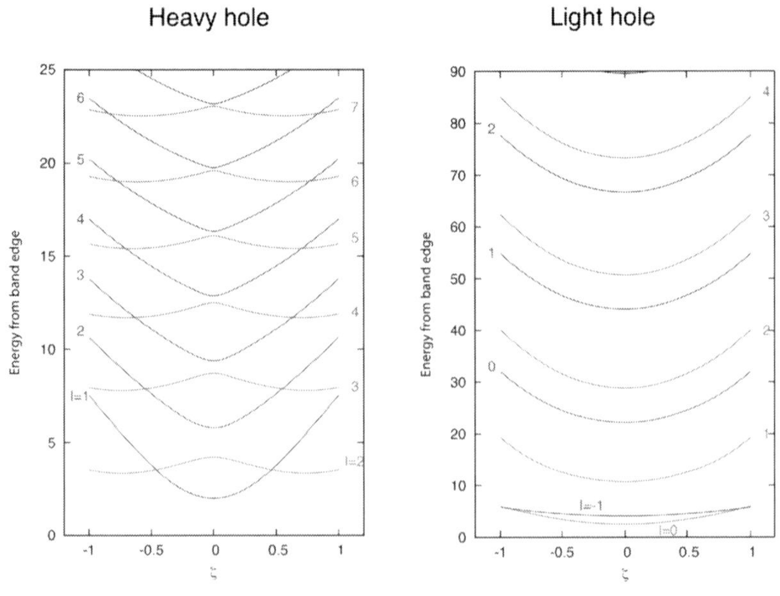

Figure 1. Eigenenergies with respect to $\zeta = k_z/\sqrt{s}$ for heavy (left) and light (right) holes under a magnetic field. The energies are measured in units of $e\hbar B/mc$. The energy axis are inverted, i.e., the hole band looks like "electron like" in this figure.

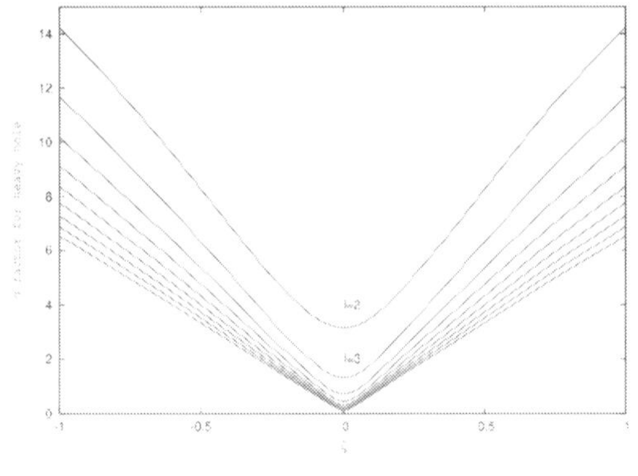

Figure 2. Effective g-factor for heavy holes as a function of $\zeta = k_z/\sqrt{s}$.

4. Matrix elements and selection rules

In order to obtain detailed information of the conductivity, it is important to investigate the selection rules for the matrix elements of the velocity operator since the conductivity can be calculated in terms of the matrix elements of the velocity operators as

$$\sigma_{\mu\nu} = \frac{1}{i\omega}\left[\Phi_{\mu\nu}(\omega + i\delta) - \Phi_{\mu\nu}(0 + i\delta)\right], \tag{16}$$

$$\Phi_{\mu\nu}(i\omega_\lambda) = -e^2 T \sum_{n,i,j} \langle i| \boldsymbol{v}_\mu |j\rangle \langle j| \boldsymbol{v}_\nu |i\rangle \mathscr{G}(i\tilde{\epsilon}_n)\mathscr{G}(i\tilde{\epsilon}_n - i\omega_\lambda) \tag{17}$$

on the basis of the Kubo formula [13, 14], where $\mathscr{G}$ is the thermal Green's function. The velocity operator along the field direction is given by

$$\begin{aligned} v_z &= \frac{\partial \mathcal{H}}{\hbar \partial k_z} \\ &= \frac{1}{m}\begin{pmatrix} (\gamma_1 - 2\bar{\gamma})\pi_z & 0 & -\sqrt{3}\bar{\gamma}\pi_- & 0 \\ 0 & (\gamma_1 + 2\bar{\gamma})\pi_z & 0 & \sqrt{3}\bar{\gamma}\pi_- \\ -\sqrt{3}\bar{\gamma}\pi_+ & 0 & (\gamma_1 + 2\bar{\gamma})\pi_z & 0 \\ 0 & \sqrt{3}\bar{\gamma}\pi_+ & 0 & (\gamma_1 - 2\bar{\gamma})\pi_z \end{pmatrix}, \end{aligned} \tag{18}$$

where $\pi_\pm = \pi_x \pm i\pi_y$. It is found by using Eqs. (11)-(14) that the matrix elements for v_z, $\langle \psi_{l\alpha}|v_z|\psi_{l'\alpha'}\rangle$ is finite only between $l = l'$. (Note that the matrix elements for v_x and v_y give contribution only between $l = l' \pm 1$.) The result for v_z is obtained as follows:

$$\langle \psi_{l\alpha}| v_z |\psi_{l\alpha'}\rangle = V_{\rm intra} + V_{\rm inter}, \tag{19}$$

$$V_{\rm intra} = \left[(c_1 c_1' + c_4 c_4')(\gamma_1 - 2\bar{\gamma}) + (c_2 c_2' + c_3 c_3')(\gamma_1 + 2\bar{\gamma})\right]\frac{\hbar k_z}{m}, \tag{20}$$

$$V_{\rm inter} = \left[(c_1 c'_3 + c_3 c'_1)\sqrt{l-1} - (c_2 c'_4 + c_4 c'_2)\sqrt{l+1}\right]\frac{\hbar\bar{\gamma}\sqrt{6s}}{m}, \tag{21}$$

where we omitted the subscript α of c_i. $V_{\rm intra}$ originates from the diagonal elements of Eq. (18) or $\mathcal{H}_1$, so it corresponds to the intraband transition. $V_{\rm inter}$ originates from the off-diagonal elements of Eq. (18) or $\mathcal{H}_2$, so it corresponds to the interband transition between light and heavy hole bands. The intraband contribution is proportional to $\hbar k_z/m$, which is just the velocity at $B = 0$, and independent form the magnetic field. This is the trivial contribution: the normal longitudinal magnetoconductivity does not depend on the magnetic field. On the other hand, the interband contribution is proportional to $\bar{\gamma}\sqrt{B}$. This magnetic field dependent term, whose origin is the same as the anomalous g-factor, would give a nontrivial contribution; the spin current proposed in the first section of this paper.

When we measure the dc magnetoconductivity, we obtain the summation of intra- and interband contributions, and we cannot distinguish between the two contributions. However, it is interesting to compare the intensity of the interband contribution to the intraband one as

$$\left|\frac{V_{\rm inter}}{V_{\rm intra}}\right|^2 = \left|\frac{(c_1 c'_3 + c_3 c'_1)\sqrt{l-1} - (c_2 c'_4 + c_4 c'_2)\sqrt{l+1}}{(c_1 c_1' + c_4 c_4')(\gamma_1 - 2\bar{\gamma}) + (c_2 c_2' + c_3 c_3')(\gamma_1 + 2\bar{\gamma})}\right|^2 \frac{6\bar{\gamma}^2 s}{k_z^2} \sim \frac{\bar{\gamma}^2}{\gamma_1^2 \zeta^2}. \tag{22}$$

The interband contributions will increase as the magnetic field is increased. The interband contribution should be amplified in the ac magnetoconductivity with an appropriate frequency. Also, if we can obtain a material that has large $\bar{\gamma}/\gamma_1$, we will have the larger interband contributions. This will be achieved by using heavier elements such as Sn and Pb.

Journal of Physics: Conference Series **603** (2015) 012022 doi:10.1088/1742-6596/603/1/012022

5. Conclusion

We have proposed the nontrivial contribution appears in the magnetoconductivity of the Luttinger Hamiltonian by investigating the effective g-factor and the selection rules for the matrix elements of the longitudinal velocity operator. The effective g-factor for every Landau levels of the Luttinger Hamiltonian is exactly obtained for the first time. It was confirmed that, for heavy hole band, $g^* \propto |k_z|$ for large quantum numbers of Landau levels, which was originally pointed out by approximate theory of Yafet [11]. This anomalous g-factor is due to the interband effect of the magnetic field between the light and heavy hole bands.

We have proposed that the k_z-dependent g-factor can generate a nontrivial spin current in the longitudinal magnetoconductivity. To confirm this idea, we have investigated the selection rules for the matrix elements of the longitudinal velocity operator. We found a nontrivial contribution actually appears in the longitudinal magnetoconductivity also due to the interband effect between light and heavy hole bands. This nontrivial contribution will generate the spin current, which can be amplified by the magnetic field or the external light with an appropriate frequency.

References

[1] Luttinger J M 1956 *Phys. Rev.* **102**(4) 1030–1041
[2] Voon L C L Y and Willatzen M 2009 *The $k \cdot p$ Method* (Springer-Verlag)
[3] Murakami S, Nagaosa N and Zhang S C 2003 *Science* **301** 1348
[4] Murakami S, Nagosa N and Zhang S C 2004 *Phys. Rev. B* **69**(23) 235206
[5] Wolff P A 1964 *J. Phys. Chem. Solids* **25** 1057
[6] Fuseya Y, Ogata M and Fukuyama H 2015 *J. Phys. Soc. Jpn.* **84** 012001
[7] Fuseya Y, Ogata M and Fukuyama H 2012 *J. Phys. Soc. Jpn* **81** 093704
[8] Fuseya Y, Ogata M and Fukuyama H 2014 *J. Phys. Soc. Jpn.* **83** 074702
[9] Cohen M H and Blount E I 1960 *Phil. Mag.* **5** 115
[10] Wallis R F and Bowlden H J 1960 *Phys. Rev.* **118** 456–461
[11] Yafet Y 1963 *Solid State Phys.* **14** 1 – 98
[12] Zhu Z, Fauqué B, Fuseya Y and Behnia K 2011 *Phys. Rev. B* **84** 115137
[13] Kubo R 1957 *J. Phys. Soc. Jpn.* **12** 570
[14] Fukuyama H, Ebisawa H and Wada Y 1969 *Prog. Theor. Phys.* **42** 494

Magnetoconductivity of Dirac electrons in bismuth

Mitsuaki Owada and Yuki Fuseya

Department of Engineering Science, University of Electro-Communications, Chofu, Tokyo 182-8585, Japan

E-mail: `oowada@kookai.pc.uec.ac.jp`

Abstract. The magnetoconductivity of bismuth is theoretically investigated on the basis of the Kubo formula to interpret the longstanding mystery of linear magnetoresistance in bismuth. First, the magnetoconductivity for the isotropic Dirac model is studied. It is found that the inverse magnetoconductivity increases quadratically at low magnetic fields and is saturated at high fields. This high field property is in contrast to that obtained by the semi-classical theory, where the inverse magnetoconductivity keeps quadratic increase. Next, the magnetoconductivity for the extended Dirac model, which is a realistic model for bismuth, is studied. The inverse magnetoconductivity so obtained is not saturated, but is reduced at high fields. Implications of present results to the linear magnetoresistivity of bismuth are discussed.

1. Introduction

In 1928, Kapitza reported that the magnetoresistance of bismuth increases linearly with respect to the external magnetic eld [1]. Next year, he further reported such a linear dependence can be observed in various materials, which is known as "Kapitza's linear law" [2–5]. The linear magnetoresistance was a mystery at that time since the all existing theory predicted a quadratic dependence at small elds and a saturating behavior at high elds. This problem was partly solved by Lifshits and Peschanskii for the case with open Fermi surface [6]. Much later, it was shown by Abrikosov that the magnetoresistance becomes linear at extremely high magnetic elds, in the so-called quantum limit, where only the lowest Landau level is occupied, by taking into account the in nite order of impurity scattering beyond the Born approximation [5, 7]. The theory of Abrikosov seems to succeed to give a reasonable explanation for the linear magnetoresistance on bismuth. However, it is well known that the e ective Hamiltonian of electrons in bismuth is given by the form equivalent to the Dirac Hamiltonian and the carrier density greatly varies as a function of the magnetic eld due to the charge neutrality condition of compensated metals [8–10]. These speci c features of bismuth are indispensable to explain various properties of bismuth, such as the angle resolved quantum oscillation [8, 9] and diamagnetism [11], but they were not taken into account in the Abrikosov's theory.

In this work, we investigate the magnetoconductivity of bismuth by taking into account the quantum e ect of whole Landau levels on the basis of the Kubo formula [12], the Dirac dispersion, the spatial anisotropy, and the eld dependence of the Fermi energy, aiming at giving a possible interpretation to the Kapitza's linear law.

2. Theory

We consider the isotropic Dirac Hamiltonian under an external magnetic eld [10, 13–15]:

$$\mathscr{H} = \begin{pmatrix} \Delta & i \ \boldsymbol{\pi} \ \boldsymbol{\sigma} \\ -i \ \boldsymbol{\pi} \ \boldsymbol{\sigma} & -\Delta \end{pmatrix}, \tag{1}$$

where 2Δ is the energy gap,   is the isotropic velocity, and $\boldsymbol{\sigma}$ is the Pauli matrix. $\boldsymbol{\pi} = -i\boldsymbol{\nabla} + (e/c)\boldsymbol{A}$, where $\boldsymbol{A}$ is the vector potential. The eigenenergy of this Hamiltonian is given by

$$E^{\rm iso}_{n,\ }(k_z) = \pm\sqrt{\Delta^2 + 2\Delta\left\{\left(n + \frac{1}{2} + \frac{\sigma}{2}\right)\omega_{\rm c} + \frac{k_z^2}{2m_{\rm c}}\right\}}, \tag{2}$$

where the cyclotron frequency is de ned as

$$\omega_{\rm c} = \frac{eB}{m_{\rm c}c} = \frac{eB\ ^2}{c\Delta}. \tag{3}$$

Even if we take into account the anisotropy of the velocity, the above isotropic model is essentially valid only with a replacement $m_{\rm c} \to m_h$ in Eq. (2) [9, 15, 16]. It is a remarkable characteristic of Dirac electron system that the lowest Landau level, $E_{\rm LLL} = E_{n=0,\ =-1}$, is xed at the initial energy even at high magnetic elds [10]. Note that the cyclotron mass $m_{\rm c} = \Delta/\ ^2$ is much less than the bare electron mass, m, in the case of bismuth, e. g., $m_{\rm c}/m \simeq 0.0019$ for the eld parallel to the bisectrix axis [9]. This is the reason why we can reach quantum limit at relatively low magnetic eld.

The complete and normalized wave function has the form

$$\psi = \sqrt{\frac{E+\Delta}{2E}}\begin{pmatrix} \chi_{\rm u} \\ \dfrac{-i\ (\boldsymbol{\pi}\ \boldsymbol{\sigma})\chi_{\rm u}}{E+\Delta} \end{pmatrix}, \tag{4}$$

where $\chi_{\rm u}$ is a wave function of free electrons under a magnetic eld [13]. The matrix elements of the velocity operator, which is given by $v_i = \partial\mathscr{H}/\partial k_i$, are calculated as

$$\begin{aligned} \langle\psi|v_i|\psi'\rangle &= \sqrt{\frac{(E+\Delta)(E'+\Delta)}{4EE'}}\begin{pmatrix} \chi_{\rm u} & \dfrac{i\ (\boldsymbol{\pi}\ \boldsymbol{\sigma})\chi_{\rm u}}{E+\Delta}\end{pmatrix}\begin{pmatrix} 0 & i\ \sigma_i \\ -i\ \sigma_i & 0\end{pmatrix}\begin{pmatrix} \chi'_{\rm u} \\ \dfrac{-i\ (\boldsymbol{\pi}\ \boldsymbol{\sigma})\chi'_{\rm u}}{E'+\Delta}\end{pmatrix} \\ &= \ ^2\sqrt{\frac{(E+\Delta)(E'+\Delta)}{4EE'}}\left(\chi_{\rm u}\ \frac{\sigma_i(\boldsymbol{\pi}\ \boldsymbol{\sigma})}{E'+\Delta} + \frac{(\boldsymbol{\pi}\ \boldsymbol{\sigma})\sigma_i}{E+\Delta}\chi'_{\rm u}\right). \end{aligned} \tag{5}$$

After some straightforward calculations, we obtain

$$\langle\psi|\boldsymbol{v}|\psi'\rangle = \frac{\ ^2}{2\sqrt{EE'(E+\Delta)(E'+\Delta)}}\langle\chi_{\rm u}|(E+E'+2\Delta)\boldsymbol{\pi} + i(E-E')(\boldsymbol{\pi}\times\boldsymbol{\sigma})|\chi'_{\rm u}\rangle. \tag{6}$$

Then the matrix elements for v_x is given by

$$\begin{aligned} \langle\psi|v_x|\psi'\rangle &= \frac{\ ^2A(E,E')}{2\sqrt{2}}\langle\chi_{\rm u}|(E+E'+2\Delta)(\pi^+ + \pi^-) \\ &\quad +(E-E')\left\{(\pi^+ - \pi^-)\sigma_z - \pi_z(\sigma^+ - \sigma^-)\right\}|\chi_{\rm u}\rangle, \end{aligned} \tag{7}$$

where $A(E,E') = [EE'(E+\Delta)(E'+\Delta)]^{-1/2}$, $\pi \quad (v_x \pm i v_y)/\sqrt{2}$, and $\sigma \quad (\sigma_x \pm i\sigma_y)/\sqrt{2}$. The operator π changes the orbital quantum number $n \to n \pm 1$, while σ changes the spin quantum number $\sigma \to \sigma \pm 1$. The only good quantum number is $j = n + 1/2 + \sigma$, and the operator v_x changes j as $j \to j \pm 1$, namely, orbital and spin quantum number cannot be changed simultaneously.

The diagonal conductivity σ_{xx} is given on the basis of the Kubo formula as [12, 17]

$$\sigma_{xx}(\omega) = \frac{e^2}{i\omega}\left[\Phi_{xx}(\omega + i\delta) - \Phi_{xx}(0 + i\delta)\right], \tag{8}$$

$$\Phi_{xx}(i\omega_\lambda) = -T\sum_{n,i,j}\langle i|v_x|j\rangle\langle j|v_x|i\rangle\mathscr{G}(i\tilde{\varepsilon}_n)\mathscr{G}(i\tilde{\varepsilon}_n - i\omega_\lambda), \tag{9}$$

where $\varepsilon_n = (2n+1)\pi T$, $\omega_\lambda = 2\pi\lambda T$ (n, λ: integer). Here we introduced the e ect of impurity scattering, Γ, as $i\tilde{\varepsilon}_n = i\varepsilon_n + i\Gamma\varepsilon_n/|\varepsilon_n|$. For the Green function part,

$$F(i\omega_\lambda) = -T\sum_n \mathscr{G}(i\tilde{\varepsilon}_n)\mathscr{G}(i\tilde{\varepsilon}_n - i\omega_\lambda) = -T\sum_n \frac{1}{i\tilde{\varepsilon}_n - E'}\frac{1}{i\tilde{\varepsilon}_n - i\omega_\lambda - E}, \tag{10}$$

the summation with respect to ε_n can be carried out as

$$F(\omega) = \frac{1}{2\pi i}\int_{-\infty}^{\infty} dx \quad f(x)\left[\frac{1}{x+\omega-E'+i\Gamma}\frac{1}{x-E+i\Gamma} - \frac{1}{x+\omega-E'+i\Gamma}\frac{1}{x-E-i\Gamma}\right.$$
$$\left.+\frac{1}{x-E'+i\Gamma}\frac{1}{x-\omega-E-i\Gamma} - \frac{1}{x-E'-i\Gamma}\frac{1}{x-\omega-E-i\Gamma}\right]. \tag{11}$$

(ω_λ was analytically continued as $i\omega_\lambda \to \omega$.) The contributions from the second and third terms of Eq. (11) correspond to the "Fermi surface term", and that from the rst and fourth terms correspond to the "Fermi sea term".

For the interband transition, we have

$$\begin{aligned}\Phi^{\rm cv}_{xx} = \frac{e^2\ {}^4 N_{\rm L}}{8}\sum_{n,k_z,}^{\prime}\Big[& \\ & F(\omega, -E_{n,}\ , E_{n+1,}\ )(A^{\rm cv}_{n+1\ ,n}\ )^2 m\omega_{\rm c}(n+1)\left[(E_{n+1,}\ - E_{n,}\ + 2\Delta) + \sigma(E_{n+1,}\ + E_{n,}\ )\right]^2 \\ & +F(\omega, -E_{n,}\ , E_{n-1,}\ )(A^{\rm cv}_{n-1\ ,n}\ )^2 m\omega_{\rm c} n\left[(E_{n-1,}\ - E_{n,}\ + 2\Delta) - \sigma(E_{n-1,}\ + E_{n,}\ )\right]^2 \\ & +2F(\omega, -E_{n,}\ , E_{n,-}\ )(A^{\rm cv}_{n-\ ,n}\ )^2 k_z^2(E_{n,-}\ + E_{n,}\ )^2\Big] \\ & +\frac{e^2\ {}^2}{2}F(\omega, -E_{0\downarrow}, E_{0\uparrow})\frac{(E_{0\uparrow}+E_{0\downarrow})}{E_{0\uparrow}},\end{aligned} \tag{12}$$

where $\sum'_{n,k_z,}$ denotes summations with respect to n, k_z and σ except for $n = 0, k_z = 0, \sigma = -1$ of the valence band, and we omitted the k_z dependences in energy. $N_{\rm L}$ is the degrees of the Landau-level degeneracy, $N_{\rm L} = eB/2\pi c = \omega_{\rm c}\Delta/2\pi\ {}^2$. $A^{\rm cv}_{n+1\ ,n}$ means $A(E_{n+1,}\ , -E_{n,}\ )$. ("c" and "v" abbreviate conduction and valence bands, respectively.) The rst and second term corresponds to the orbital transition, and the third term to the spin transition, which appear also in the Wol 's theory [13]. The fourth term is a new term only appears in the interband transition (not appear in the Wol 's theory). The total conductivity is calculated both from the intra- and inter-band contributions.

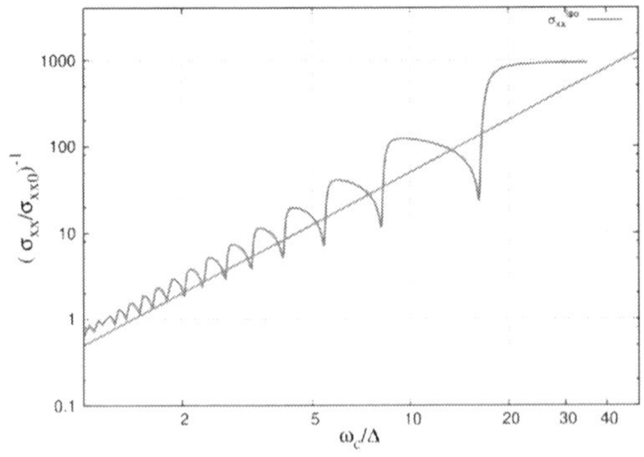

Figure 1. Inverse magnetoconductivity of the isotropic Dirac model as a function of magnetic eld, ω_c/Δ, for $E_F/\Delta = 5.8$, $\Gamma/\Delta = 0.02$. The straight line shows B^2-dependence.

3. Results

3.1. Isotropic Dirac model

The results of the inverse magnetoconductivity for the isotropic Dirac model is shown in Fig. 1 as a function of ω_c/Δ. Here we xed the Fermi energy to be $E_F/\Delta = 5.8$, where E_F is measured from the center of the band gap. There are clear quantum oscillations, which cannot be obtained by the semi-classical theory. When the Fermi energy E_F touches the minimum of the sub bands, $\sigma_{xx}^{\rm iso}$ shows a peak structure (appears as a dip in the plot of $1/\sigma_{xx}^{\rm iso}$). The peak width is widened by the impurity scattering Γ. $1/\sigma_{xx}^{\rm iso}$ exhibits B^2 dependence with the exception of the quantum oscillation in the weak eld region, $\omega_c/\Delta \lesssim 15$. This is consistent with the semi-classical transport theory for free electrons[18],

$$\sigma_{xx}^{\rm semi}(B) = \frac{ne^2\tau}{m^*}\frac{1}{1+\omega_c^2\tau^2}, \tag{13}$$

where, $\tau = 1/2\Gamma$. The system reaches the quantum limit for $\omega_c/\Delta \gtrsim 17$, where only the lowest Landau level is occupied. Beyond the quantum limit, $1/\sigma_{xx}^{\rm iso}$ is saturated by the eld. This is in contrast to the semi-classical result, where $1/\sigma_{xx}$ keeps quadratic increase.

3.2. Model for bismuth

Next, we consider the anisotropy and the change of the Fermi energy E_F according to the "extended Dirac model", whose energy is given as [9, 19]

$$E_{n,}^{\rm ex}(k_z) = \sqrt{\Delta^2 + 2\Delta\left\{\left(n + \frac{1}{2} + \frac{\sigma}{2}\right)\omega_c + \frac{k_z^2}{2m_h}\right\}} + \frac{g'\sigma}{2}\frac{{}_0}{2}B, \tag{14}$$

where m_h is the e ective mass along the magnetic eld, and ${}_0 = e/mc$. This model also takes into account the spin splitting and the shift of the lowest Landau level by introducing the g'-term in Eq. (14) [9, 20, 21]. This model has succeeded in giving good agreements with experiments on bismuth [9, 19]. The anisotropy is taken into account in ω_c and m_h. The eld dependences of E_F (shown in the left panel of Fig. 2) is calculated so as to satisfy the charge neutrality condition [9].

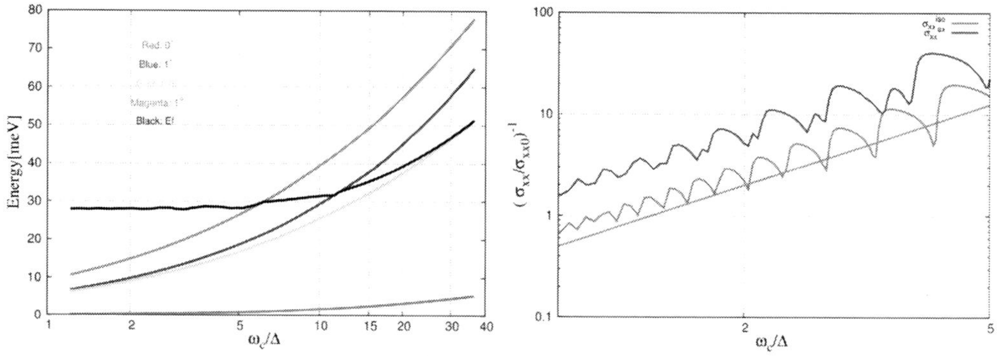

Figure 2. (left) Field dependence of the Fermi energy and Landau levels $E_{0,}^{\mathrm{ex}}$ and $E_{1,}^{\mathrm{ex}}$ for the extended Dirac model. (right) Inverse magnetoconductivity for the extended and isotropic Dirac models at low elds.

The result of the inverse magnetoconductivity so obtained is shown in the right panel of Fig. 2 for the magnetic eld parallel to the trigonal axis. Since E_{F} does not vary so much at low elds, the properties of $\sigma_{xx}^{\mathrm{ex}}$ is basically the same as that of $\sigma_{xx}^{\mathrm{iso}}$. At high elds, on the other hand, E_{F} changes drastically as $E_{\mathrm{F}} \propto B$ in bismuth due to the charge neutrality condition [8, 9] and the e ect of the g'-term becomes more relevant. Thus the properties of $\sigma_{xx}^{\mathrm{ex}}$ becomes di erent from those of $\sigma_{xx}^{\mathrm{iso}}$ as shown in Fig. 3. $1/\sigma_{xx}^{\mathrm{ex}}$ is not saturated but greatly reduced at high elds in contrast to the saturated $1/\sigma_{xx}^{\mathrm{iso}}$. This reduction in $1/\sigma_{xx}^{\mathrm{ex}}$ is a precursor to the quantum limit. For the eld parallel to the trigonal axis, the second lowest Landau level, $E_{0,+}$, approaches E_{F}, but cannot cross E_{F} even at high elds since both E_{F} and $E_{0,+}$ increases in almost the same way [9]. The system keeps the state just before entering the quantum limit, namely, $1/\sigma_{xx}^{\mathrm{ex}}$ keeps the dip structure at high elds. Consequently, the quantum e ect and the change of E_{F} make the behavior of conductivity completely di erent from that of semi-classical one.

In order to understand the behavior of magnetoresistance, we need to calculate the magnetic eld dependence of Hall conductivity, $\sigma_{xy}(B)$. Moreover, we also need to estimate the contributions not only from electrons, but also from holes for bismuth. According to the semi-classical theory, the total magnetoresistance exhibits unsaturated $\rho_{xx}^{\mathrm{semi}} \propto B^2$ behavior at high elds when the carrier numbers of electrons and holes are equal. This consequence is obtained from $1/\sigma_{xx}^{\mathrm{semi}} \propto B^2$ at high elds. Our result of $1/\sigma_{xx}^{\mathrm{ex}}$ exhibit the drastic reduction in $1/\sigma_{xx}^{\mathrm{ex}}$ at high elds. Therefore, it is naively expected that the ρ_{xx}^{Bi} will be reduced from the value o $\rho_{xx}^{\mathrm{semi}}$ at high elds and become close to $\rho_{xx} \propto B$ behavior.

4. Conclusion

We have investigated the magnetoconductivity of Dirac electrons in bismuth on the basis of the Kubo formula. For the isotropic Dirac model, $1/\sigma_{xx}^{\mathrm{iso}}(B) \propto B^2$ at low elds, which is consistent with the semi-classical results for free electrons. In the quantum limit, on the other hand, $1/\sigma_{xx}^{\mathrm{iso}}$ is saturated. This is contrast to the semi-classical result. We have also calculated the magnetoconductivity for electrons in bismuth based on the extended Dirac model. We have took into account the spacial anisotropy, the spin splitting, and the eld dependences of the Fermi energy E_{F}. $1/\sigma_{xx}^{\mathrm{ex}} \propto B^2$ at low elds, whereas $1/\sigma_{xx}^{\mathrm{ex}}$ is drastically reduced at high elds. This reduction in $1/\sigma_{xx}^{\mathrm{ex}}$ is signi cantly contrast to the semi-classical result, which keeps quadratic

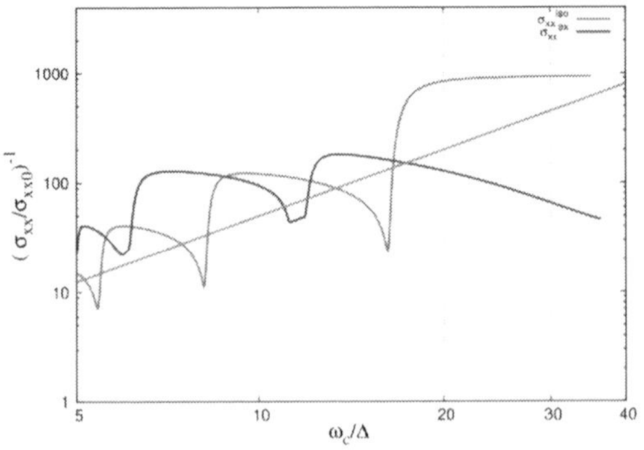

Figure 3. Inverse magnetoconductivity for the extended and isotopic Dirac models at high elds.

increase. Consequently, the magnetoconductivity of bismuth will be reduced from the semi-classical value $\rho_{xx}^{\rm semi} \propto B^2$ toward $\rho_{xx} \propto B$. We believe such a conclusion is useful to interpret the Kapitza's linear law in bismuth.

References

[1] Kapitza P 1928 *Proc. Roy. Soc. A* **119** 358
[2] Kapitza P 1929 *Proc. Roy. Soc. A* **123** 292
[3] Peierls R E 1980 *Proc. R. Soc. Lond. A* **371** 28
[4] Shoenberg D 1985 *Biogr. Mems Fell. R. Soc.* **31** 326–374
[5] Abrikosov A A 2003 *J. Phys. A: Math. Gen.* **36** 9119
[6] Lifshits I M and Peschanskii V G 1959 *Sov. Phys. JETP* **8** 875
[7] Abrikosov A A 1969 *Sov. Phys. JETP* **29** 746
[8] Smith G E, Bara G A and Rowell J M 1964 *Phys. Rev.* **135** A1118
[9] Zhu Z, Fauqué B, Fuseya Y and Behnia K 2011 *Phys. Rev. B* **84** 115137
[10] Fuseya Y, Ogata M and Fukuyama H 2015 *J. Phys. Soc. Jpn.* **84** 012001
[11] Fukuyama H and Kubo R 1970 *J. Phys. Soc. Jpn.* **28** 570
[12] Kubo R 1957 *J. Phys. Soc. Jpn.* **12** 570
[13] Wol P A 1964 *J. Phys. Chem. Solids* **25** 1057
[14] Fuseya Y, Ogata M and Fukuyama H 2012 *J. Phys. Soc. Jpn* **81** 013704
[15] Fuseya Y, Ogata M and Fukuyama H 2012 *J. Phys. Soc. Jpn* **81** 093704
[16] Fuseya Y, Ogata M and Fukuyama H 2014 *J. Phys. Soc. Jpn.* **83** 074702
[17] Fukuyama H and Kubo R 1969 *J. Phys. Soc. Jpn.* **27** 604
[18] Grosso G and Parravicini G 2000 *Solid State Physics* (Elsevier Science)
[19] Zhu Z, Fauqué B, Malone L, Antunes A B, Fuseya Y and Behnia K 2012 *Proc. Natl. Acad. Sci. U. S. A.* **109** 14813
[20] Bara G A 1965 *Phys. Rev.* **137** A842
[21] Vecchi M P, Pereira J R and Dresselhaus M S 1976 *Phys. Rev. B* **14** 298

Ferrimagnetism and single-particle excitations in a periodic Anderson model on the honeycomb lattice

Kazuhiro Seki[1], **Tomonori Shirakawa**[1,2], **Qinfang Zhang**[3], **Tao Li**[4], and **Seiji Yunoki**[1,2,5]

[1] Computational Condensed Matter Physics Laboratory, RIKEN, Wako, Saitama 351-0198, Japan

[2] Computational Materials Science Research Team, RIKEN Advanced Institute for Computational Science (AICS), Kobe Hyogo 650-0047, Japan

[3] Key Laboratory for Advanced Technology in Environmental Protection of Jiangsu Province, Yancheng Institute of technology,Yancheng 224051, China

[4] Department of Physics, Renmin University of China, Beijing, 100872, China

[5] Computational Quantum Matter Research Team, RIKEN Center for Emergent Matter Science (CEMS), Wako, Saitama 351-0198, Japan

E-mail: `kazuhiro.seki@riken.jp`

Abstract. By using the variational cluster approximation and cluster perturbation theory, we investigate the magnetism and single-particle excitations of a periodic Anderson model on the honeycomb lattice as an effective model for the single-side hydrogenated graphene, namely, graphone. We calculate the magnetic moment as a function of U (Coulomb interaction on impurity sites) with showing that the ground state is ferrimagnetic for any $U > 0$. We then calculate the single-particle excitations and show that the single-particle excitations are gapless and exhibit quadratic dispersion relation near the Fermi energy.

1. Introduction

Recently, a series of hydrogenated graphene [1, 2] attracts increasing attention because of its potential rich physics. A first-principles calculation based on density functional theory (DFT) has predicted that the single-side hydrogenated graphene, called graphone, becomes a ferromagnetic semiconductor with a small indirect gap [3]. Other DFT based study has suggested that the single-side hydrogenated and fluorinated graphenes can be a quantum spin liquid [4]. Possible increase of the spin-orbit coupling due to the sp^3 lattice distortion has been also discussed [5, 6]. However, electron correlation effects on the hydrogenated graphene beyond the DFT based calculations have not been investigated so far.

Here, we employ the variational cluster approximation (VCA) [7] and the cluster perturbation theory (CPT) [8] to investigate electron correlation effects on graphone by modeling it as a half-filled periodic Anderson model on the honeycomb lattice. We find that the ground state is ferrimagnetic (FM). The calculation of the magnetic moment as a function of electron correlation U at zero temperature shows that the magnetic moment in the weak-coupling region is naturally related to the flat-band FM. We also calculate the single-particle excitations and show that the low-energy excitations display the gapless quadratic dispersion.

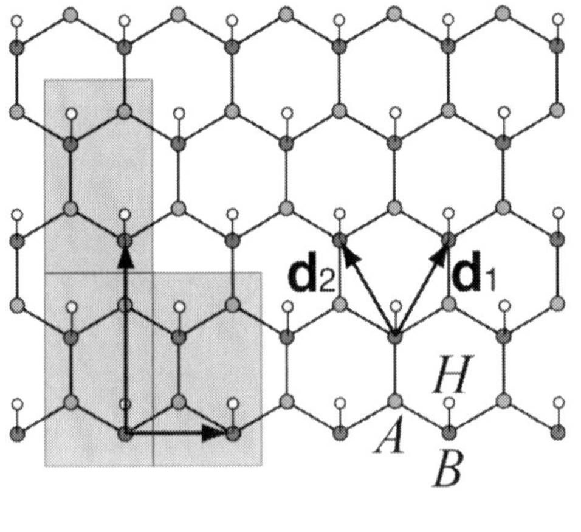

Figure 1. The periodic Anderson model on the honeycomb lattice. Carbon sites on A- and B-sublattices are denoted as green and red circles, respectively, and hydrogen sites are denoted as yellow circles. The shaded regions represent the 4-site (6-orbital) clusters used in the VCA and the CPT calculations. The arrows indicate primitive translational vectors for the cluster. The primitive translational vectors of the honeycomb lattice are also denoted by $\mathbf{d}_1$ and $\mathbf{d}_2$.

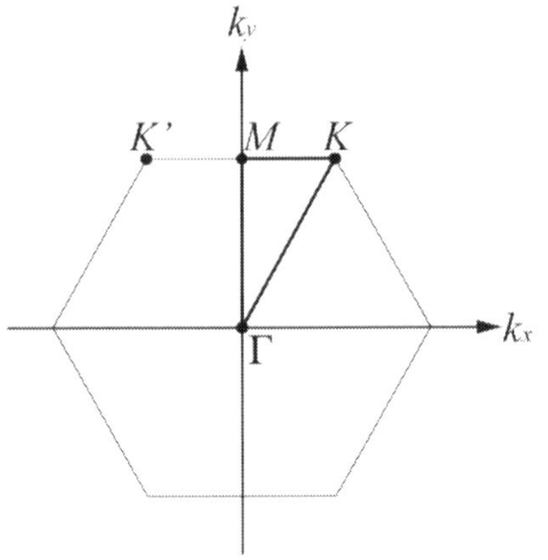

Figure 2. The Brillouin zone of the honeycomb lattice. Γ, M, K, and K' points are denoted as dots. Here $\Gamma = (0,0)$ and $M = 4\pi/3(0,\sqrt{3}/2)$. The thick triangle connecting Γ, M, and K points represents the momentum path used in Figs. 3(a) and (b).

2. Model

2.1. Periodic Anderson model

We consider a periodic Anderson model on the honeycomb lattice defined as

$$\mathcal{H} = \mathcal{H}_0 + \epsilon_H \sum_{i,\sigma} n_{i\sigma H} + U \sum_i n_{i\uparrow H} n_{i\downarrow H}, \tag{1}$$

where

$$\mathcal{H}_0 = -t \sum_{\langle i,j\rangle,\sigma} c^\dagger_{i\sigma A} c_{j\sigma B} - t \sum_{i,\sigma} c^\dagger_{i\sigma A} c_{i\sigma B} + t_{sp} \sum_{i,\sigma} c^\dagger_{i\sigma B} c_{i\sigma H} + \text{H.c.}, \tag{2}$$

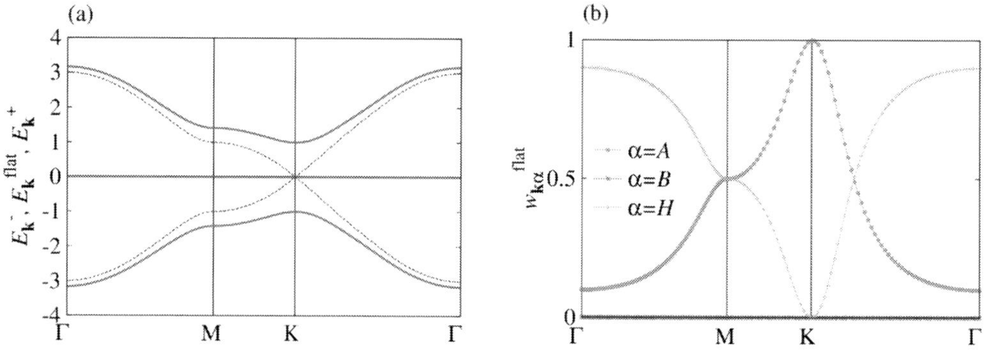

Figure 3. (a) The non-interacting tight-binding dispersion relations $E_{\mathbf{k}}^{-}$, $E_{\mathbf{k}}^{\mathrm{flat}}$, and $E_{\mathbf{k}}^{+}$ with $t_{sp}/t = 1$ (red solid lines). The non-interacting tight-binding dispersion relations for pure graphene model are also shown (blue dashed lines). (b) The orbital-resolved spectral weights $w_{\mathbf{k}A}^{\mathrm{flat}}$, $w_{\mathbf{k}B}^{\mathrm{flat}}$, and $w_{\mathbf{k}H}^{\mathrm{flat}}$ for the flat band $E_{\mathbf{k}}^{\mathrm{flat}}$ in the non-interacting limit with $t_{sp}/t = 1$. The results are shown along the path in the momentum space given in Fig. 2.

$c_{i\sigma\alpha}^{\dagger}$ is the electron creation operator with spin $\sigma\,(=\uparrow,\downarrow)$ and orbital $\alpha\,(=A,B,H)$ in the i-th unit cell, and $n_{i\sigma\alpha} = c_{i\sigma\alpha}^{\dagger}c_{i\sigma\alpha}$. Here, orbital A (B) denotes carbon p_z orbital on A (B)-sublattice of the honeycomb lattice and orbital H indicates hydrogen s orbital [see Fig. 1(a)]. The sum in the first term of Eq. (2), indicated by $\langle i,j\rangle$, runs over all pairs of nearest-neighboring unit cells. The first and second terms represent the hopping between the nearest-neighboring carbon sites with the hopping integral t, forming the conduction bands. t_{sp} is the hybridization between the hydrogen "impurity" site and the carbon site on B-sublattice. ϵ_H is the on-site potential energy and U is the on-site Coulomb repulsion of the hydrogen sites. This is the simplest model for graphone by implicitly assuming the hopping integral t in Eq. (2) as the renormalized one due to electron correlations in carbon sites [9, 10]. We consider the particle-hole symmetric case with $\epsilon_H = -U/2$, in which the electron density n is exactly 1 for any U values. We also set $t = t_{sp} = 1$ and $\hbar = k_B = 1$.

2.2. Non-interacting limit

In the non-interacting case with $\epsilon_H = 0$, the Hamiltonian in the momentum space can be written as

$$\mathcal{H}_0 = \sum_{\mathbf{k},\sigma} \left(c_{\mathbf{k}\sigma A}^{\dagger} c_{\mathbf{k}\sigma B}^{\dagger} c_{\mathbf{k}\sigma H}^{\dagger} \right) \begin{pmatrix} 0 & \gamma_{\mathbf{k}} & 0 \\ \gamma_{\mathbf{k}}^{*} & 0 & t_{sp} \\ 0 & t_{sp} & 0 \end{pmatrix} \begin{pmatrix} c_{\mathbf{k}\sigma A} \\ c_{\mathbf{k}\sigma B} \\ c_{\mathbf{k}\sigma H} \end{pmatrix}, \tag{3}$$

where $c_{\mathbf{k}\sigma\alpha}^{\dagger}$ is the Fourier transform of the real-space creation operator and

$$\gamma_{\mathbf{k}} = -t\left(1 + e^{i\mathbf{k}\cdot\mathbf{d}_1} + e^{i\mathbf{k}\cdot\mathbf{d}_2}\right) \tag{4}$$

is the matrix element between carbon p_z orbitals on A and B sublattices. Here,

$$\mathbf{d}_1 = \left(\frac{1}{2}, \frac{\sqrt{3}}{2}\right), \quad \mathbf{d}_2 = \left(-\frac{1}{2}, \frac{\sqrt{3}}{2}\right), \tag{5}$$

are primitive translational vectors of the honeycomb lattice (Fig. 1). Notice that, at $\mathbf{k} = K$ and K' points defined as (see also Fig. 2)

$$K = \frac{4\pi}{3}\left(\frac{1}{2}, \frac{\sqrt{3}}{2}\right), \quad K' = \frac{4\pi}{3}\left(-\frac{1}{2}, \frac{\sqrt{3}}{2}\right), \tag{6}$$

$\gamma_{\mathbf{k}}$ becomes zero, i.e., $\gamma_K = \gamma_{K'} = -t(1 + e^{2\pi i/3} + e^{4\pi i/3}) = 0$. The non-interacting tight-binding band dispersion relations are given as

$$E_{\mathbf{k}}^{\mathrm{flat}} = 0, \tag{7}$$

$$E_{\mathbf{k}}^{\pm} = \pm\sqrt{|\gamma_{\mathbf{k}}|^2 + t_{sp}^2}. \tag{8}$$

We thus find that $\mathcal{H}_0$ has a flat band just at the Fermi energy $E_{\mathrm{F}} = 0$. It also should be noted that the massless Dirac electronic dispersion is absent in the non-interacting limit [see Fig. 3(a)]. The orbital-resolved spectral weights for the flat band are given as

$$w_{\mathbf{k}A}^{\mathrm{flat}} = \langle\phi_{\mathbf{k}}^{\mathrm{flat}}|c_{\mathbf{k}A\sigma}^{\dagger}c_{\mathbf{k}A\sigma}|\phi_{\mathbf{k}}^{\mathrm{flat}}\rangle = \frac{t_{sp}^2}{|\gamma_{\mathbf{k}}|^2 + t_{sp}^2} \tag{9}$$

$$w_{\mathbf{k}B}^{\mathrm{flat}} = \langle\phi_{\mathbf{k}}^{\mathrm{flat}}|c_{\mathbf{k}B\sigma}^{\dagger}c_{\mathbf{k}B\sigma}|\phi_{\mathbf{k}}^{\mathrm{flat}}\rangle = 0 \tag{10}$$

$$w_{\mathbf{k}H}^{\mathrm{flat}} = \langle\phi_{\mathbf{k}}^{\mathrm{flat}}|c_{\mathbf{k}H\sigma}^{\dagger}c_{\mathbf{k}H\sigma}|\phi_{\mathbf{k}}^{\mathrm{flat}}\rangle = \frac{|\gamma_{\mathbf{k}}|^2}{|\gamma_{\mathbf{k}}|^2 + t_{sp}^2}, \tag{11}$$

where $|\phi_{\mathbf{k}}^{\mathrm{flat}}\rangle$ is the eigenstate of the flat band. Equations (9)-(11) indicate that the flat band consists of A and H orbitals, but not B orbital. In particular, the flat band at K (and K') point is solely consists of A orbital [see Fig. 3(b)].

The existence of the flat band Eq. (7) can be understood by recalling the Lieb's argument on bipartite lattices [11]. Namely, because of the imbalance of the number of sublattices, $\mathcal{H}_0$ has $|A| + |H| - |B|$ ($|\alpha|$: the number of α orbitals) zero eigenvalues, forming the flat band. This also explains why the wave functions with zero eigenvalues are contributed only from A and H orbitals, but not from B orbital. It should be also noted here that the tight-binding model considered here captures the characteristic features of the band structures for graphone obtained by a spin-unpolarized DFT calculation [4].

3. Methods

3.1. Variational cluster approximation

In order to investigate a possible symmetry breaking state, we employ the VCA [7] which is a cluster method formulated based on the self-energy-functional theory [12]. In the VCA, the original lattice is divided into disconnected finite-size clusters, and then variational parameters are introduced to examine possible symmetry breaking states. Here the two-body interaction term must be the same as the original one, whereas the one-body term can be changed [12]. The collection of these disconnected clusters is called "reference system" and its self-energy is used as a trial function for the grand-potential functional to be optimized based on the self-energy functional theory.

Here, we introduce, as a variational parameter, a uniform field h' on the hydrogen (impurity) sites [13] described as

$$\mathcal{H}_{h'} = h' \sum_i \left(n_{i\uparrow H} - n_{i\downarrow H}\right). \tag{12}$$

The reference system $\mathcal{H}_{\rm ref}$ considered is thus composed of a collection of disconnected finite size clusters as shown in Fig. 1(a), each cluster being described by $\mathcal{H}$ with no hopping terms between clusters (denoted as $\mathcal{H}_{\rm c}$), and $\mathcal{H}_{h'}$, i.e., $\mathcal{H}_{\rm ref} = \mathcal{H}_{\rm c} + \mathcal{H}_{h'}$.

In the VCA, the grand-potential functional per unit cell at zero temperature is given as [14]

$$\Omega = \frac{1}{L_c}\Omega' - \frac{1}{NL_c\pi}\int_0^\infty {\rm d}x \sum_\sigma \sum_{\tilde{\mathbf{k}}} \ln \left| \det \left(\boldsymbol{I} - \boldsymbol{V}_\sigma(\tilde{\mathbf{k}}) \boldsymbol{G}'_\sigma(ix) \right) \right| . \quad (13)$$

where Ω' is the exact grand potential of the single cluster, $\tilde{\mathbf{k}}$ is the wave vector defined in the reduced Brillouin zone of the reference system, N is the number of clusters and L_c is the number of unit cells in a cluster. The exact single-particle Green's function of $\mathcal{H}_{\rm ref}$ is denoted as $\boldsymbol{G}'_\sigma(ix)$ and $\boldsymbol{V}_\sigma \equiv \boldsymbol{G}'^{-1}_{\sigma 0} - \boldsymbol{G}^{-1}_{\sigma 0}$ represents the difference between the one-body terms of $\mathcal{H}$ and $\mathcal{H}_{\rm ref}$. We employ the exact diagonalization technique to calculate $\boldsymbol{G}'_\sigma(ix)$ and Ω'. The FM state is obtained when a saddle point $\partial\Omega/\partial h'|_{h'=h^*} = 0$ with the lowest Ω is at $h^* \neq 0$. Notice that in this paper we focus on the physics at zero temperature.

3.2. Cluster perturbation theory

The CPT [8] is employed to obtain the translational invariant Green's function of the infinite systems. In the CPT, the single-particle Green's function $\mathcal{G}^{\alpha\beta}_\sigma(\mathbf{k},\omega)$ is given as

$$\mathcal{G}^{\alpha\beta}_\sigma(\mathbf{k},\omega) = \frac{1}{3L_c}\sum_{i,j} \left(\boldsymbol{G}'^{-1}_\sigma(\omega) - \boldsymbol{V}_\sigma(\mathbf{k}) \right)^{-1}_{i\alpha,j\beta} e^{-i\mathbf{k}\cdot(\mathbf{r}_i - \mathbf{r}_j)}, \quad (14)$$

where $\mathbf{r}_i$ denotes the location of the i-th unit cell in the cluster. Here, the momentum $\mathbf{k}$ can take arbitrary value in the momentum space, and the frequency ω can take arbitrary value in the complex-frequency space. The single-particle spectral function for α orbital with spin σ is given as

$$A^{\alpha\alpha}_\sigma(\mathbf{k},\omega) = -\frac{1}{\pi}\Im \mathcal{G}^{\alpha\alpha}_\sigma(\mathbf{k},\omega + i\eta), \quad (15)$$

where η gives the Lorentzian broadening of the spectra.

4. Results

4.1. Ferrimagnetism

By calculating the grand-potential functional Ω as a function of h', we find that the FM state is stabilized for $U > 0$. We then calculate the magnetic moment per unit cell defined as

$$m_z = \sum_\alpha m_{z\alpha} \quad (16)$$

where $m_{z\alpha}$ is the magnetic moment of α orbital per unit cell which is given as [14]

$$m_{z\alpha} = \frac{1}{NL_c\pi}\int_0^\infty {\rm d}x \sum_\sigma \sum_{\tilde{\mathbf{k}}} \sum_i \sigma_z \Re \left[\left(\boldsymbol{G}'^{-1}_\sigma(ix) - \boldsymbol{V}_\sigma(\tilde{\mathbf{k}}) \right)^{-1} \right]_{i\alpha,i\alpha} \quad (17)$$

with σ_z being the z component of the Pauli matrices, i.e., $\sigma_z = 1(-1)$ for $\sigma = \uparrow (\downarrow)$. Figure 4 shows the calculated results of m_{zA}, m_{zB}, m_{zH}, and m_z as a function of U/t at zero temperature. We find that $m_z = 1$ within the numerical accuracy for any U. We also find that $m_{zA} > 0$, $m_{zB} < 0$, and $m_{zH} > 0$ as expected from the bipartite structure of the lattice. In the small U region, m_{zB} approaches to zero, whereas m_{zA} and m_{zH} approach to finite values in the non-interacting limit. Now let us consider the magnetic moment in the non-interacting limit. Since

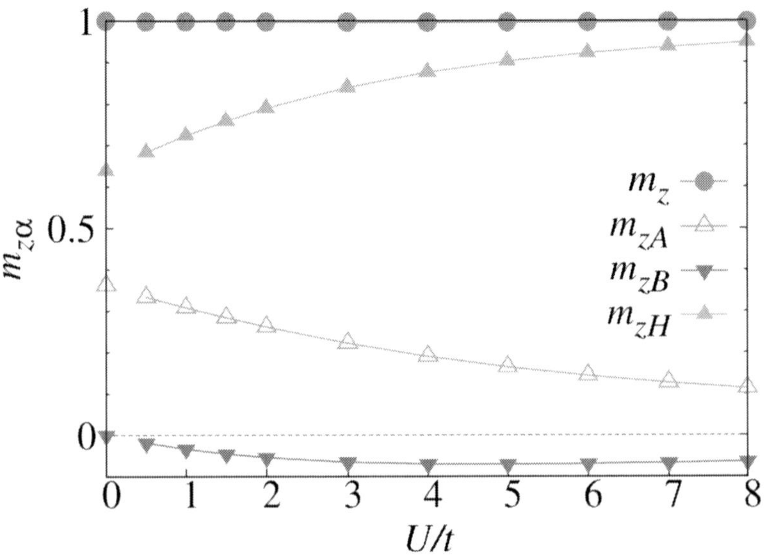

Figure 4. U-dependence of the orbital-resolved magnetic moments m_{zA}, m_{zB}, and m_{zH}, and the total magnetic moment m_z per unit cell. The magnetic moments in the non-interacting limit are also shown as symbols on the vertical axis at $U/t = 0$.

there exists the flat band at $E_{\rm F}$, the system is unstable against the FM long-range order for any $U > 0$. Assuming that the flat band is fully polarized in the non-interacting limit, the magnetic moments in the non-interacting limit can be calculated as [see Eqs. (9), (10), and (11)]

$$\lim_{U\to 0} m_{zA} = \frac{1}{L}\sum_{\mathbf{k}} \frac{t_{sp}^2}{|\gamma_{\mathbf{k}}|^2 + t_{sp}^2} \simeq 0.362, \tag{18}$$

$$\lim_{U\to 0} m_{zB} = 0, \tag{19}$$

$$\lim_{U\to 0} m_{zH} = \frac{1}{L}\sum_{\mathbf{k}} \frac{|\gamma_{\mathbf{k}}|^2}{|\gamma_{\mathbf{k}}|^2 + t_{sp}^2} \simeq 0.638, \tag{20}$$

where L is the number of unit cells and the numerical values are for $t = t_{sp} = 1$. The magnetic moments calculated by the VCA indeed approach to those values with decreasing U. Notice that a similar model on the square lattices has been recently studied in Ref. [15].

In the large U region, the magnetic moment becomes dominated by H orbital. Note that, in the strong coupling limit, an electron in each hydrogen site is completely localized and the RKKY interaction [16, 17, 18] between these localized spins is ferromagnetic [19], which naturally induces the FM ground state.

4.2. Single-particle excitations

Figures 5(a)-(d) show the single-particle excitation spectra $A_\sigma^{\alpha\alpha}(\mathbf{k},\omega)$ for momentum $\mathbf{k}$ around K point at $U/t = 4$. The low-energy single-particle excitations show quadratic dispersion relation around K point, indicating massive Dirac quasi-particle excitations. Although the dispersion

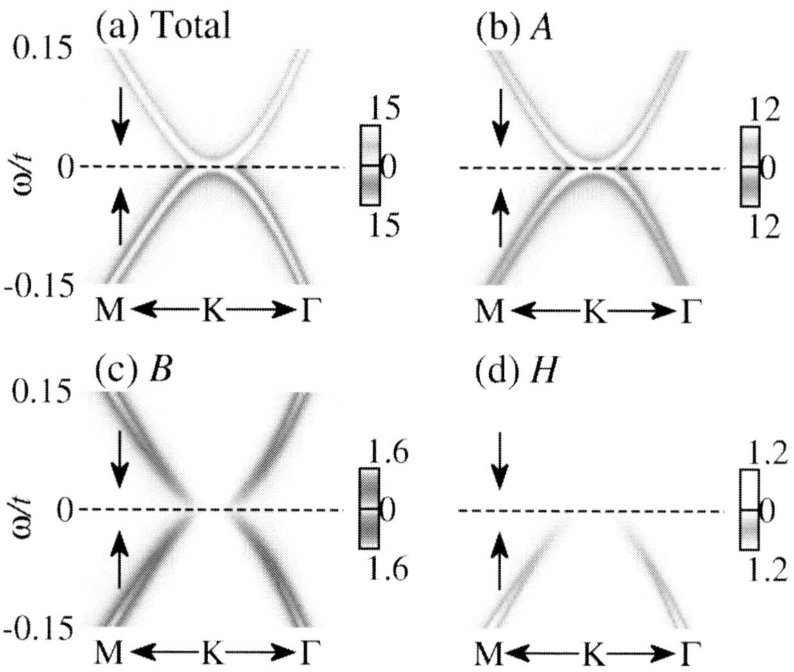

Figure 5. The single-particle excitations for $U/t = 4$ at the vicinity of K point near $E_{\rm F}$. (a) $A_\sigma(\mathbf{k},\omega) = \sum_\alpha A_\sigma^{\alpha\alpha}(\mathbf{k},\omega)$, (b) $A_\sigma^{AA}(\mathbf{k},\omega)$, (c) $A_\sigma^{BB}(\mathbf{k},\omega)$, and (d) $A_\sigma^{HH}(\mathbf{k},\omega)$. Here $\eta/t = 0.01$ is used. $A_\uparrow^{\alpha\alpha}(\mathbf{k},\omega)$ and $A_\downarrow^{\alpha\alpha}(\mathbf{k},\omega)$ are indicated with different colors. The region of momenta taken in the horizontal axis is 0.2π in the K-M (K-Γ) direction from K.

relation is massive, the single-particle gap is absent at K point. We also find that the A-orbital spectral weight at K point at $E_{\rm F}$ is finite even there exists the electron correlation U in H orbital, which also forms the flat band in the non-interacting limit. This is because the A orbital is completely decoupled from the H orbital at K point in the non-interacting limit [see Fig. 3(b)], thereby the electron correlation U on H orbital does not affect the spectrum of A orbital at K point. Therefore, the single-particle gap is absent at K point even at finite U. The same discussion holds for K' point. The single-particle excitation spectra in the FM state thus exhibit point-contact massive Dirac dispersion.

5. Summary

Using the VCA and the CPT, we have studied the magnetism and single-particle excitations of a half-filled periodic Anderson model at zero temperature. We have calculated the spontaneous magnetization as a function of electron correlation U. In the small U region, the calculated magnetic moments approach to those estimated from the flat-band ferrimagnetism. In the large U region, the H orbital gives the dominant contribution to the magnetic moment. Interestingly, the single-particle excitations in the FM state show a point-contact quadratic dispersion relation around the K and K' points. So far our study has been done at zero temperature in the magnetic state. Further study at finite temperatures in which the magnetic long-range order is destabilized will be reported elsewhere [20].

Acknowledgments
The computations have been done using the RIKEN Integrated Cluster of Clusters (RICC) facility. This work has been supported in part by Grant-in-Aid for Scientific Research from MEXT Japan under the grant Nos. 24740269 and 26800171 and by RIKEN iTHES Project and Molecular Systems. Q.Z. also acknowledges the National Natural Science Foundation of China (11204265), the Natural Science Foundation of Jiangsu Province (BK2012248) and the College Natural Science Research Project of Jiangsu Province (13KJ430007).

References
[1] Sofo J O, Chaudhari A S and Barber G D 2007 *Phys. Rev. B* **75** 153401
[2] Elias D C, Nair R R, Mohiuddin T M G, Morozov S V, Blake P, Halsall M P, Ferrari A C, Boukhvalov D W, Katsnelson M I, Geim A K and Novoselov K S 2009 *Science* **323** 610
[3] Zhou J, Wang Q, Sun Q, Chen X X, Kawazoe Y and Jena P 2009 *Nano Lett.* **9** 3867
[4] Rudenko A N, Keil F J, Katsnelson M I and Lichtenstein A I 2013 *Phys. Rev. B* **88** 081405(R)
[5] Castro Neto A H and Guinea f 2009 *Phys. Rev. Lett.* **103** 026804
[6] Gmitra M, Kochan D and Fabian J 2013 *Phys. Rev. Lett.* **110** 246602
[7] Potthoff M, Aichhorn M and Dahnken C 2003 *Phys. Rev. Lett.* **91** 206402
[8] Sénéchal D, Perez D and Pioro-Ladriere M 2000 *Phys. Rev. Lett.* **84** 522
[9] Sorella S, Otsuka Y and Yunoki S 2012 *Sci. Rep.* **2** 992
[10] Schuler M, Rosner M, Wehling T O, Lichtenstein A I and Katsnelson M I 2013 *Phys. Rev. Lett.* **111** 036601
[11] Lieb E H 1989 *Phys. Rev. Lett.* **62** 1201
[12] Potthoff M 2003 *Eur. Phys. J. B* **32** 429
[13] Horiuchi S, Kudo S, Shirakawa T and Ohta Y 2008 *Phys. Rev. B* **78** 155128
[14] Sénéchal D *e-print arXiv:0806.2690*
[15] Titvinidze I, Schwabe A and Potthoff M 2014 *Phys. Rev. B* **90** 045112
[16] Ruderman M A and Kittel C 1954 *Phys. Rev.* **96** 99
[17] Kasuya T 1956 *Prog. Theor. Phys.* **16** 45
[18] Yoshida K 1958 *Phys. Rev.* **106** 893
[19] Saremi S 2007 *Phys. Rev. B* **76** 184430
[20] Seki K, Shirakawa T, Zhang Q, Li T and Yunoki S *submitted*

International Workshop on Dirac Electrons in Solids 2015 IOP Publishing
Journal of Physics: Conference Series **603** (2015) 012025 doi:10.1088/1742-6596/603/1/012025

Band structure of superconducting Dirac electron systems

Takahiro Kuraya and Yuki Fuseya

Department of Engineering Science, University of Electro-Communications, Chofu, Tokyo 182-8585, Japan

E-mail: `kuraya@kookai.pc.uec.ac.jp`

Abstract. The band structures of PbTe and SnTe are theoretically investigated. First-principles calculation is carried out on the basis of the density functional theory within the generalized gradient approximations. Then, in order to study the fine band structures in the low energy region, the tight-binding model is introduced based on the most localized Wannier functions derived from the first-principles calculations. It is found that the topology of the equal energy surface is di erent between PbTe and SnTe due to the spin-orbit interaction. The information of the topologically di erent Fermi surface will be important to study the superconducting mechanism on doped PbTe and SnTe.

1. Introduction

The narrow gap IV-VI semiconductors, such as PbTe and SnTe, are famous for their highly efficient thermoelectricity [1]. Although they are semiconductors stoichiometrically, they exhibit clear evidence for the existence of Fermi surfaces without any doping [2]. Moreover, PbTe and SnTe become superconductors by doping. The latter exhibits superconductivity by self-doping, As and Sb counter doping [3] and In doping [4]. The former exhibits superconductivity by Tl doping [5], though it has not exhibited superconductivity by self-doping.

The band structures of these compounds have been investigated repeatedly. Some of them are done by rst-principles calculations [6, 7, 8] and some of them by linear combination of atomic orbital (LCAO) calculations [9]. Since the energy gap of IV-VI semiconductors are very small ($E_{\text{gap}} = 0.186$ eV for PbTe [10] and $E_{\text{gap}} \simeq 0.3$ eV for SnTe [11]), the accuracy of rst-principles calculations are not enough and it is hard to obtain information of ne band structure. On the other hand, LCAO calculations can tune the parameters so as to t with experimental data and can investigate ne band structures, but theoretical grounds for the parameters are not so reliable.

In the present work, we carry out rst-principles calculations for PbTe and SnTe, and then lead to the tight-binding model on a basis of the most localized Wannier functions derived from the rst-principles calculations. With this scheme, we can investigate the ne band structure with parameters determined ab initio, namely, our scheme has both strong points of rst-principles and LCAO calculations.

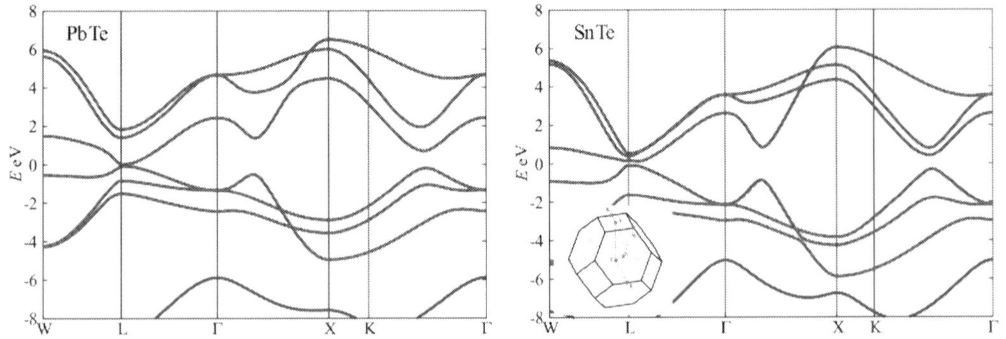

Figure 1. Band structures for PbTe (left) and SnTe (right). The solid lines are the results by the rst-principles calculation, and the dotted lines are those by the tight-binding model on the basis of the most localized Wannier functions derived from rst-principles calculations.

2. Band structures

We carried out the rst-principles calculations based on the density-functional theory within the generalized gradient approximation (GGA) using the OpenMX code [12, 13, 14], where the core Coulomb potential is replaced by norm-conserving pseudopotentials and pseudo-atomic orbitals (PAOs) are used as basis functions to expand one-particle Kohn-Sham wave functions. In the following, the PAO basis functions are chosen as Pb10s1p1d1, Sb9s1p1d1, Te9s1p1d1, where the number implies the cuto radius of the PAO functions and s1p1d1 means that one primitive orbitals for each s-, p-, and d-orbitals are used for PAO basis functions. The k-point sampling for the st-principles calculation is $12 \times 12 \times 12$ and that for the tight-binding calculation is $200 \times 200 \times 200/2$ in the rst Brillouin zone.

The band structures obtained by the rst-principles calculation for PbTe and SnTe are shown as the solid lines in Fig. 1. The results obtained by the tight-binding model on the basis of the most localized Wannier functions derived from the rst-principle calculations are also shown as the dotted lines in Fig. 1. The tight-binding results agree quite well with those by the rst-principles calculations.

The global features are consistent with previous works [6, 7, 8, 9]. The minimum of the conduction band and the maximum of the valence band locate at the L-point. The energy gaps are $\Delta_{\rm Pb} = 0.115$ eV and $\Delta_{\rm Sn} = 0.280$ eV, which are close to experimental values. The second maximum of the valence band appear close to the midpoint of the Γ-K lines. Here we denote the position of the second maximum of the valence band in the Γ-K line as Σ_0. Basically, the upper three bands originate from three p-orbitals of anion (Te) and the lower three bands originate from three p-orbitals of cation (Pb, Sn). However, they are hybridized due to the strong spin-orbit interactions. It should be noted that, for PbTe, the lowest conduction band and the highest valence band are strongly coupled and isolated from the other bands. In such a case, the so-called Wol Hamiltonian, which is essentially equivalent to the Dirac Hamiltonian but with spatial anisotropy, should be a good e ective Hamiltonian [15, 16]. On the other hand, for SnTe, the other two conduction bands are close to the lowest conduction bands. Thus there should be relevant couplings between four bands besides the coupling between the lowest conduction and the highest valence bands. This "tug-of-war" between bands leads to the relatively flat band for the lowest conduction band of SnTe.

In the low energy region, $|E| < 1$ eV, there is an important di erence between PbTe and SnTe; the energy di erence between the L- and Σ_0-point of the highest valence band. The energy di erence for SnTe, $E_L - E_{\Sigma_0} = 201$ meV, is about twice larger than that of PbTe, $E_L - E_{\Sigma_0} = 104$ meV. This is a quantitative di erence, but can generate a qualitative di erence for the Fermi surface as discussed later.

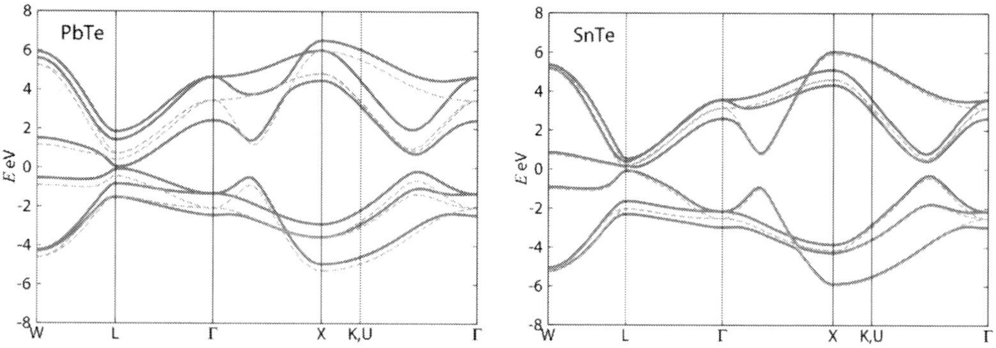

Figure 2. Band structures for PbTe (left) and SnTe (right) obtained by the tight-binding model on the basis of the most localized Wannier functions derived from rst-principles calculations.

In Fig. 2, the band structures with (solid lines) and without (dotted lines) the spin-orbit interaction (SOI) are shown. It is clearly seen that the band structure of PbTe is modi ed a great deal by the SOI, while that of SnTe is not so much. These di erences can be easily understood if we consider the atomic SOI of Pb (~ 1.35 eV) and Sn (~ 0.45 eV). It is interesting that the band gap at the L-point of PbTe is greatly reduced by the SOI from 0.840 eV to 0.115 eV, while that of SnTe is slightly enhanced from 0.0409 eV to 0.289 eV. According to Ref. [6], the highest valence (lowest conduction) band has L_6^+ (L_6^-) symmetry both for PbTe and SnTe when the SOI is absent. Basically, the SOI lowers the L_6^- band and lifts up the L_6^+ band [17], so that the band gap decreases as is seen in PbTe. In SnTe, however, since the gap without SOI in small, the gap rst decreases, and them increases after the band inversion between L_6^+ and L_6^-.

An important di erence appears at the Σ_0 point. The energy at the Σ_0 of PbTe is enhanced from -0.645 eV to -0.179 eV, while that of SnTe is not changed so much.

The di erence between PbTe and SnTe can be seen more clearly from the virtual crystal approximation (VCA) for $Pb_{1-x}Sn_x$, where the tight-binding parameters are linearly extrapolated between PbTe and SnTe, as is shown in Fig. 3. At the L-point, the top of the valence band hardly changes as Sn content (x) increases in the case with the SOI (the left panel of Fig. 3). The energy decrease is less than 20 meV. In the case without the SOI, on the other hand, the energy at the L-point rapidly increases by x for more than 400 meV. Therefore, the top of the valence band is pinned due to the SOI. In contrast, at the Σ_0-point, the top of the band rapidly decreases as x decreases in the case with the SOI and it increases without the SOI. The energy decrease is about 114 meV for the case with SOI, and the energy increase is about 207 meV for the case without SOI. Hence, the top of the band at the Σ_0-point decreases about 6 times faster than that at the L-point due to the SOI.

Journal of Physics: Conference Series **603** (2015) 012025 doi:10.1088/1742-6596/603/1/012025

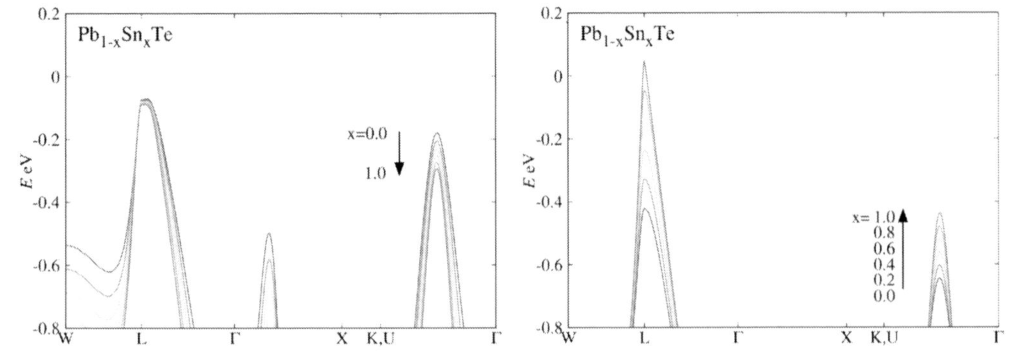

Figure 3. Low energy band structures for $Pb_{1-x}Sn_xTe$ with (left) and without (right) the SOI obtained by the VCA.

Figure 4. Equal energy surfaces for PbTe (left, $\Delta E_F = 126$meV) and SnTe (right, $\Delta E_F = 228$meV).

3. Fermi surfaces

Figure 4 shows the equal energy surfaces of PbTe and SnTe. This equal energy surfaces should correspond to the Fermi surfaces of the doped PbTe and SnTe. We have also calculated the Fermi surface of Tl doped PbTe on the basis of the VCA, and have checked the equal energy surface corresponds to the doped Fermi surface as far as the Tl content is less than 1%

In the case of PbTe, the equal energy surface at the L-point grows as decreasing the energy (doping) for $\Delta E_F < 100$ meV. The shape of the equal energy surface at the L-point is not simply ellipsoidal, but is like a "barrel" [18]. Then, a second pocket appears at around the Σ_0-point for $\Delta E_F \geq 100$ meV. The shape of the second pocket is almost ellipsoidal but warped as is shown in the left panel of Fig. 2. Both the rst and second pocket grow by lowering the energy, and they are connected with each other for $\Delta E_F \geq 140$ meV. The energy region where the second pocket exists is 100 meV $< \Delta E_F <$ 140 meV.

In the case of SnTe, on the other hand, the equal energy surface at the L-point grows solely until $\Delta E_F < 200$ meV. The shape of the rst pocket is similar to that of PbTe. Then, the second pocket appears at around $\Delta E_F \simeq 200$ meV, but is soon connected to the rst pocket as is shown in the right panel of Fig. 2. The energy region where the second pocket is not connected to the rst one is only 200 meV $< \Delta E_F <$ 210 meV. Therefore, the doping region where the second pocket exists is very narrow, suggesting that the second pocket is hard to observe practically for SnTe.

4. Conclusion

In the present work the low energy band structures of PbTe and SnTe have been investigated by ab initio calculations. The rst-principles calculations within the GGA have been carried out by using the OpenMX code. The tight-binding model has been derived based on the most localized Wannier functions obtained by the rst-principles calculations. The direct energy gaps at the L-point have been evaluated to be $\Delta_{\rm Pb} = 0.115$ eV and $\Delta_{\rm Sn} = 0.280$ eV, which are close to those obtained by experiments. There are maxima of the highest valence band near the midpoint (Σ_0) of the Γ-K line both for PbTe and SnTe. The energy di erences between the L- and Σ_0-points are $E_L - E_{\Sigma_0} = 201$ meV for SnTe and $E_L - E_{\Sigma_0} = 104$ meV for PbTe. This di erence between *PbTe* and *SnTe* is mainly due to the SOI and brings the topological di erence in the Fermi surfaces of doped PbTe and SnTe. For doped PbTe, the second pocket appears at around the Σ_0-point for wide doping levels. On the other hand, for SnTe, the doping level where the second pocket appears is very narrow, so that the second pocket will not appear in practice. These topologically di erent Fermi surfaces due to the SOI can yield di erent properties of superconductivity on doped PbTe and SnTe.

References

[1] Mahan G D 1997 Good thermoelectrics *SOLID STATE PHYSICS* vol 51 (ACADEMIC PRESS INC.) p 81
[2] Doiron-Leyraud N 2011 *La Physique au Canada* **67** 89
[3] Hulm J 1969 *Phys. Rev.* **169** 388
[4] Erickson A S, Chu J H, Toney M F, Geballe T H and Fisher I R 2009 *Phys. Rev. B* **79** 024520
[5] Nakayama K, Sato T, Takahashi T and Murakami H 2008 *Physical Review Letters* **100** 227004
[6] Martinez G, Schluter M and Cohen M L 1975 *Phys. Rev. B* **11**(2) 651–659
[7] Singh D 2010 *Phys. Rev. B* **81** 195217
[8] Littlewood P B, Mihaila B and Schulze R K 2010 *Phys. Rev. Lett.* **105** 086404
[9] Lent C S, Bowen M A, Dow J D, Allgaier R S, Sanky O F and Ho E S 1986 *Superlattices Microstruct.* **2** 491–499
[10] Mitchell D L, Palik E D and Zemel J N 1964 *Proc. Seventh Int. Conf. Phys. Semicond.* p 325
[11] Dimmock J O, Melngaillis I and Strauss A J 1966 *Phys. Rev. Lett.* **16** 1193
[12] Ozaki T 2003 *Phy. Rev. B* **67** 155108
[13] Ozaki T, Kino H, Yu J and Han M J 2014 Openmx package URL `http://www.openmx-square.org`
[14] Weng H, Ozaki T and Terakura K 2009 *Phys. Rev. B* **79** 235118
[15] Fuseya Y, Ogata M and Fukuyama H 2014 *J. Phys. Soc. Jpn.* **83** 074702
[16] Fuseya Y, Ogata M and Fukuyama H 2015 *J. Phys. Soc. Jpn.* **84** 012001
[17] Niu C, Dai Y, Ma Y, Yu L and Huang B 2013 *Materials Express* **3** 159
[18] Buchauer L F 2014 Master's thesis Technishe Universitat Darmstadt

AUTHOR INDEX

Ahmadi, E 131
Chiu, Ching-Kai 16
Ezawa, Motohiko 48
Fuchs, J-N 1
Fujimoto, A 164
Fukuda, A. 164
Fuseya, Yuki 26, 172, 178, 192
Jafari, S A 40, 124, 131
Jalali, Z. 40
Jiang, Y 164
Jiang, Z. 164
Joiner, C A 164
Kanao, Taro 110
Kariyado, Toshikaze 62
Kubo, K 104
Kuraya, Takahiro 192
Li, Tao 184
Mahyaeh, I 124
Mashkoori, M 124
Matsuura, Hiroyasu 110, 151, 158
Mizoguchi, Tomonari 33
Montambaux, G. 1
Morinari, T 104
Morinari, Takao 54, 80
Nomura, Kentaro 137
Ogata, Masao 33, 68, 94, 110, 144, 151, 158
Okuma, Nobuyuki 144
Owada, Mitsuaki 178
Piechon, F 1
Piechon, Frederic 80
Proskurin, Igor 68, 94
Raoux, A 1
Schnyder, Andreas P 16
Seki, Kazuhiro 184
Sekine, Akihiko 137
Shirakawa, Tomonori 184
Suemasa, Riki 172
Suetsugu, Shota 158
Suzumura, Yoshikazu 68, 80, 94
Terasawa, D 164
Vogel, E M 164
Yanagisawa, Takashi 120
Yoshioka, Nobuyuki 151
Yunoki, Seiji 184
Zhang, Qinfang 184

Institute of Physics
Dirac House, Temple Back
Bristol BS1 6BE UK